Linear Difference Equations with Discrete Transform Methods

Mathematics and Its Applications

Volume 363

Linear Difference Equations with Discrete Transform Methods

by

Abdul J. Jerri
Clarkson University

KLUWER ACADEMIC PUBLISHERS
DORDRECHT / BOSTON / LONDON

A C.I.P. Catalogue record for this book is available from the Library of Congress.

ISBN 0-7923-3940-1

Published by Kluwer Academic Publishers,
P.O. Box 17, 3300 AA Dordrecht, The Netherlands.

Kluwer Academic Publishers incorporates
the publishing programmes of
D. Reidel, Martinus Nijhoff, Dr W. Junk and MTP Press.

Sold and distributed in the U.S.A. and Canada
by Kluwer Academic Publishers,
101 Philip Drive, Norwell, MA 02061, U.S.A.

In all other countries, sold and distributed
by Kluwer Academic Publishers Group,
P.O. Box 322, 3300 AH Dordrecht, The Netherlands.

Printed on acid-free paper

Printed in the Netherlands

To my youngest daughter Huda.

Contents

List of Figures

xii

List of Tables

PREFACE

This book covers the basic elements of difference equations and the tools of difference and sum calculus necessary for studying and solving, primarily, ordinary linear difference equations. Examples from various fields are presented clearly in the first chapter, then discussed along with their detailed solutions in Chapters 2-7. The book is intended mainly as a text for the beginning undergraduate course in difference equations, where the "operational sum calculus" of the direct use of the discrete Fourier transforms for solving boundary value problems associated with difference equations represents an added new feature compared to other existing books on the subject at this introductory level. This means that in addition to the familiar methods of solving difference equations that are covered in Chapter 3, this book emphasizes the use of discrete transforms. It is an attempt to introduce the methods and mechanics of discrete transforms for solving ordinary difference equations. The treatment closely parallels what many students have already learned about using the operational (integral) calculus of Laplace and Fourier transforms to solve differential equations. As in the continuous case, discrete operational methods may not solve problems that are intractable by other methods, but they can facilitate the solution of a large class of discrete *initial and boundary value* problems. Such operational methods, or what we shall term "operational sum calculus," may be extended easily to solve partial difference equations associated with initial and/or boundary value problems.

Difference equations are often used to model "an approximation" of differential equations, an approach which underlies the development of many numerical methods. However, there are many situations, for example, recurrence relations and the modeling of discrete processes such as traffic flow with finite number of entrances and exits in which difference equations arise naturally, as we shall illustrate in Chapters 1 and 7. This further justifies the use of the operational sum calculus of discrete transforms. One of the goals of this book is to present various classes of difference equations and then introduce the discrete transforms which are compatible with them.

The first chapter starts with examples that illustrate how difference equations are the natural setting for problems that range from forecasting population to the electrical networks. These are in addition to the typical use of difference equations as an "approximation" of differential equations. In the rest of the chapter we present some of the fundamental difference operators along with their basic properties and their inverses as "sum" operators, which are necessary for modeling difference equations as well as developing pairs for the basic discrete transforms. Chapter 2 gives a clear introduction to difference equations, including their general classification, which hopefully will help establish the structure of the remaining chapters. Chapter 2 concludes with a very brief introduction and illustration of how the Laplace and Fourier transforms are used for simplifying the solution of differential equations, i.e., how the "operational integral calculus" method helps in algebraizing the differential operator, and at the same time involves the prescribed auxiliary conditions. This is complemented by a brief account of the discrete Fourier transform method, as the direct "sum" analog of the "integral" transforms method, for solving difference equations with boundary conditions. Thus we may coin the phrase "operational sum calculus or algebra," since the operation of the discrete Fourier transforms is no more than multiplying a column matrix by a square matrix. Chapter 3 discusses in detail the typical methods of solving linear difference equations, along with the most basic theorems and a variety of illustrative examples. This includes constant as well as variable coefficients, and homogeneous and nonhomogeneous equations. In addition to the main topic here, which is linear difference equations of one variable, or "ordinary linear difference equations," a clear introduction to difference equations of several variables, or "partial difference equations" is also presented, which is supported by a number of interesting examples. Chapter 3 concludes with a brief elementary presentation of the important topics of convergence and stability of the equilibrium solution of difference equations with initial conditions. Chapter 4, a relatively new chapter for introductory courses in difference equations, presents the most important transform, the discrete Fourier transform (DFT), along with its basic properties. The DFT represents the main operational sum calculus tool for solving difference equations associated with boundary conditions in the most direct way. The efficient algorithm of computing the DFT, the fast Fourier transform (FFT), is also discussed. A number of discrete Fourier transform pairs are deduced, which are needed for illustrating the solution of boundary value problems by this direct discrete transform method. Chapter 5 builds upon the work of the preceding chapter to derive the discrete sine and cosine transforms. The boundary value problems with which these transforms are compatible are also discussed and illustrated. Chapter 6 presents the z-transform which is used to solve difference equations associated with initial values, or "initial value problems," and which can be encompassed under the "operational sum calculus" methods. This is in parallel to how the Laplace transform is used to solve linear differential equations associated with initial conditions. The basic properties and pairs of the z-transform are established, which are very necessary for illustrating this direct transform method for solving difference equations associated with initial value problems. These solutions are compared with those obtained by the usual methods of Chapter 3. Chapter 7, a main chapter, presents examples that sample the many kinds of practical problems which give rise, mostly in a natural way, to difference equations. This chapter is considered to be the culmination of the preceding chapters with the typical methods of solution in Chapter 3, as well as the discrete transforms method of "operational sum calculus" in Chapters 4-6. Of course, these problems can be distributed over the different sections of the book, where their appropriate methods of solution are covered. To accomplish such a purpose, we shall, at the end of each of these sections, make a clear reference to the corresponding examples of Chapter 7. The applications include chemical mixing, compounding interest and investments, population growth, traffic flow, coupled springs and masses system, evaluating determinants and integrals, and diffusion processes. For each section, there are many related exercises with answers to most of the exercises, which are found at the end of the book.

Acknowledgements

The project which is the subject of this book started over a decade ago with the idea of the author to use discrete Fourier transforms for solving difference equations in parallel to the use of Fourier transforms for solving differential equations, where it was termed "Operational Difference Calculus." The author then invited Professor William L. Briggs to join him in writing a short monograph. When the time came last year for writing a textbook by expanding the monograph, Prof. Briggs declined sharing the efforts due to his heavy writing commitments. Not to let this chance escape, he consented that the author go it alone and kindly agreed to read the initial manuscript. To my friend and former colleague Prof. Briggs, I owe sincere thanks and appreciation and hope that this new and complete project with its new flavor of the direct "operational sum method" will meet his expectations as well as those in the field.

I would like to give thanks to Professor Michiel Hazewinkel, the managing editor of the Kluwer Academic Publishers series on Mathematics and its Applications, for inviting me to write this book as well as a forthcoming book on "The Gibbs Phenomenon – Analysis and Remedies." I would like to also thank Professors Saber Elaydi, Ronald E. Mickens, and Johnny L. Henderson for their careful reading of the manuscript and for their detailed constructive suggestions. Professors M. Zuhair Nashed, Allan C. Peterson, Gerry Ladas, John R. Graef, and Russel L. Herman also made many helpful suggestions; they deserve my thanks. I would like also to thank many of the students who read the manuscript and made various remarks. I have seriously attempted to accommodate almost all these valuable suggestions. Mrs. Joyce Worley was of great assistance in typesetting and finalizing this manuscript, and she deserves my special thanks. Thanks are also due Mr. Joseph Hruska for producing the drawings with care.

Some topics and examples of the original monograph were used in the author's book "Integral and Discrete Transforms with Applications and Error Analysis," (Marcel Dekker Inc., 1992), which are used now in this book, with due thanks to Marcel Dekker Inc. Mr. John Martindale of Kluwer Academic Publishers showed a great understanding and was most helpful with both projects, and he deserves my sincere thanks. Ms. Arlene Apone, of Kluwer, who reviewed the final manuscript and made constructive editorial suggestions, deserves my thanks.

My very special thanks go to my wife Suad and my daughter Huda for being so supportive and very patient during the preparation of this book.

COURSE ADOPTION

This book may be incorporated into a one semester introductory course for students of engineering, mathematics or science. We assume that students have had an elementary exposure to functions of a complex variable. While a course in differential equations or numerical analysis might increase the appreciation of the material in this book, neither course is truly a prerequisite for an understanding of this book. However, the basic sophomore course of differential equations is most helpful, as we shall often refer to the parallel with difference equations. For such an introductory one semester course, we suggest Chapters 1 to 6 and problems of interest to the particular class from Chapter 7. Section 2.3 is optional for those who are familiar with Laplace and Fourier integral transforms, and Fourier series.

Of course, the book, with its added direct discrete transforms method flavor, represents a readable reference to students as well as to researchers in the applied fields.

Chapter 1

SEQUENCES AND DIFFERENCE OPERATORS

1.1   Sequences

Our first experience with operators comes from infinitesimal calculus. In that setting, we are familiar with the differentiation and integration operators as they are applied to functions. As we shall see shortly, the operators of the difference calculus are applied not to functions, but to *sequences* of numbers. Therefore, before investigating the difference calculus, it might be useful to investigate the ways in which the sequences which underlie the difference calculus might arise. We will denote a sequence of numbers by $\{u_k\}$, where the index k takes integer values. The precise indexing of a particular sequence (that is, the first and last values of k) depends upon the problem in which the sequence occurs, but is usually apparent from the problem. The notation $\{u_k\}$ may represent a finite set of numbers or an infinite (singly or doubly) set of numbers such as $\{u_k\}_{k=0}^{\infty}$ or $\{w_{k,l}\}_{k,l=0}^{\infty}$. The index k here in $\{u_k\}$ is only a dummy variable, and we often use n, m, j, k, l, etc.

The following examples are types of problems that might produce the sequences we will be studying. Usually they occur as solutions of

1

recurrence **relations** (or what are presently termed difference equations.)

Sequences–Various Examples of Producing a Sequence

Example 1.1   Sampling (Discretizing or Digitizing)

A function u defined on an interval I, is sampled at the equally spaced grid points $\{x_0, x_1, x_2, \ldots\ldots, x_k\}$ which partition the interval I. This process of sampling (or discretizing or digitizing) a function u produces a sequence $\{u_k\}$, where $u_k = u(x_k)$ for $0 \le k \le K$ as shown in Fig. 1.1.

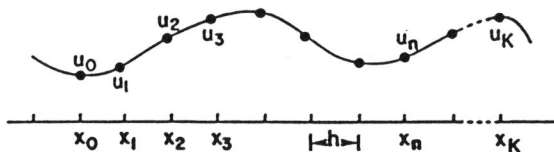

Figure 1.1: Sampling a function

In problems which involve sampling, the spacing of the grid points (or sampling frequency) is often important. We shall denote this quantity $h = x_{k+1} - x_k$ and will return to its significance shortly. An every day example that involves sampling is the way we receive our phone calls. Our voice is first recorded as a signal $s(t)$, which is a continuous function of time t. However, such a signal cannot be transmitted on a practical channel, since as a continuous function of time t, it has infinite information even when it is taken in a finite interval of time $a < t < b$. The way around this is to first sample such a signal at time intervals $t_n = \frac{n\pi}{W}$, where W is the highest frequency for the transmission channel. The samples spacing here is $h = \frac{\pi}{W}$. The finite sequence of the samples of the signal $s_n = s(\frac{n\pi}{W})$, $n = 1, 2, \cdots, N$, becomes now practical to transmit. At the receiver end, (assuming no noise) this sampled message of our voice $s(\frac{n\pi}{W})$, $n = 1, 2, \cdots, N$, can then be interpolated to generate our continuous voice signal $s(t), \frac{\pi}{W} \le t \le \frac{N\pi}{W}$. Indeed, this vital process of sampling may sometimes be noticed in some long distance conversations.

Example 1.2   Collecting Data at (Discrete) Time Intervals

A sequence $\{u_k\}$ may occur as the result of collecting data. For example, if temperature readings are recorded at hourly intervals, with u_k denoting the temperature at time t_k, $0 \leq k \leq K$, then a finite sequence of readings $\{u_k\}$ is generated. This also applies when we speak about forecasting a population after each interval of time, which may be a year. Let us consider a population where its size is known at discrete set of time intervals $\{t_n = n\Delta t\}$ for a fixed interval Δt, n is a positive integer. We may denote the population in the nth time interval as the sequence $P_n = P(n\Delta t)$, the population in the next interval $t_{n+1} = (n+1)\Delta t$ becomes $P_{n+1} = P((n+1)\Delta t)$. Now if we make the reasonable assumption that the population P_{n+1} in the next interval is proportional to the population P_n in the preceding interval, we have

$$P_{n+1} = \alpha P_n, \tag{1.1}$$

where α is a (positive) proportionality constant, and the (recurrence) relation in (1.1) generates the sequence $\{P_o, P_1, P_2, \cdots\}$ provided that the *initial* population $P_0 = P(t_0)$ at the initial time t_0 is given. We will elaborate more on recurrence relations in the next example.

Example 1.3   A Difference Equation (Recurrence Relation)–
                   The Fibonacci Relation

In the preceding examples, the sequence $\{u_k\}$ was closely related to a second (independent) variable. Often there is no apparent independent variable, except for the index k itself. This is the situation in *recurrence relations* , in which sequences are generated by an explicit rule. For example, the *Fibonacci relation* $u_{k+1} = u_k + u_{k-1}$, together with the *initial conditions* $u_0 = 0, u_1 = 1$ produce the sequence $\{0, 1, 1, 2, 3, 5, 8, \cdots\}$.

Next we present two familiar examples, where recurrence relations appear in generating important polynomials as well as in the process of solving differential equations. The reader who is not familiar with the elementary course of differential equations may skip Examples 1.4 and 1.5.

Example 1.4   Recurrence Relations for Polynomials

A very familiar example in mathematics, for the appearance of recurrence relations, is in the properties of some important polynomials. For example, the *Legendre polynomials* $P_n(x)$, $-1 < x < 1$,
$n = 0, 1, 2, \cdots$ as functions of the *continuous variable* x are governed
by the well known *Legendre* differential equation in $P_n(x)$,

$$(1 - x^2)\frac{d^2 P_n}{dx^2} - 2x\frac{dP_n}{dx} + n(n+1)P_n(x) = 0, \quad -1 < x < 1. \quad (E.1)$$

However, $P_n(x)$, as a sequence in the *discrete variable* $n = 2, 3, \cdots$,
is governed by the recurrence relation:

$$nP_n(x) = (2n - 1)xP_{n-1}(x) - (n - 1)P_{n-2}(x), \quad n = 2, 3, \cdots. \quad (E.2)$$

We note that this recurrence relation (E.2), just like differential equations, needs *initial values* of the sequence P_n, namely P_0 and P_1, to
be used in the right hand side of (E.2) in order to generate the next
member of the sequence $P_2(x)$ on the left hand side. Such P_2 and
P_1, then, can be used again in the same recurrence relation (E.2) to
generate P_3 and so on. In this case it is known that $P_0(x) = 1$ and
$P_1(x) = x$, whence (E.2) with n=2 gives

$$2P_2(x) = 3xP_1(x) - P_0(x) = 3x^2 - 1,$$

$$P_2(x) = \frac{3}{2}x^2 - \frac{1}{2}. \quad (E.3)$$

With these $P_2(x)$, $P_1(x)$ and n=3 in (E.2) we can easily find $P_3(x)$,

$$3P_3(x) = 5xP_2(x) - 2P_1(x) = 5x(\frac{3}{2}x^2 - \frac{1}{2}) - 2x = \frac{15}{2}x^3 - \frac{9}{2}x,$$

$$P_3(x) = \frac{1}{2}(5x^3 - 3x). \quad (E.4)$$

This process may be continued to easily generate all needed Legendre
polynomials (see exercises 2).

There are many more very useful polynomials that have their own
recurrence relations (see exercises 3).

Example 1.5 Recurrence Relations(Difference Equations) in the Power Series Solution of Differential Equations

One of our first exposures to a serious recurrence relation in the beginning courses of calculus and differential equations, was in determining the coefficients $\{c_n\}$ in the power series expansion of the solution $y(x)$,

$$
y(x) = \sum_{n=0}^{\infty} c_n (x-a)^n \qquad (E.1)
$$
$$
= c_0 + c_1(x-a) + \cdots + c_n(x-a)^n + \cdots
$$

about a *regular* point $x_0 = a$ of a (homogeneous) differential equation. For example, the differential equation

$$
\frac{d^2 y}{dx^2} + y = 0 \qquad (E.2)
$$

is homogeneous, linear, and with constant coefficients. Hence, for the case of constant coefficients, $x_0 = a = 0$ is a regular point, and the solution $y(x)$ of (E.2) may assume the following power series expansion about the (regular) point $x_0 = 0$,

$$
y(x) = \sum_{n=0}^{\infty} c_n x^n \qquad (E.3)
$$
$$
= c_0 + c_1 x + c_2 x^2 + \cdots + c_n x^n + \cdots
$$

Such power series method of solving the differential equation (E.2) consists of substituting the above power series for $y(x)$ in (E.2), then collect the terms of the same powers of x in the resulting (E.2), and finally equate their coefficients to zero in order to arrive at a recurrence relation for the general term of the coefficient c_n in the power series solution (E.3).

For now, and in order not to lose track of our aim for illustrating various recurrence relations, we give such resulting recurrence relation as,

$$
n(n-1)c_n + c_{n-2} = 0, \qquad (E.4)
$$

$$c_n = -\frac{1}{n(n-1)}c_{n-2}, \quad n \neq 0, 1,$$

and we leave its detailed derivation for the exercises, (see exercise 5). We may observe again that this recurrence relation, just as in the case of a differential equation, needs auxiliary conditions, namely the values of c_0 and c_1, for the complete determination of all the coefficients c_n, $n = 2, 3, \cdots$. In this case we need the initial values c_0, and c_1, whence the recurrence relation (E.4) can generate all the *even* indexed coefficients c_{2n} in terms of c_0, and all the *odd* indexed coefficients c_{2n+1} in terms of c_1,

$$c_2 = -\frac{1}{2}c_0,$$

$$c_4 = -\frac{1}{12}c_2 = \frac{1}{24}c_0 = \frac{1}{4!}c_0,$$

$$c_3 = -\frac{1}{6}c_1 = -\frac{1}{3!}c_1, \qquad (E.5)$$

$$c_5 = -\frac{1}{20}c_3 = \frac{1}{120}c_1 = \frac{1}{5!}c_1,$$

and if we continue this process we can arrive at general even and odd solutions

$$c_{2n} = \frac{(-1)^n}{(2n)!}c_0, \quad n = 1, 2, \cdots, \qquad (E.6)$$

$$c_{2n+1} = \frac{(-1)^n}{(2n+1)!}c_1, \quad n = 1, 2, \cdots, \qquad (E.7)$$

of the recurrence relation (E.4) in terms of the given initial terms c_0, and c_n, respectively. This means that the recurrence relation (E.4) has two solutions c_{2n}, and c_{2n+1}, which remind us of the general solution of second order (homogeneous) differential equations. Indeed, the recurrence relation (E.4) is what we term a "difference equation," which, as we shall see from the classification in Chapter 2, is of second order. Also from the methods of solving linear difference equations discussed in Chapter 3, we can arrive at the above two (linearly independent) solutions of (E.6), (E.7) of the difference equation (E.4) in c_n. For the definition of linear independence of sequences, see Definition 3.1, Examples 3.4, 3.5, and Theorem 3.5 in Chapter 3.

So, we may conclude here that solving a difference equation is a necessary supporting step in the finding of the power series solution (E.3) of the (homogeneous second order) differential equation in (E.2).

Now, if we substitute c_n from (E.6) and (E.7) in (E.3), we have a solution to the differential equation (E.2),

$$
\begin{aligned}
y(x) &= c_0 \sum_{n=0}^{\infty} \frac{(-1)^n x^{2n}}{(2n)!} + c_1 \sum_{n=0}^{\infty} \frac{(-1)^n x^{2n+1}}{(2n+1)!} \\
&= c_0 \cos x + c_1 \sin x,
\end{aligned}
$$

after recognizing the above two series as the Maclaurin series for $\sin x$ and $\cos x$, respectively. This is the expected general solution of (E.4) as the linear combination of its two linearly independent solutions $\cos x$ and $\sin x$, which can be verified very easily.

Example 1.6   Stirred Tank Problems

Sequences often arise in describing a discrete process or a system which evolves in "jumps." For example, consider a tank which contains a brine solution of a known concentration. At regular intervals, one drop of brine of another known concentration is added to the tank while one drop is emitted from the tank. The volume of solution in the tank remains constant, yet at each stage of the process, the concentration of the solution in the tank changes. Letting u_k denote the concentration of the solution after k drops have been added, this process generates the sequence $\{u_0, u_1, u_2, \cdots\}$. In this case the index k plays the role of "time-like" variable. Increasing k corresponds to the system advancing in time. We will return to the modeling of this system in Chapter 7 (Example 7.1, see Fig 7.1).

Example 1.7   Electrical Network

Sequences and difference equations often appear in the description of networks. A familiar example is that of electrical networks in which resistors and voltage sources are connected by segments of wire in some prescribed way. Labeling the wire segments and letting

i_k denote the current passing through segment k, we have a sequence $\{i_k\}$ of unknown currents as shown in Fig. 1.2.

Using a set of algebraic relations (Kirchoff's Laws) relating the currents to the (known) resistances and voltages, it is possible to determine the terms of $\{i_k\}$ as we shall illustrate next for the circuit in Fig. 1.2. In this problem, there is no time evolution of the system. The sequence $\{i_k\}$ gives the *steady state* currents in the network and the index k plays the role of a "space-like" variable. Changing k corresponds to moving spatially through the network.

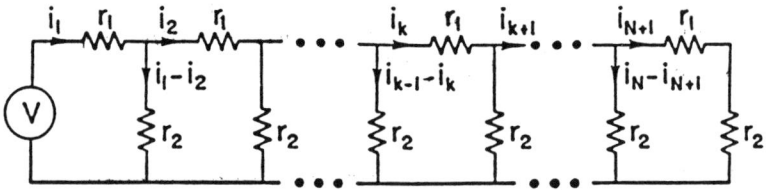

Figure 1.2: An electric network with N Loops

With two constant resistances r_1, and r_2 as shown in Fig. 1.2, the currents in the kth loop can be shown easily, with the help of Kirchoff's law, to satisfy the following recurrence relation

$$i_k r_1 + (i_k - i_{k+1})r_2 - (i_{k-1} - i_k)r_2 = 0, \quad k = 1, 2, \ldots, N+1. \quad (1.2)$$

What remains now is to solve this recurrence relation or difference equation for i_k, which is the subject of Chapter 3 for finding the solution, and Chapter 7 for this and other relevant applications.

Partial Difference Equations

The above sequences $\{c_k\}$ and their recurrence relations, or difference equations are dependent on only one variable, for example x_k in $u(x_k) \equiv c_k$, $u(x_k + \Delta x) \equiv c_{k+1}$, $\cdots$, etc. We may also have dependence on two or more variables. For example, consider the

function $u(x,t)$, where $u(x_k, t_j)$ is a sequence that has two independent variables, and hence we write it as a sequence with two indices
$u(x_k, t_j) \equiv c_{k,j}$. It is easily seen that $u(x_k + \Delta x, t_j) = c_{k+1,j}$ and
$u(x_k + \Delta x, t_j + \Delta t) = c_{k+1,j+1}$. In this case the recurrence relation, or difference equation, is called *partial* difference equation. For
example, the equation,

$$u(x_k + \Delta x, t_j) + 3u(x_k, t_j + \Delta t) = 0 \tag{1.3}$$

can be written as the following partial differential equation in $c_{k,j}$.

$$c_{k+1,j} + 3c_{k,j+1} = 0. \tag{1.4}$$

We will have a simple introduction to linear ordinary difference equations in Chapter 2, their basic methods of solutions in Chapter 3,
where the partial difference equations are covered in Section 3.5.

Having seen how sequences may arise in several fairly realistic
kinds of problems, we will return in the next section to the operators
of the difference calculus. In working with a general sequence $\{u_k\}$ it
will be helpful to remember the many ways in which such a sequence
might arise.

Exercises 1.1

1. The (N+1) sample points $x_k = \frac{2\pi k}{N}$ are placed on the interval
   $[0, 2\pi]$. Find the sequences which are produced when the functions $f(x) = \sin(\frac{Nx}{3})$ and $g(x) = \sin(\frac{2Nx}{3})$ are sampled at these
   points. How do they compare? How do you explain this?

   Hint: Graph the sample points and observe the frequencies of
   the two periodic sequences.

2. (a) For the Legendre polynomials $P_n(x)$ and their recurrence
   relation in (E.2) of Example 1.4, use the first three polynomials $P_0(x), P_1(x)$ and $P_2(x)$ in that example to generate
   the next two members of the sequence, i.e., $P_3(x)$ and
   $P_4(x)$ as

   $$P_3(x) = \frac{1}{2}(5x^3 - 3x)$$

and
$$P_4(x) = \frac{1}{8}(35x^4 - 30x^2 + 3).$$

(b) For $P_0(x) = 1$, $P_1(x) = x$, and the fact that $P_2(0) = -\frac{1}{2}$, use the following recurrence relation to find $P_2(x)$,

$$\frac{dP_{n+1}}{dx} = (2n+1)P_n + \frac{dP_{n-1}}{dx}. \qquad (E.1)$$

Hint: Let n=1 to have $\frac{dP_2}{dx} = 3x$, then use $P_2(0) = -\frac{1}{2}$ to determine the constant of integration.

(c) As in part (b), use the following recurrence relation to show that $P_2(x) = \frac{3}{2}x^2 - \frac{1}{2}$.

$$\frac{dP_{n+1}}{dx} = (n+1)P_n + x\frac{dP_n}{dx} \qquad (E.2)$$

(d) If the recurrence relation in P_n as given in (E.2) of Example 1.4

$$nP_n(x) = (2n-1)xP_{n-1}(x) - (n-1)P_{n-2}(x),$$

$$n = 2, 3, \cdots \qquad (E.3)$$

is called a *difference equation*, as far as the variation in the index n is concerned, what would we call the recurrence relations in (E.1) and (E.2)?

Hint: Note that both recurrence relations involve differencing in n for $P_n(x)$ as a function of n, and also differentiation of $P_n(x)$ as a function of the other continuous variable x.

3. In parallel to the Legendre polynomial $P_n(x)$ and its recurrence relation in Example 1.4 and problem 2, the following Tchebychev Polynomial of the first kind $T_n(x)$ of degree n satisfies the following differential equation in the function $f(x) = T_n(x)$

$$(1 - x^2)\frac{dT_n}{dx} - x\frac{dT_n}{dx} + n^2 T_n(x) = 0, \qquad (E.1)$$

as well as a recurrence relation for T_n as a sequence in n,

$$T_{n+1}(x) = 2xT_n(x) - T_{n-1}(x). \qquad (E.2)$$

Also $T_n(x)$ happens to have the (important) simple expression in terms of a trigonometric function,

$$T_n(x) = \cos(n\arccos x), \quad -1 < x < 1. \qquad (E.3)$$

(a) Use the explicit expression in (E.3) to show that $T_0(x) = 1$ and $T_1(x) = x$, then use the recurrence relation (E.2) to show that $T_2(x) = 2x^2 - 1$ and $T_3(x) = 4x^3 - 3x$.

(b) Use the definition in (E.3) to prove the recurrence relation (E.2).

Hint: Find $T_{n+1}(x)$ and $T_{n-1}(x)$ from (E.3), remembering the simple trigonometric formulas, $\cos(\alpha \mp \beta) = \cos\alpha\cos\beta \pm \sin\alpha\sin\beta$, and that $\cos(\arccos x) \equiv x$.

4. We know that a line divides the plane into two regions, and two lines divide it into four regions. For three lines, if we insist that no more than two lines intersect in the same point, then the plane is divided into 7 regions. Generalize this problem in n such restricted lines to find how many separate regions R_n the plane will be divided into? Show that R_n satisfies the recurrence relation

$$R_{n+1} = n + 1 + R_n.$$

Hint:
$$R_2 = 4 = 2 + 2 = R_1 + 2,$$
$$R_3 = 7 = 4 + 3 = R_2 + 3,$$

noting that the $n + 1st$ line produces $n + 1$ extra regions.

5. Derive the recurrence relation

$$n(n-1)c_n + c_{n-2} = 0, \quad n = 2, 3, \cdots. \qquad (E.1)$$

in (E.4) of Example 1.5.

Hint:

$$y(x) = \sum_{n=0}^{\infty} c_n x^n, \qquad (E.2)$$

$$y'(x) = c_1 + 2c_2 x + 3c_3 x^2 + \cdots + n c_n x^{n-1} + \cdots$$
$$= \sum_{n=1}^{\infty} n c_n x^{n-1}, \qquad (E.3)$$

$$y''(x) = 2c_2 + 6c_3 + 12c_4 x^2 + 20c_5 x^3 + \cdots$$
$$+ n(n-1)c_n x^{n-2} + \cdots \qquad (E.4)$$
$$= \sum_{n=2}^{\infty} n(n-1)c_n x^{n-2}.$$

Then substitute $y(x)$ from (E.1) and $y''(x)$ from (E.4) in

$$\frac{d^2 y}{dx^2} + y = 0, \qquad (E.5)$$

and collect terms with equal powers of x to have

$$(2c_2 + c_0) + (6c_3 + c_1)x + (12c_4 + c_2)x^2 + (20c_5 + c_3)x^3$$
$$+ \cdots + [n(n-1)c_n + c_{n-2}]x^{n-2} + \cdots = 0. \qquad (E.6)$$

where the vanishing of the coefficient of the general power x^{n-2} gives the desired recurrence relation (E.1),

$$n(n-1)c_n + c_{n-2} = 0, \quad n \geq 2. \qquad (E.1)$$

6. Consider the differential equation with variable coefficients,

$$\frac{d^2 y}{dx^2} - x \frac{dy}{dx} + 2y = 0.$$

Since $x = 0$ is a regular point of this equation, we can use a power series expansion about $x_0 = 0$,

$$y(x) = \sum_{n=0}^{\infty} c_n x^n$$

for a solution. Show that the recurrence relation that governs the sequence c_n of the coefficients is

$$(n+2)(n+1)c_{n+2} - (n-2)c_n = 0, \quad n = 2, 3, \cdots.$$

Hint: See the detailed step in the hint of problem 5.

7. Consider the differential equation,

$$\frac{d^2 y}{dx^2} + x\frac{dy}{dx} + 3y = 0.$$

Use the power series expansion,

$$y(x) = \sum_{n=0}^{\infty} c_n x^n$$

for its solution about its regular point $x = 0$, to show that the recurrence relation for its coefficients is,

$$n(n-1)c_n + (n+1)c_{n-2} = 0, \quad n = 2, 3, \cdots.$$

1.2 Difference Operators

The simplest operator that can be applied to a sequence is the *identity operator* I which leaves a sequence unchanged. This may be written

$$Iu_k = u_k, \tag{1.5}$$

where u_k is an arbitrary element of the sequence. Almost as simple is the *(forward) shift operator* E which takes each element of a sequence into its successor. This may be expressed as

$$Eu_k = u_{k+1} \tag{1.6}$$

for all relevant k. We will also encounter the *averaging operator* A which may be defined as

$$Au_k = \frac{1}{2}(u_k + u_{k+1}). \tag{1.7}$$

These three operators applied to the sequence $\{u_k\} = \{k\} = \{1,2,3,4,\cdots\}$ have the following effects:

$$I\{k\} = \{k\}, \quad \text{for } k \geq 1 ,$$

$$E\{k\} = \{k+1\}, \quad \text{for } k \geq 1 ,$$

$$A\{k\} = \{\frac{1}{2}(k + k + 1)\} = \{k + \frac{1}{2}\}, \quad \text{for } k \geq 1 .$$

Of more interest to our present work is the *forward difference operator* Δ defined by

$$\Delta u_k = (E - I)u_k = u_{k+1} - u_k \tag{1.8}$$

In situations in which the sequence $\{u_k\}$ arises from a function u defined on a set of grid points $\{x_k\}$ (as in Examples 1.1 and 1.2 in Section 1.1), the forward difference operator may also be expressed as

$$\Delta u_k = u_{k+1} - u_k = u(x_k + h) - u(x_k), \quad h = \Delta x \tag{1.9}$$

Our first correspondence with the infinitesimal calculus may be made by noting that the *difference quotient*

$$\frac{1}{h}\Delta u_k = \frac{u(x_k + h) - u(x_k)}{h} \tag{1.10}$$

is an *approximation* to $u'(x_k)$, the derivative $\frac{du}{dx}$ at x_k. This approximation improves as h decreases. The difference calculus which will be developed in the coming pages will reflect many familiar properties of the calculus of derivatives.

The second order difference operator is $\Delta^2 = \Delta(\Delta)$, whose action on the sequence u_k is the twice repeated operation of the first order operator Δ, in exact analogy of the second order differential operator $\frac{d^2}{dx^2} = \frac{d}{dx}(\frac{d}{dx})$ when it acts on the continuous function $f(x)$ as $\frac{d^2 f}{dx^2}$,

$$\begin{aligned}
\Delta^2 u_k &= \Delta(\Delta u_k) = \Delta(u_{k+1} - u_k) \\
&= (u_{k+1+1} - u_{k+1}) - (u_{k+1} - u_k), \\
\Delta^2 u_k &= u_{k+2} - 2u_{k+1} + u_k.
\end{aligned} \tag{1.11}$$

Higher order difference operators Δ^n can be defined in the same manner, where it is easily seen that this operation is commutative, i.e.,

$\Delta^m(\Delta^n) = \Delta^n(\Delta^m) = \Delta^{m+n}$ as we shall illustrate in the following example.

Example 1.8 Commutivity of the Difference Operator Δ

We will show first that $\Delta(\Delta^2) = \Delta^3$,

$$
\begin{aligned}
\Delta^3 u_k &= \Delta(\Delta^2 u_k) = \Delta(u_{k+2} - 2u_{k+1} + u_k) \\
&= (u_{k+3} - 2u_{k+2} + u_{k+1}) - (u_{k+2} - 2u_{k+1} + u_k) \\
&= u_{k+3} - 3u_{k+2} + 3u_{k+1} - u_k. \tag{1.12}
\end{aligned}
$$

after using (1.11) for Δ^2 then (1.9) for Δ. Next, we show that $\Delta^3 = \Delta^2(\Delta)$ acting on u_k gives the same answer above,

$$
\begin{aligned}
\Delta^3 u_k = \Delta^2(\Delta u_k) &= \Delta^2(u_{k+1} - u_k) \\
&= u_{k+3} - u_{k+2} - 2(u_{k+2} - u_{k+1}) + (u_{k+1} - u_k) \tag{1.12} \\
&= u_{k+3} - 3u_{k+2} + 3u_{k+1} - u_k.
\end{aligned}
$$

after using (1.9) for Δ then (1.11) for Δ^2. From the above two equal results as (1.12) we conclude that $\Delta^2(\Delta) = \Delta(\Delta^2) = \Delta^3$.

Basic Properties of the Forward Difference Operator Δ

Just like we showed above, the difference operator is commutative, which is the same property of the (linear) differential operator. Many of its other properties have very close parallels. For example, we will show next that the difference operator Δ is *linear*, and we will derive the formulas of the difference of the product and quotient of two sequences. The simple proofs of these results follow an exact parallel to what is done for the derivative of the product and the quotient of two differentiable functions, except that the final result may look a bit different. As expected, it will be easily seen that the results of such differencing formulas do approach that of the differentiation as h (or Δx) in (1.10) approaches zero.

We show here that the difference operator Δ is a linear operator, since

(i) $\Delta(c u_k) = c(u_{k+1}) - (c u_k) = c(u_{k+1} - u_k) = c\Delta u_k$, (1.13) and

(ii) For the two sequences u_k and v_k,

$$
\begin{aligned}
\Delta(u_k + v_k) &= (u_{k+1} + v_{k+1}) - (u_k + v_k) \\
&= (u_{k+1} - u_k) + (v_{k+1} - v_k) \\
&= \Delta u_k + \Delta v_k,
\end{aligned} \tag{1.14}
$$

i.e., the difference of the sum of two sequences u_k and v_k is the sum of the differences Δu_k and Δv_k. If we use (1.13) and (1.14) we can prove the general property of defining a linear operator,

$$
\Delta(b_1 u_k + b_2 v_k) = b_1 \Delta u_k + b_2 \Delta v_k, \tag{1.15}
$$

for b_1 and b_2 as arbitrary constants. The proof is done by simply replacing u_k and v_k on the left side of (1.14) by $b_1 u_k$ and $b_2 v_k$, respectively, then employ the first of the linear properties (1.13) in the last step to arrive at (1.15).

In analogy to the derivative of the product $f(x)g(x)$ and the quotient $f(x)/g(x), g(x) \neq 0$, of two differentiable functions,

$$
\frac{d}{dx}\left(f(x)g(x)\right) = \frac{df}{dx}g(x) + f(x)\frac{dg}{dx}, \tag{1.16}
$$

$$
\frac{d}{dx}\left(\frac{f(x)}{g(x)}\right) = \left(g(x)\frac{df}{dx} - f(x)\frac{dg}{dx}\right)/g^2(x), \quad g(x) \neq 0. \tag{1.17}
$$

we state here their parallels for the difference operator Δ. For the two sequences u_k and v_k, we will show that

$$
(i) \qquad \Delta(u_k v_k) = u_{k+1}\Delta v_k + v_k \Delta u_k, \tag{1.18}
$$

which can also be written in terms of the difference operator Δ (on v_k and u_k only) as

$$
\begin{aligned}
\Delta(u_k v_k) &= u_{k+1}\Delta v_k + v_k \Delta u_k \\
&= (\Delta + I)u_k \Delta v_k + v_k \Delta u_k \\
&= (\Delta u_k + u_k)\Delta v_k + v_k \Delta u_k \\
&= \Delta u_k \Delta v_k + u_k \Delta v_k + v_k \Delta u_k
\end{aligned}
$$

$$(ii) \qquad \Delta(\frac{u_k}{v_k}) = \frac{v_k \Delta u_k - u_k \Delta v_k}{v_k v_{k+1}} \qquad (1.19)$$

The proofs of (1.18) and (1.19) are simple, since for (1.18) we use the definition of Δ in (1.14) for $\Delta(u_k v_k)$ to have

$$\Delta(u_k v_k) = u_{k+1} v_{k+1} - u_k v_k,$$

and as was done in the case of differentiation, we subtract and add the term $u_{k+1} v_k$, to obtain,

$$\begin{aligned}
\Delta(u_k v_k) &= (u_{k+1} v_{k+1} - u_{k+1} v_k) + (u_{k+1} v_k - u_k v_k) \\
&= u_{k+1}(v_{k+1} - v_k) + (u_{k+1} - u_k) v_k \qquad (1.18) \\
&= u_{k+1} \Delta v_k + v_k \Delta u_k
\end{aligned}$$

The same method is followed for (1.19), where the term $u_k v_k$ needs to be subtracted and added in the numerator of the expression resulting from applying the definition of Δ in $\Delta \frac{u_k}{v_k}$, i.e.,

$$\begin{aligned}
\Delta(\frac{u_k}{v_k}) &= \frac{u_{k+1}}{v_{k+1}} - \frac{u_k}{v_k} = \frac{u_{k+1} v_k - v_{k+1} u_k}{v_k v_{k+1}} \\
&= \frac{(u_{k+1} v_k - u_k v_k) + (u_k v_k - v_{k+1} u_k)}{v_k v_{k+1}} \\
&= \frac{v_k(u_{k+1} - u_k) - u_k(v_{k+1} - v_k)}{v_k v_{k+1}} \qquad (1.19) \\
&= \frac{v_k \Delta u_k - u_k \Delta v_k}{v_k v_{k+1}}
\end{aligned}$$

The nth order differencing of the product of two sequences $\Delta^n(u_k v_k)$ comes out parallel to that of the following *Leibnitz Rule* for $\frac{d^n}{dx^n}(f(x)g(x))$ in differential calculus,

$$\frac{d^n}{dx^n}[f(x)g(x)] \equiv D^n[f(x)g(x)],$$

$$\frac{d^n}{dx^n}[f(x)g(x)] = D^n[f(x)g(x)]$$

$$= (f)(D^n g) + \binom{n}{1}(Df)(D^{n-1}g) +$$

$$\binom{n}{2}(D^2 f)(D^{n-2}g) + \cdots + \binom{n}{n}(D^n f)(g)$$

$$= \sum_{j=0}^{n} \binom{n}{j}(D^j f)(D^{n-j}g) \tag{1.20}$$

where

$$\binom{n}{j} \equiv \frac{n!}{(n-j)!j!}, \quad 0! \equiv 1. \tag{1.21}$$

The analogous Leibnitz formula for the difference operator $\Delta^n(u_k v_k)$ is

$$\Delta^n(u_k v_k) = u_k \Delta^n(v_k) + \binom{n}{1}\Delta(u_k)\Delta^{n-1}(v_{k+1})$$

$$+ \binom{n}{2}(\Delta^2 u_k)(\Delta^{n-2}v_{k+2}) + \cdots$$

$$+ \binom{n}{n}(\Delta^n u_k)(v_{k+n})$$

$$= \sum_{j=0}^{n} \binom{n}{j}\Delta^j(u_k)\Delta^{n-j}(v_{k+j}) \tag{1.22}$$

We can easily see that (1.18) is a special case of this Leibnitz formula for $n = 1$,

$$\Delta(u_k v_k) = \binom{1}{0}\Delta^0(u_k)\Delta v_k + \binom{1}{1}\Delta u_k \Delta^0 v_{k+1}$$

$$= \frac{1!}{1!0!}u_k \Delta v_k + \frac{1!}{0!1!}v_{k+1}\Delta u_k$$

$$= u_k \Delta v_k + v_{k+1}\Delta u_k \tag{1.23}$$

after using $0! \equiv 1$, and $\Delta^0 u_k = I u_k = u_k$, and noting that (1.23) is the same as (1.18), since u_k and v_k can be exchanged in (1.22), and hence (1.23) to produce (1.18).

If we look closely at (1.18) for the difference of the product of two sequences, we notice that it is not very symmetric like the one we had for the derivative of the product of two functions in (1.16)

$$\frac{d}{dx}\left(f(x)g(x)\right) = \frac{df}{dx}g(x) + f(x)\frac{dg}{dx}. \tag{1.16}$$

Also, our attempt in (1.18a) for writing the result (1.18) in terms of the Δ operator (on u_k and v_k only) did not come out in a simple form. On the other hand, if the *(forward) shift operator* E of (1.6) is used, its formulas are very simple,

$$E(u_k v_k) = u_{k+1}v_{k+1} = Eu_k Ev_k. \tag{1.24}$$

It is easy to see that

$$
\begin{aligned}
E(u_k^n) &= E(\overbrace{u_k \cdot u_k \cdots u_k}^{n}) \\
&= \overbrace{Eu_k \cdot Eu_k \cdots Eu_k}^{n} \\
&= (Eu_k)^n \\
&= (u_{k+1})^n \tag{1.25}
\end{aligned}
$$

Also,

$$E^n(u_k v_k) = u_{k+n}v_{k+n} = E^n u_k E^n v_k \tag{1.26}$$

which is simpler than the Leibnitz formula (1.22) of the difference operator Δ.

Example 1.9 Differencing the Sequence $\{u_k\} = \{a^k\}$

Let us consider the sequence $u_k = a^k$, which is the discrete version of $u(x) = a^x$, whose derivative is

$$\frac{d}{dx}a^x = a^x \ln a. \tag{1.27}$$

The difference Δa^k of a^k is

$$\Delta a^k = a^{k+1} - a^k = a^k(a-1), \tag{1.28}$$

which parallels the result of differentiating $u(x) = a^x$ in (1.27). Now we may check $\Delta(a^k b^k)$, where according to the product formula (1.18), we have,

$$
\begin{aligned}
\Delta(a^k b^k) &= a^{k+1}\Delta b^k + b^k \Delta a^k \\
&= a^{k+1}b^k(b-1) + b^k a^k(a-1) \\
&= a^k b^k \left[a(b-1) + a - 1 \right] \\
&= a^k b^k (ab-1) \qquad\qquad (1.29)
\end{aligned}
$$

after using (1.28) in the second line. On the other hand, we can also use (1.28) to have the same result (1.29) in a faster way.

$$
\begin{aligned}
\Delta(a^k b^k) &= \Delta \left[(ab)^k \right] \\
&= (ab)^k (ab-1) \qquad\qquad (1.29) \\
&= a^k b^k (ab-1).
\end{aligned}
$$

As we shall see later, the sequence a^k will play an important role in difference calculus, and obviously, in the solution of the homogeneous constant coefficient difference equations since, as seen in (1.28), it is the sequence that springs back under differencing. This is analogous to the role played by $y = e^{mx} = (e^m)^x = a^x$, $a \equiv e^m$ in the solution of homogeneous constant coefficient differential equations.

The Backward and Central Difference Operators ∇, δ

It is also possible to define a *backward* difference operator as follows:

$$
\nabla u_k = u_k - u_{k-1} \quad \text{or} \quad \nabla u_k = u(x_k) - u(x_k - h). \qquad (1.30)
$$

Again we may note that the quotient

$$
\frac{1}{h}\nabla u_k = \frac{u(x_k) - u(x_k - h)}{h} \qquad\qquad (1.31)
$$

is an *approximation* to $u'(x_k)$ which also improves as h becomes small. The asymmetry of the forward and backward difference operators Δ and ∇ may be eliminated by defining a *central difference operator* δ,

$$
\delta u_k = u_{k+1/2} - u_{k-1/2}
$$

or

$$\delta u_k = u(x_k + \frac{h}{2}) - u(x_k - \frac{h}{2}).$$ (1.32)

The quotient

$$\frac{1}{h}\delta u_k = \frac{u(x_k + \frac{h}{2}) - u(x_k - \frac{h}{2})}{h}$$ (1.33)

is yet another *approximation* which converges to $u'(x_k)$ as $h \to 0$.

It is now possible to apply combinations of the above operators in succession to generate higher order differences. For example, the *second order* difference $\delta^2 u$ is given by:

$$\begin{aligned}
\delta^2 u_k &= \delta(\delta u_k) \\
&= \delta(u_{k+1/2} - u_{k-1/2}) \\
&= (u_{k+1} - u_k) - (u_k - u_{k-1}) \\
&= u_{k+1} - 2u_k + u_{k-1},
\end{aligned}$$ (1.34)

or if we wish to show the independent variable x_k

$$\delta^2 u_k = u(x_k + h) - 2u(x_k) + u(x_k - h).$$ (1.35)

In the latter case where $\{u_k\}$ is derived from a function $u(x)$, it is possible to show that the quotient $\frac{1}{h^2}\delta^2 u_k$ is an approximation to $u''(x_k)$ which improves as h becomes smaller, and we shall leave as an exercise.

There are countless ways in which to construct such higher order differences and there is a multitude of identities which relate these differences. This topic is usually covered in classical numerical analysis texts, but is not central to the ideas which will be developed in the coming pages.

It will be useful to become familiar with sequences and the effect of difference operators through some examples.

Example 1.10   Differencing (a simple power) Sequence $\{u_k\} = \{k\}$

Returning to the linear sequence $\{u_k\} = \{k\} = \{1, 2, 3, 4, \cdots\}$, and applying the above difference operators, (in (1.9), (1.30), (1.32), (1.34)), we have:

$$\Delta(k) = (k + 1) - k = 1,$$ (1.36)

$$\nabla(k) = k - (k-1) = 1, \tag{1.37}$$

$$\delta(k) = (k + \frac{1}{2}) - (k - \frac{1}{2}) = 1, \tag{1.38}$$

$$\delta^2(k) = k + 1 - 2k + k - 1 = 0. \tag{1.39}$$

Compare these results with the effect of applying the first and second derivative operator to the linear function $f(x) = x$.

────────────

The following example illustrates the differencing of the trigonometric sequence $\{u_k\} = \{\sin \alpha k\}$.

Example 1.11   Differencing a Trigonometric Sequence $\{u_k\} = \{\sin \alpha k\}$

$$
\begin{aligned}
\Delta(\sin \alpha k) &= \sin(\alpha(k+1)) - \sin(\alpha k) \\
&= \sin(\alpha k)\cos \alpha + \cos(\alpha k)\sin \alpha - \sin(\alpha k) \\
&= \sin(\alpha k)[\cos \alpha - 1] + \cos(\alpha k)\sin \alpha
\end{aligned}
$$

after using the sine addition law. We can also write

$$
\begin{aligned}
\Delta(\sin(\alpha k)) &= \sin(\alpha(k+1)) - \sin(\alpha k) \\
&= 2\cos\left(\alpha(k + \frac{1}{2})\right)\sin(\frac{\alpha}{2}). \tag{1.40}
\end{aligned}
$$

after using the identity,

$$\sin A - \sin B = 2\cos\left(\frac{A+B}{2}\right)\sin\left(\frac{A-B}{2}\right).$$

────────────

The Factorial Polynomial Sequence $\{k^{(m)}\}$

While the results of these difference operators are simple for the sequence $\{u_k\} = \{k\}$, they are much more complicated when it comes to the *power sequence* $\{u_k\} = \{k^m\}$. This is an extreme contrast with the simple result of the derivative of the power function $f(x) = x^m$,

$$\frac{dx^m}{dx} = mx^{m-1}, \tag{1.41}$$

To show this rather involved result for Δk^m, we have

$$\Delta k^m = (k+1)^m - k^m, \tag{1.42}$$

where, obviously, we need the following *binomial expansion* to use for $(k+1)^m$,

$$
\begin{aligned}
(a+b)^m &= a^m + ma^{m-1}b + \binom{m}{2} a^{m-2}b^2 + \cdots \\
&\quad + \binom{m}{m-1} a^{m-1}b^1 + b^m \\
&= \sum_{j=0}^{n} \binom{m}{j} a^{m-j}b^j.
\end{aligned} \tag{1.43}
$$

with

$$\binom{m}{j} \equiv \frac{m!}{(m-j)!j!}. \tag{1.44}$$

If we use (1.43) for the expansion of $(k+1)^m$ in (1.42), we have

$$
\begin{aligned}
\Delta k^m &= (k+1)^m - k^m \\
&= \left[k^m + mk^{m-1} + \binom{m}{2} k^{m-2} + \cdots \right. \\
&\quad \left. + \binom{m}{m-1} k + k + 1 \right] - k^m \\
&= mk^{m-1} + \binom{m}{2} k^{m-2} + \cdots \\
&\quad + \binom{m}{m-1} k + k + 1,
\end{aligned} \tag{1.45}
$$

a rather *cumbersome* formula compared to its parallel of $\frac{dx^m}{dx} = mx^{m-1}$ in (1.41). This difficulty may be stated in another way, namely, that this power sequence $\{u_k\} = \{k^m\}$ is *not compatible* with the difference operator Δ. So it is difficult to work with, and we must search for another sequence that may play a more natural role with Δ in much the same way x^m does for the differentiation operator $D \equiv \frac{d}{dx}$, and thus have a much simpler formula than the one in (1.45). It turns out that the following *factorial type polynomial*

$$
\begin{aligned}
u_k &= k^{(m)} = k(k-1)(k-2)\cdots(k-(m-1)) \\
&= \frac{k!}{(k-m)!} \equiv k^{(m)} \text{ for } m \geq 0, \ k^{(0)} = 1,
\end{aligned}
\tag{1.46}
$$

does exactly that, where

$$
\Delta k^{(m)} = mk^{(m-1)}, \quad m \geq 0.
\tag{1.47}
$$

In (1.46) we note how the symbol $k^{(m)}$ is used to indicate the parallel role this sequence plays to the power function x^m. $k^{(m)}$ is read "k, m factorial." We shall leave the derivation of (1.47), for an exercise (see exercise 11), and be satisfied, for now, with a few simple illustrations. First we see that $k^{(k)} = k!$,

$$
\begin{aligned}
k^{(k)} &= k(k-1)(k-2)\cdots(k-k+1) \\
&= k(k-1)(k-2)\cdots 1 = k!.
\end{aligned}
\tag{1.48}
$$

or from (1.46),

$$
k^{(k)} = \frac{k!}{0!} = \frac{k!}{1} = k!,
$$

since $0! = 1$. Also,

$$
k^{(1)} = \frac{k!}{(k-1)!} = k,
$$

and,

$$
k^{(m)} = 0 \quad \text{for } m > k.
\tag{1.49}
$$

since $(k-m)!$ in the denominator of (1.46) is *unbounded* for negative $k - m$ when $m > k$.

Example 1.12 Differencing the Factorial Polynomial Sequence $\{u_k\} = \{k^{(m)}\}$

For a few special cases, we will use the definition of the difference operator as well as the formula (1.47) for the difference of the factorial polynomial. For $u_k = k(k-1) = k^{(2)}$, we have the direct use of the definition of Δ in (1.9).

$$
\begin{aligned}
\Delta u_k = \Delta k^{(2)} &= \Delta \left(k(k-1) \right) = (k+1)k - k(k-1) \\
&= k^2 + k - k^2 + k = 2k,
\end{aligned}
\tag{E.1}
$$

while the formula in (1.47) gives $\Delta k^{(2)} = 2k^{(1)} = 2k$.
For $\Delta k^{(3)}$ we have

$$
\begin{aligned}
\Delta k^{(3)} &= \Delta\left[k(k-1)(k-2)\right] \\
&= (k+1)k(k-1) - k(k-1)(k-2) \\
&= 3k^2 - 3k = 3k(k-1)
\end{aligned} \tag{E.2}
$$

and (1.47) gives the same result

$$
\Delta k^{(3)} = 3\Delta k^{(2)} = 3k(k-1). \tag{E.2a}
$$

The important result (1.47) for the difference of the factorial polynomials $k^{(m)}$ can be extended for *negative* m as

$$
\begin{aligned}
k^{(-m)} &\equiv \frac{1}{(k+m)^{(m)}} \\
&\equiv \frac{1}{(k+m)(k+m-1)\cdots(k+1)}, \quad \text{for } m > 0.
\end{aligned} \tag{1.50}
$$

and we can show (see exercise 12) that

$$
\Delta k^{(-m)} = -mk^{(-m-1)}. \tag{1.51}
$$

Therefore, in general, we have that

$$
\Delta k^{(m)} = mk^{(m-1)}. \tag{1.52}
$$

While $k^{(m)}$ is a special polynomial of degree m, it is nevertheless a polynomial, and we expect that any polynomial sequence of degree m may be written in terms of a finite number m of such factorial polynomials as we shall illustrate in the next example.

Example 1.13 Polynomial Sequences Expressed in Terms of the Factorial Polynomials

We can verify easily that the familiar simple polynomial sequence k^3 of degree 3 can be expressed in terms of the factorial polynomials $k^{(3)}$, $k^{(2)}$ and $k^{(1)}$ as

$$
k^3 = k^{(3)} + 3k^{(2)} + k^{(1)}, \tag{E.1}
$$

when we substitute for the factorial polynomials on the right hand side, using the definition of $k^{(m)}$ in (1.46) for positive m,

$$
\begin{aligned}
k^{(3)} + 3k^{(2)} + k^{(1)} &= k(k-1)(k-2) + 3k(k-1) + k \\
&= k^3 - 3k^2 + 2k + 3k^2 - 3k + k \qquad\qquad (E.2) \\
&= k^3 .
\end{aligned}
$$

Exercises 1.2

1. The function $f(x) = \sin x$ is sampled at intervals of h. Let x be any of the sample points.

   (a) Find $\Delta(\sin x)$ in terms of x and h.

   Hint: See Example 1.11 and the trigonometric identity used for (1.40),

   $$
   \sin A - \sin B = 2 \cos \left(\frac{A+B}{2} \right) \sin \left(\frac{A-B}{2} \right).
   $$

   (b) Evaluate

   $$
   \lim_{h \to 0} \Delta \left(\frac{\sin x}{h} \right),
   $$

   using

   $$
   \lim_{\theta \to 0} \frac{\sin \theta}{\theta} = 1.
   $$

   (c) Find $\delta^2(\sin x)$ in terms of x and h.

   Hint: See (1.35).

   (d) Evaluate

   $$
   \lim_{h \to 0} \delta^2 \left(\frac{\sin x}{h^2} \right),
   $$

   using

   $$
   \lim_{h \to 0} 4 \frac{1 - \cos h}{h^2} = 1.
   $$

   Hint: See part (c).

2. Verify the following identities

    (a) $E\Delta = \Delta E$

    (b) $\Delta^2 = (E-1)^2 = E^2 - 2E + 1$

    (c) $\delta^2 = \Delta \nabla$

3. Evaluate Δu_k when

    (a) $u_k = c,\;\; c$ is a real number

    (b) $u_k = k^{(4)}$

        Hint: See (1.47).

    (c) $u_k = k^{(-2)}$

        Hint: See (1.51) and (1.50).

    (d) $u_k = k^2 + 3k + 1$

        Hint: See the definition of Δ in (1.9).

    (e) $u_k = \sin \frac{\pi k}{4}$

        Hint: See Example 1.11.

    (f) $u_k = \cos \alpha k, c$ is a real number

        Hint: You may need the trigonometric identity

$$\cos C - \cos D = 2 \sin \left(\frac{D+C}{2} \right) \sin \left(\frac{D-C}{2} \right).$$

    (g) $u_k = \ln k$

    (h) $u_k = (-1)^k$

        Hint: See (1.9).

    (i) $u_k = (-1)^k k$

        Hint: See (1.9).

4. (a) Show that the factorial polynomial $k^{(m)}$ of (1.46) can be written in terms of the familiar factorials as

$$k^{(m)} = \frac{k!}{(k-m)!} \qquad (E.1)$$

    (b) Show that

i. $k^{(k)} = k!$.

ii. $k^{(n)} = 0$, for $n > k$.

Hint: for (a), note that $0! \equiv 1$, and for (b) that $(k - n)!$ is infinite for negative integer $k - n$ which is the case when $n > k$.

5. We understand the word "factorial" for $k^{(m)}$, however the other word "polynomial," in calling it "factorial polynomial" may appear more clear if we generalize $k^{(m)}$ to $x^{(m)}$ with $\Delta x = h$ instead of $\Delta k = 1$ in $k^{(m)}$,

$$x^{(m)} = x(x-h)(x-2h)\cdots(x-mh+h), \quad m = 1, 2, 3, \cdots \quad (E.1)$$

(a) Generate $x^{(1)}$, $x^{(2)}$ and $x^{(3)}$ as polynomials in x of degree 1, 2 and 3, respectively. Note $x^{(0)} = 1$.

(b) To illustrate that the polynomial $P_n(x)$ can be expressed in terms of the above factorial polynomial $x^{(n)}$, show that

$$x^3 = x^{(3)} + 3x^{(2)}h + x^{(1)}h^2. \qquad (E.2)$$

Hint: start with the right hand side, and substitute for $x^{(m)}$ as defined in (E.1) and part (a).

(c) With $\Delta f(x) = f(x + h) - f(x)$, use the result in part (a) to show that $\Delta x^{(3)} = 3x^{(2)}h$, then attempt to prove the result,

$$\Delta x^{(m)} = m x^{(m-1)}h, \qquad (E.3)$$

which is a generalization of $\Delta k^{(m)} = mk^{(m-1)}$ in (1.47) for $\Delta x = h$ instead of $\Delta k = 1$.

(d) In parallel to what was done in Example 1.11 of the special case

$$\Delta(\sin \alpha k) = 2 \cos\left[\alpha(k + \frac{1}{2})\right] \sin \frac{\alpha}{2},$$

Show that

i. $\Delta(\sin \alpha x) = 2 \cos\left[\alpha(x + \frac{h}{2})\right] \sin \frac{\alpha h}{2}$

ii. $\Delta(\cos \alpha x) = -2 \sin\left[\alpha(x + \frac{h}{2})\right] \sin \frac{\alpha h}{2}$

(e) Prove that

    i. $\Delta(\ln x) = \ln\left(1 + \frac{h}{x}\right)$

    ii. $\Delta(b^x) = b^x(b^h - 1)$

    iii. $\Delta(e^{\alpha x}) = e^{\alpha x}(e^{\alpha h} - 1)$

    iv. $\Delta\left[(px + q)^{(m)}\right] = mph(px + q)^{(m-1)}$

(f) Attempt to write the generalization of the factorial polynomial $x^{(m)}$ of x to that which corresponds to any function $f(x)$, i.e., write the expression for $[f(x)]^{(m)}$.

6. (a) With Δ as in the problem 5, find

    i. $\Delta(2x^2 + 3x)$

    ii. $(\Delta + 1)(2\Delta - 1)(x^2 + 2x + 1)$

    iii. $\Delta^2(x^3 - x^2)$

  (b) With $E = \Delta + 1$, find

    i. $E(4x - x^2)$

    ii. $E^3(3x - 2)$

    iii. $(E - 2)(E - 1)(2^{\frac{x}{h}} + x)$

7. With Δ as in problem 5(c) prove the linearity properties of Δ,

  (a) $\Delta[f(x) + g(x)] = \Delta f(x) + \Delta g(x)$

  (b) $\Delta[\alpha g(x)] = \alpha \Delta[g(x)]$

8. Evaluate

  (a) $\Delta(7x^{(4)})$

  (b) $\Delta(9x^{(-2)})$

  (c) $\Delta(3x^{(-2)} - 2x^{(2)} - 5x^{(-1)})$

9. As a generalization to the (forward) shift operator E, $Eu_k = u_{k+1}$ in (1.6), we define

$$Ef(x) = f(x + h),$$

$$E^{\frac{1}{2}}f(x) = f\left(x + \frac{h}{2}\right),$$

$$E^{-1}f(x) = f(x - h),$$

and

$$E^{-\frac{1}{2}}f(x) = f(x - \frac{h}{2}).$$

Also, we generalize the central and the backward difference operators δ and ∇, where

$$\delta f(x) = f(x + \frac{h}{2}) - f(x - \frac{h}{2}),$$

and

$$\nabla f(x) = f(x) - f(x - h).$$

   (a) Show that

      i. $\delta = E^{\frac{1}{2}} - E^{-\frac{1}{2}}$

      ii. $\nabla = 1 - E^{-1}$

      iii. $\delta = \nabla E^{\frac{1}{2}}$

   (b) Find

      i. $\Delta(x^2 + 2x)$

      ii. $\delta((x^2 + 2x)$

10. Consider the sequence

$$u_k = P_k = a_0 k^3 + a_1 k^2 + a_2 k + a_3 \qquad (E.1)$$

as a polynomial of degree 3 in k

   (a) Show that

$$\Delta^3 P_k = a_0 3!. \qquad (E.2)$$

Hint:

$$
\begin{aligned}
\Delta P_k &= a_0(k+1)^3 + a_1(k+1)^2 + a_2(k+1) + a_3 \\
&\quad - [a_0 k^3 + a_1 k^2 + a_2 k + a_3] \\
&= 3a_0 k^2 + \text{terms of degree less than 2,}
\end{aligned}
$$

$$\Delta^2 P_k = \Delta(\Delta P_k) = 6a_0 k + \text{terms of degree less than 1}$$

$$\Delta^3 P_k = \Delta(\Delta^2 P_k) = 6a_0 = a_0 \cdot 3!.$$

(b) Use the method employed in part (a) to prove the general result for P_k as a polynomial of degree n in k, $P_k \equiv P_n(k)$,

$$P_k \equiv P_n(k) = a_0 k^n + a_1 k^{n-1} + \cdots + a_{n-1} k + a_n, \quad (E.3)$$

i.e., show that

$$\Delta^n P_k = a_0 n!. \quad (E.4)$$

11. Prove the result that

$$\Delta(k^{(m)}) = m k^{(m-1)} \text{ for } m > 0,$$

where the factorial polynomial $k^{(m)}$ is given by (1.46) as

$$k^{(m)} = k(k-1)(k-2) \cdots (k-m+1), \quad \text{and } k^{(0)} = 1.$$

Hint: Write $\Delta k^{(m)} = (k+1)^{(m)} - k^{(m)}$, and use the definition (1.46) for both terms; then simplify.

12. Extend the result of problem 11 to negative integers by proving that

$$\Delta k^{(-m)} = -m k^{(-m-1)} \text{ for } m > 0$$

where

$$k^{(-m)} = \frac{1}{(k+m)^{(m)}} = \frac{1}{(k+m)(k+m-1) \cdots (k+1)},$$

as given in (1.50), hence conclude that

$$\Delta k^{(m)} = m k^{(m-1)} \text{ for all integer } m.$$

Hint: Write

$$\Delta k^{(-m)} = \frac{1}{(k+m+1)^{(m)}} - \frac{1}{(k+m)^{(m)}},$$

and use the definition (1.50) for both terms, then simplify.

13. (a) Express $\Delta^r k^n$ in terms of the E operator to show that

$$\Delta^r k^n = \{\sum_{j=0}^{r} \binom{r}{j} E^{r-j}\}k^n$$

$$= (E^r k^n) - \binom{r}{1} (E^{r-1}k^n) + \cdots + (-1)^r k^n.$$

(E.1)

Hint: Write $\Delta^r = (E-1)^r$, then use the binomial expansion (1.43) for $(E-1)^r$ and the definition of $E^j u_k = u_{k+j}$.

(b) Use the result in part (a) to show that

$$\sum_{j=0}^{r} \binom{r}{j} (k+r-j)^n = \begin{cases} 0, & 0, \text{ for } r > n \\ n!, & \text{ for } r = n \end{cases}$$

Hint: For $r = n$ the first term in (E.1) is $(k+r)^n$, noting that $\Delta^n k^n = \Delta^n(k+r)^n = n!$, since n is a constant and that $\Delta^{n+j}k^n = 0$ for $j \geq 1$.

14. (a) Show that $k^{(n)}(k-n)^{(j)} = k^{(n+j)}$, where $k^{(0)} = 1$.

(b) Given the gamma function $\Gamma(x)$, $x > 0$ with its important property

$$x\Gamma(x) = \Gamma(x+1),$$ (E.1)

where x is a positive number, which is not restricted to integers. Attempt to give a general definition of the factorial polynomial $k^{(n)}$ to that of $x^{(n)}$, and show that

$$x^{(n)} = \frac{\Gamma(x+1)}{\Gamma(x+1-n)}, \quad x+1-n > 0.$$

Hint: Note that $k\Gamma(k) = \Gamma(k+1)$ makes $\Gamma(k+1) = k!$

(c) Write $k^{(4)}$ the factorial polynomial of degree 4 explicitly.

(d) Write $k^{(3)}$, $k^{(2)}$ and $k^{(1)}$ explicitly as polynomials, then attempt to express k^3 in terms of them, i.e., as a linear combination of them.

15. Show that

(a) $\Delta(a+bx)^{(n)} = bn(a+bx)^{(n-1)}$

(b) $\Delta^2(a+bx)^{(n)} = bn(n-1)(a+bx)^{(n-2)}$.

1.3 The Inverse Difference Operator as a Sum Operator

We shall briefly discuss Δ^{-1}, the *inverse* of the difference operator Δ,

$$(\Delta^{-1}\Delta)(u_k) = u_k, \qquad (1.53)$$

and which, as may be expected, will amount to *summation* as opposed to the differencing of Δ. Let $\Delta^{-1}u_k = y_k$, we shall show that y_k, aside from an arbitrary constant y_1, is a finite summation of the original sequence u_k as we will develop next in (1.55). First, we have

$$\Delta(\Delta^{-1}u_k) = \Delta(y_k) = y_{k+1} - y_k.$$

Now, if we write this equation for $u_1, u_2, u_3, \cdots, u_k$,

$$\Delta(\Delta^{-1}u_1) = y_2 - y_1$$

$$\Delta(\Delta^{-1}u_2) = y_3 - y_2$$

$$\Delta(\Delta^{-1}u_3) = y_4 - y_3$$

$$\Delta(\Delta^{-1}u_4) = y_5 - y_4$$

$$\vdots$$

$$\Delta(\Delta^{-1}u_{k-1}) = y_k - y_{k-1}$$

$$\Delta(\Delta^{-1}u_k) = y_{k+1} - y_k$$

then add, the left hand sides will add up to $u_1 + u_2 + \cdots + u_k$ by using the definition $\Delta(\Delta^{-1}u_j) = u_j$, $j = 1, 2, 3, \cdots, k$ for each of the above k equations. Adding the right hand side, with the telescoping of y_2 to y_k, yields $y_{k+1} - y_1$,

$$u_1 + u_2 + \cdots + u_k = y_{k+1} - y_1. \qquad (1.54)$$

To have $y_k = \Delta^{-1}u_k$, we use k instead of $k+1$ in (1.54) above to have

$$u_1 + u_2 + \cdots + u_{k-1} = y_k - y_1,$$

$$y_k = \Delta^{-1}u_k = u_1 + u_2 + \cdots + u_{k-1} + y_1,$$

$$y_k = \Delta^{-1}u_k = y_1 + \sum_{j=1}^{k-1} u_j \qquad (1.55)$$

as the desired finite *summation* operation of Δ^{-1}. With this result we can interpret the action of the inverse difference operator Δ^{-1} on the sequence u_k, namely $y_k = \Delta^{-1}u_k$, as the $k-1$ *partial sum* of the sequence u_j plus an *arbitrary constant* y_1. The latter may remind us of the spirit of the indefinite integration of function of continuous variable $f(x)$, along with its arbitrary constant of integration

$$\int f(x)dx = F(x) + C$$

where $F(x)$ is the antiderivative of $f(x)$, i.e. $\frac{dF}{dx} = f(x)$. Often $\Delta^{-1}u_k$ is called the *antidifference* or *sum* operator, and for obvious reasons!

With the inverse operator Δ^{-1} being defined in (1.55) and interpreted as a finite sum, it should be very easy to show that it is also a linear operator, just like what we showed for Δ, i.e., for the two sequences u_k and v_k, and c_1 and c_2 constants we have

$$\Delta^{-1}(c_1u_k + c_2v_k) = c_1\Delta^{-1}u_k + c_2\Delta^{-1}v_k \qquad (1.56)$$

In the following example we will derive the following two simple results, and that of the *geometric series* (E.7) of part (c).

$$\Delta^{-1}(1) = k, \qquad (1.57)$$

$$\Delta^{-1}(a^k) = \frac{1}{a-1}a^k, \quad a \neq 1 \qquad (1.58)$$

which are among many other useful summation formulas of the inverse difference (or sum) operator Δ^{-1}, which are listed in (1.59)-(1.69).

Example 1.14   Antidifferencing Sequences with Δ^{-1}

(a). It is easy to illustrate that the antidifference operation on a constant sequence $u_k = 1$ is $y_k = k$ if we assume the (special consistent) condition that the arbitrary constant y_1 in (1.55) is 1,

$$\Delta^{-1}(1) = y_1 + \sum_{j=1}^{k-1} 1 = \overbrace{1 + 1 + \cdots + 1}^{k-1} \qquad (E.1)$$

$$= 1 + k - 1 = k$$

where y_1 is given as 1, and that the above $k - 1$ partial sum of constants 1 adds up to $k - 1$.

(b). Also, we had

$$\Delta(a^k) = (a - 1)a^k \qquad (E.2)$$

in (1.28), now we can very easily show that

$$\Delta^{-1}(a^k) = \frac{a^k}{a - 1}, \quad a \neq 1 \qquad (E.3)$$

a formula that is the parallel to what we get for the integration of a^x,

$$\int a^x dx = \frac{a^x}{\ln a}, \quad a \neq 1. \qquad (E.4)$$

From (E.2) we have

$$a^k = \frac{1}{a - 1} \Delta a^k, \quad a \neq 1 \qquad (E.5)$$

and if we apply the antidifference operator Δ^{-1} on both sides of (E.5), we have

$$\Delta^{-1} a^k = \Delta^{-1}\left(\frac{1}{a - 1} \Delta a^k\right) = \frac{1}{a - 1} \Delta^{-1}(\Delta a^k)$$

$$= \frac{1}{a - 1} a^k, \quad a \neq 1. \qquad (E.6)$$

after using the linearity property of Δ^{-1} for taking the constant factor $\frac{1}{a-1}$ outside the operation of Δ^{-1} in the first step, then using the definition $(\Delta^{-1}\Delta)(a^k) = Ia^k = a^k$ in the following step.

(c). Now we will use the definition of the antidifference operator Δ^{-1} as the sum in (1.55)

$$y_k = \Delta^{-1} u_k = y_1 + \sum_{j=1}^{k-1} u_j, \qquad (1.55)$$

to derive the formula for the sum of the *geometric series*

$$\sum_{j=0}^{k} a^j = \frac{a^{k+1} - 1}{a - 1}, \quad a \neq 1. \qquad (E.7)$$

From (1.55) and (E.6) we have

$$\sum_{j=1}^{k-1} a^j = y_k = \frac{a^k}{a - 1} + C. \qquad (E.8)$$

The constant C can be determined from letting $k = 2$ on the left hand side of (E.8) to have,

$$a = \frac{a^2}{a - 1} + C, \quad C = a - \frac{a^2}{a - 1} = -\frac{a}{a - 1}. \qquad (E.9)$$

With this value of C, (E.8) becomes

$$\sum_{j=0}^{k} a^j = 1 + \sum_{j=1}^{k-1} a^j + a^k$$

$$= 1 + \frac{a^k}{a - 1} + C + a^k$$

$$= 1 + \frac{a^k}{a - 1} - \frac{a}{a - 1} + a^k \qquad (E.7)$$

$$= \frac{a - 1 + a^k - a + a^{k+1} - a^k}{a - 1}$$

$$= \frac{a^{k+1} - 1}{a - 1}, \quad a \neq 1.$$

which is the geometric series (E.7).

_____

Example 1.15   Basic Summation Operator Pairs

Besides the summation formulas (1.57), (1.58) of the inverse operator, we list here a few more that we shall need in the following chapters. In the following Example 1.16 we will illustrate the use of one of these pairs, namely (1.64), and leave the illustration of the rest for the exercises.

$$\Delta^{-1}k = \frac{1}{2}k(k-1) + C \tag{1.59}$$

$$\Delta^{-1}k^2 = \frac{1}{6}k(k-1)(2k-1) + C \tag{1.60}$$

$$\Delta^{-1}k^3 = \frac{1}{4}(k-1)^2k^2 + C \tag{1.61}$$

$$\Delta^{-1}(-1)^k = \frac{1}{2}(-1)^{k+1} + C \tag{1.62}$$

$$\Delta^{-1}ka^k = \frac{a^k}{a-1}\left(k - \frac{a}{a-1}\right) + C, \quad a \neq 1 \tag{1.63}$$

$$\Delta^{-1}k^{(m)} = \frac{k^{(m+1)}}{m+1} + C, \quad m \neq -1 \tag{1.64}$$

$$\Delta^{-1}(ak+b)^{(m)} = \frac{(ak+b)^{m+1}}{a(m+1)} + C, \quad m \neq -1 \tag{1.65}$$

$$\Delta^{-1}e^{\alpha k} = \frac{e^{\alpha k}}{e^{\alpha}-1} + C, \quad \alpha \neq 0 \tag{1.66}$$

$$\Delta^{-1}\sin(ak+b) = -\frac{\cos(ak+b-\frac{a}{2})}{2\sin\frac{a}{2}} + C \tag{1.67}$$

$$\Delta^{-1}\cos(ak+b) = \frac{\sin a(k+b-\frac{a}{2})}{2\sin\frac{a}{2}} + C \tag{1.68}$$

$$\Delta^{-1}\alpha^k F(k) = \frac{\alpha^k}{\alpha-1}\left[1 - \frac{\alpha\Delta}{\alpha-1} + \frac{\alpha^2\Delta^2}{(\alpha-1)^2}\right.$$
$$\left. - \frac{\alpha^3\Delta^3}{(\alpha-1)^3} + \cdots\right] F(k) + C, \quad \alpha \neq 1 \tag{1.69}$$

where if $F(k) = P_m(k)$, a polynomial of degree m, then this series terminates with the term involving the operator Δ^m, since $\Delta^{m+j}P_m(k) = 0$ for $j \geq 1$.

Example 1.16 Δ^{-1} for the Summation of Series

In this example we illustrate the use of the important pair (1.64),

$$\Delta^{-1}k^{(m)} = \frac{k^{(m+1)}}{m+1} + C, \quad m \neq -1 \qquad (E.1)$$

to find the finite sum $\sum_{j=1}^{n} j^4$, where we recognize that

$$\sum_{j=1}^{k-1} j^4 = \Delta^{-1}k^4,$$

$$\sum_{j=1}^{n} j^4 = \Delta^{-1}(n+1)^4 \qquad (E.2)$$

The latter involves powers of n up to 4, and the best way to get to it is via the pair in (E.1). Hence, we must expand $(n+1)^4$ in terms of the factorial polynomials of degree up to 4,

$$\begin{aligned}
(n+1)^4 &= c_0 + c_1 n^{(1)} + c_2 n^{(2)} + c_3 n^{(3)} + c_4 n^{(4)} \\
&= c_0 + c_1 n + c_2 n(n-1) + c_3 n(n-1)(n-2) \qquad (E.3) \\
&\quad + c_4 n(n-1)(n-2)(n-3)
\end{aligned}$$

To find the constants c_0, c_1, c_2, and c_3 of the expansion, we set $n = 0$, 1, 2, and 3 in (E.3), which results in four simple simultaneous equations in c_0, c_1, c_2, c_3 and c_4, whose solution is $c_0 = 1$, $c_1 = 15$, $c_2 = 25$, $c_3 = 10$ and $c_4 = 1$, whence (E.3) becomes

$$(n+1)^4 = n^{(4)} + 10n^{(3)} + 25n^{(2)} + 15n^{(1)} + 1 \qquad (E.4)$$

Now we use Δ^{-1} on both sides of (E.4),

$$\Delta^{-1}(n+1)^4 = \Delta^{-1}n^{(4)} + 10\Delta^{-1}n^{(3)} + 25\Delta^{-1}n^{(2)} + 15\Delta^{-1}n^{(1)} + \Delta^{-1}1,$$

$$= \sum_{j=1}^{n} j^4 = \frac{1}{5}n^{(5)} + 10\frac{n^{(4)}}{4} + 25\frac{n^{(3)}}{3} + 15\frac{n^{(2)}}{2} + n^{(1)} + C \qquad (E.5)$$

To determine the arbitrary constant C, we let $n = 1$ in both sides of (E.5), noting that $n^{(m)} = 0$ for $m > n$, to have $1 = 1 + C$, $C = 0$.

Now, if we use the definition of the factorial polynomial $n^{(j)}$ on the right hand sides of (E.5), the result with $C = 0$ simplifies to

$$\sum_{j=1}^{n} j^4 = \frac{n(6n^4 + 15n^3 + 10n^2 - 1)}{30} \qquad (E.6)$$

We shall return in Section 2.2 to the subject of summing finite series, with the help of the above Δ^{-1} sum operator pairs, when we cover the important tool of *summation by parts* and the *Fundamental Theorem of Sum Calculus* (Theorem 2.2).

Exercises 1.3

1. Find the sum (or antidifference) $\Delta^{-1} u_k$ of the following sequences u_k:

   (a) $u_k = 0$

   (b) $u_k = 1$
       Hint: See Example 1.10.

   (c) $u_k = k$
       Hint: $k = k^{(1)}$.

   (d) $u_k = k^2$
       Hint: $k^2 = k^{(2)} + k^{(1)}$.

   (e) $u_k = 2k + 4$

   (f) $u_k = 2^k$

   (g) $u_k = 2(-1)^{k+1}$

   (h) $u_k = (-1)^{k+1}(2k + 1)$

   (i) $u_k = \ln \frac{k+1}{k}$
       Hint: Check $\Delta \ln k = \ln(k + 1) - \ln k = \ln \frac{k+1}{k}$.

   (j) $u_k = 2\left(\sin \frac{\alpha}{2}(2k + 1)\right) \sin \frac{\alpha}{2}$

   (k) $u_k = 2\left(\cos \frac{\pi}{4}(k + \frac{1}{2})\right) \sin \frac{\pi}{4}$

   (l) $u_k = 2\left(\cos \alpha(k + \frac{1}{2})\right) \sin \frac{\alpha}{2}$

2. Use the sum operator Δ^{-1} to show that

$$\sum_{k=1}^{n} k(k+1) = \frac{1}{3}(n+2)(n+1)n$$

Hint: Note that $k(k+1) = (k+1)[(k+1) - 1] = (k+1)^{(2)}$,
$\sum_{k=1}^{n-1}(k+1)^{(2)} = \Delta^{(-1)}(k+1)^{(2)} = \frac{1}{3}(k+1)^{(3)}$
$= \frac{1}{3}(k+1)(k)(k-1)|_{k=1}^{n} = \frac{1}{3}(n+1)(n)(n-1)$.

3. Show that

$$\sum_{j=1}^{n} j \cdot 2^j = 2^{n+1}(n-1) + 2$$

Hint: See (1.63), and use $n = 1$ to find the arbitrary C of the summation.

4. Prove the identity in (1.69) for $F(k) = P_m(k)$, a polynomial of degree m.

Hint: See problems 13 and 14 in the Exercises of Section 3.3, and their detailed hints.

Chapter 2

SUM CALCULUS AND THE DISCRETE TRANSFORMS METHODS

2.1 Difference Equations

In this chapter we will present some very basic elements of difference equations in Section 2.1. This is followed by a detailed discussion of the sum calculus in Section 2.2, which includes the *summation by parts*, and the *fundamental theorem of sum calculus*. These topics are essential for the sum calculus that we shall use in solving difference equations in Chapter 3, and they parallel the integration by parts and the fundamental theorem of calculus in *integral* calculus. Just as the integration by parts is used to derive the properties (and pairs) of the Laplace or Fourier transforms, for example, the summation by parts is essential for developing the operational properties of the *discrete* Fourier transforms for solving difference equations associated with boundary conditions, which is termed "the operational sum method." This method will be introduced and illustrated in Section 2.4. Since this method parallels the Fourier integral transforms method for solving differential equations with boundary conditions, we designate Section 2.3 for a brief review of such well known integral transforms, or operational (integral) calculus method.

We now begin an investigation of the difference equations which can be generated from the few difference operators which we have already seen. A *difference equation* is simply a prescribed relationship between the elements of a sequence. Having posed a difference equation, the usual aim is to solve it, which means finding a sequence that satisfies the relationship. It is best to proceed by example and we may also proceed by analogy to differential equations. Indeed, much of the terminology used for differential equations carries over directly to difference equations. An analog of the *first order, constant coefficient, non-homogeneous* differential equation

$$u'(x) + au(x) = f(x) \tag{2.1}$$

would be

$$\Delta u_k + au_k = f_k$$

or

$$u_{k+1} + (a-1)u_k = f_k. \tag{2.2}$$

In both cases a is a constant. In (2.1) the function $f(x)$ is given. In (2.2), the sequence $\{f_k\}$ is also given. A complete specification of the problem associated with (2.1) would include the interval on which the problem is to be solved (for example $x \geq 0$) and also the value of the solution at one point in that interval (for example $u(0) = A$). In a similar way, a complete specification of the problem associated with (2.2) would include the interval of the index k on which the problem is to be solved (for example $k \geq 0$) and the value of one element in the solution sequence $\{u_k\}$ (for example, $u_0 = A$). In the case that $f(x) = 0$ or $f_k = 0$ for all k, the corresponding equations are called *homogeneous*. The term *first order*, of course, refers to the appearance of a first order derivative in (2.1) and a first order difference in (2.2). While a first order of the differential equation (2.1) means the highest order of the derivative of one, the first order of the difference equation (2.2) is indicated by the highest difference in the indices of the unknown sequence u_k, which is $(k+1) - k = 1$ as seen on the left hand side of (2.2).

In analogy with the *second order, constant coefficient, nonhomogeneous* differential equation

$$u''(x) + au'(x) + bu(x) = f(x). \tag{2.3}$$

we could write the corresponding difference equation

$$\Delta^2 u_k + a\Delta u_k + bu_k = f_k$$

or

$$(u_{k+2} - 2u_{k+1} + u_k) + a(u_{k+1} - u_k) + bu_k = f_k$$

or

$$u_{k+2} + cu_{k+1} + du_k = f_k, \tag{2.4}$$

where $c = a - 2$ and $d = b - a + 1$. The use of δ^2 or ∇^2 instead of Δ^2 would result in slightly different, but equally analogous difference equations. A complete specification of a problem associated with (2.3) would include an interval of solution and two pieces of information about the solution $u(x)$ (for example, $u(0) = A$, $u'(0) = B$). In a like manner, a typical problem associated with (2.4) would specify an interval in the index k and the value of two elements of the sequence u_k (for example, $u_0 = A, u_1 = B$). Such specified (two) values of the sequence u_0 and u_1 at the initial end of the sequence, and called *initial values*. Other specifications of two values at (or around) the ends of the sequence $\{u_k\}_{k=0}^K$, for example, u_0 and u_K are termed *boundary values*.

The concept of *order* generalizes easily in the case of difference equations. A linear difference equation in u_k of the form

$$\Delta^m u_k + \Delta^{m-1} u_k + ... + \Delta u_k + u_k = f_k$$

or equivalently,

$$a_m u_{m+k} + a_{m-1} u_{m-1+k} + ... + a_1 u_{k+1} + a_0 u_k = f_k, \tag{2.5}$$

is said to be of *order* m. Again the order is determined from the largest (positive) difference of the indices of the unknown sequence u_k, which for (2.5) is $|(m + k) - k| = m$. An example is the difference equation in c_k,

$$k(k - 1)c_k + 3(k - 1)c_{k-2} = 0, \tag{2.6}$$

which is a difference equation of order 2 since $|k - (k - 2)| = 2$. In the case that the coefficients $a_0, a_1, \ldots, a_m$ in (2.5) are constants (independent of k and u_k), the equation is *constant coefficient* and *linear*. If at least one of the coefficients is a function of k, the equation is *linear* and *variable coefficient*. Clearly the above second order difference equation in (2.6) is linear, homogeneous and variable coefficient. Finally, if at least one of the coefficients actually involves an element of the unknown sequence $\{u_k\}$, then the equation is *nonlinear*, for example $u_{k-1}u_k + ku_k = 0$. This book is mainly concerned with linear difference equations. Another example is the difference equation

$$u_{k+1} + k^2 u_k = 2 \tag{2.7}$$

which is first order, linear, nonhomogeneous and variable coefficient, while

$$u_{k+2} + u_{k+1} + u_{k-1} - u_k = k^3 \tag{2.8}$$

is linear, nonhomogeneous, constant coefficient of third order. The order is 3 since the largest difference in the indices of u_k is

$$|k + 2 - (k - 1)| = 3.$$

The order of a difference equation may become more clear when the difference equation is written with the aid of the notation of the *forward shift operator* E of (1.6), where (2.7) and (2.8) can be written as

$$(E + k^2)u_k = 2, \tag{2.9}$$

and

$$(E^2 + E + E^{-1} - 1)u_k = k^3. \tag{2.10}$$

respectively. It is now clear from the largest difference in the powers of the operator E that these equations are of order 1 and 3, respectively. Of course E^{-1} in (2.10) stands for the *backward* shift operation $E^{-1}u_k = u_{k-1}$.

Initial Versus Boundary Value Problems

The above parallel classification of differential and difference equations, as we have just seen, applies to the equation itself. But, as we mentioned, such an equation does not give the complete description of the system. In other words, it does not result in the desired *unique* solution without well prescribed *auxiliary conditions*. From our experience in solving differential equations such as that of the freely falling body, we needed the initial displacement and the initial velocity as the two natural auxiliary initial conditions. These were used to determine the two arbitrary constants resulting from twice integrating the associated second order differential equation that describes Newton's Second Law of motion.

When the auxiliary conditions for a differential equation are given at an initial point, then the differential equation along with such initial conditions is called an *initial value problem*, or *one point value problem*. For example,

$$\frac{d^2u}{dt^2} = -g, \quad t > 0, \tag{2.11}$$

$$u(0) = A \tag{2.12}$$

$$u'(0) = B \tag{2.13}$$

is an initial value problem. In the same fashion, the difference equation (2.4) along with the initial values $u_0 = A$ and $u_1 = B$,

$$u_{k+2} + cu_{k+1} + du_k = f_k \tag{2.14}$$

$$u_0 = A \tag{2.15}$$

$$u_1 = B \tag{2.16}$$

constitute an initial value problem associated with the difference equation (2.14). We may note that we are given the initial value of the sequence $u_0 = A$, and also its *initial difference* $u_1 - u_0 = B - A$.

The other main class that is determined by the type of auxiliary conditions is that of *boundary value problems*. The differential equation here may be defined on a finite interval $a < x < b$, and the auxiliary conditions are given at the *two end (or boundary) points* $x = a$ and $x = b$, as in the case of the two fixed ends at $x = a, b$ of a vibrating string of length $b - a$. For example,

$$\frac{d^2y}{dx^2} + b^2y = 0, \quad 0 < x < \pi \tag{2.17}$$

$$y(0) = 0 \qquad\qquad (2.18)$$

$$y(\pi) = 0 \qquad\qquad (2.19)$$

represents a *boundary* or *two points* problem compared to the *initial* or *one point* problem in (2.11)-(2.13). These two points boundary conditions are not limited to the value of the function, but can cover a linear combination of the function and its first derivative at either one of the two points for the case of a second order differential equation. For example,

$$y(\pi) + cy'(\pi) = 0 \qquad\qquad (2.20)$$

is an acceptable boundary condition at the second point $x = \pi$, and the same can be done at the point $x = a = 0$.

We mention here the subject of the boundary value problem because we will cover difference equations that are defined on a finite number of points, for example, $u_k, 0 \le k \le N$ instead of most of what we used above, namely $u_k, k \ge 0$. The latter obviously goes with initial value problems. For the desired finite sequence $u_k, 0 \le k \le N$ of the difference equation, we must supply auxiliary conditions that are the value of the sequence at the two end points $k = 0$ and $k = N$, namely $u_0 = A$ and $u_N = B$. An example would be

$$u_k - u_{k-1} = -f_k, \quad 0 \le k \le N - 1 \qquad\qquad (2.21)$$

$$u_0 = u_N \qquad\qquad (2.22)$$

where the two boundary values happen to be equal, and such boundary condition is termed *periodic* boundary condition. We shall revisit this problem in Example 2.11 of Section 2.4, and also with the details necessary to find the solution to a particular illustration (with a specific f_k) in Example 4.8 of Section 4.4, where such a boundary value problem turns out to be a representation of a *traffic flow* in a loop of N intersections and N streets (arcs). The sequence u_k represents the number of vehicles in the kth loop, while f_k represents the traffic flow into the kth loop from the kth entrance (see Figs. 4.6, 4.7).

For the solution of such boundary value problems associated with difference equations, we will, in Chapters 4 and 5, and briefly in Section 2.4, use the direct discrete Fourier transforms method in parallel to the use of the Fourier transforms for solving boundary value problems associated with differential equations. This represents an added feature of this book for its use of the various discrete Fourier transforms in solving boundary value problems. Such an operational (sum) calculus method also parallels that of using Laplace transform for solving initial value problems associated with differential equations. The Laplace transform is recalled for its applications in solving many familiar problems such as that of falling bodies, spring oscillations and circuit theory. For initial value problems associated with difference equations, we will use the discrete analog of the Laplace transform, namely, the z-transform, which will be discussed and applied to solving initial value problems in Chapter 6. The basic methods of solving difference equations are covered in Chapter 3.

On the Solution of Difference Equations

Let us return for a moment to the second order difference equation (2.4)

$$u_{k+2} + cu_{k+1} + du_k = f_k \tag{2.4}$$

and make a few remarks about what it means to solve a difference equation. We will assume that c, d and the sequence $\{f_k\}$ are given and that the solution sequence $\{u_k\}$ is to be found for $k \geq 0$. Rewriting (2.4) as

$$u_{k+2} = f_k - cu_{k+1} - du_k, \quad \text{for } k \geq 0 \tag{2.23}$$

we note that this is really a set of relations

$$u_2 = f_0 - cu_1 - du_0 \tag{1}$$

$$u_3 = f_1 - cu_2 - du_1 \tag{2}$$

$$\vdots$$

A closer look also shows that if the (initial) values of u_0 and u_1 are given, then $u_2, u_3, u_4, \ldots$ may be found successively (recursively) from equations (1), (2), $\ldots$ of this set. In other words, the difference equation is actually a "formula" for generating the solution sequence $\{u_k\}$. It also becomes clear why a second order difference equation requires two pieces of information (in this case u_0 and u_1) to specify a solution. In general, for linear problems at least, an mth order problem should be accompanied by the value of m elements of the solution sequence if a specific solution is to be found. If these m pieces of information are not given, it is usually possible to find a *general solution* of the difference equation which possesses m arbitrary constants (in exact analogy with the corresponding differential equation), as we shall discuss and illustrate in Section 3.1.

But have we really solved the difference equation (2.4) by simply rewriting it in the form (2.23)? Returning to the *Fibonacci* sequence of Example 1.3,

$$u_{k+1} = u_k + u_{k-1} \qquad (2.24)$$

we might ask for the 1000th Fibonacci number u_{1000}, It could be found by computing $u_0, u_1, u_2, \ldots, u_{1000}$ successively using the difference equation. This means that in order to find u_{1000}, we must evaluate *all* the preceding numbers of the sequence, which does not sound very efficient. So, we will not regard such an explicit computation as a solution method. In all that follows, we will consider solving a difference equation to mean producing an expression which allows any element of the solution sequence $\{u_k\}$ to be determined *independently* of the other elements. For example, the solution of (2.24) with the conditions $u_0 = 0, u_1 = 1$ is given by:

$$u_k = \frac{1}{\sqrt{5}}\left(\frac{1+\sqrt{5}}{2}\right)^k - \frac{1}{\sqrt{5}}\left(\frac{1-\sqrt{5}}{2}\right)^k, \quad \text{for } k \geq 0. \qquad (2.25)$$

With this expression u_{1000} can be computed directly without finding all of its predecessors. The methods for determining such solution will occupy us in all that follows. In the meantime we refer to the outline of the solution (2.25) of (2.24) in exercise 4 of Section 4.4.

There is another reason for disallowing explicit calculation as a method for solving difference equations. Sometimes it does not work! The example given above involving the Fibonacci sequence is an example of *initial value problem*. We are asked to find a solution for values of the index $k \geq 0$ and we are given the first two elements (the initial values) of the sequence u_0 and u_1. As mentioned before, this corresponds closely to initial value problems with ordinary differential equations in which the initial configuration of a system is specified (for example, position and velocity of a pendulum) and the resulting solution gives the time evolution of the system. However, as mentioned earlier, the required conditions may not always be given at the initial end of the sequence. In the case of a *boundary value problem*, information about the solution sequence is given at opposite ends of the sequence. For example the second order difference equation

$$u_{k+1} + au_k + bu_{k-1} = f_k, \quad \text{for } 1 \leq k \leq 99 \qquad (2.26)$$

with the conditions $u_0 = A$ and $u_{100} = B$ would be very difficult to solve by computing $u_0, u_1, u_2, ...$, explicitly. We would need to guess at a value of u_1 to begin the process and then hope that we finished with the required value of u_{100}. However, we will see that it is possible to find solutions to boundary value problems that can be expressed in a *closed form* as we shall discuss in Section 2.4, Chapters 4 and 5 by using the discrete Fourier transforms, and briefly in Chapter 3.

We may conclude this section by writing the most general nth order difference equation in u_k, which is the equation

$$F(k, u_k, u_{k+1}, u_{k+2}, \ldots, u_{k+n}) = 0, \quad k = 0, 1, 2, \ldots \qquad (2.27)$$

or

$$u_{k+n} = g(k, u_k, u_{k+1}, u_{k+2}, \ldots, u_{k+n-1}), \quad k = 0, 1, 2, \ldots \qquad (2.28)$$

Its very special case, the *linear nth* order difference equation is

$$a_0(k)u_{k+n} + a_1(k)u_{k+n-1} + a_2(k)u_{k+n-2} + a_{n-1}(k)u_{k+1}$$
$$+ a_n(k)u_k = f_k, \quad k = 0, 1, 2, \ldots \qquad (2.29)$$

where $a_j(k)$ is the *variable coefficient*, and $f(k)$ is the *nonhomogeneous* term. With the aid of the forward shift operator E in (1.6), this linear equation (2.29) can be written as

$$\left[a_0(k)E^n + a_1(k)E^{n-1} + \ldots + a_{n-1}(k)E + a_n(k) \right] u_k = f_k, \qquad (2.30)$$

where the order n is clearly indicated as the highest power of the operator E. This is in parallel to writing the linear nth order differential equation in $u(x)$,

$$a_0(x)\frac{d^n u}{dx^n} + a_1(x)\frac{d^{n-1}u}{dx^{n-1}} + a_2(x)\frac{d^{n-2}u}{dx^{n-2}} + \ldots$$
$$+ a_{n-1}(x)\frac{du}{dx} + a_n(x)u = f(x) \qquad (2.31)$$

with the help of the differential operator $D \equiv \frac{d}{dx}$ as

$$[a_0(x)D^n + a_1(x)D^{n-1} + a_2(x)D^{n-2} + \ldots$$
$$+ a_{n-1}(x)D + a_n(x)]u(x) = f(x) \qquad (2.32)$$

It may be appropriate here to state a theorem concerning the *existence* of a *unique* solution to the general nth order difference equation (2.28) with its prescribed conditions $u_0, u_1, u_2, \ldots, u_{n-1}$.

Theorem 2.1   On the Existence of a Unique Solution

Consider the general nth order difference equation,

$$u_{k+n} = g(k, u_k, u_{k+1}, \ldots, u_{k+n-1}), \quad n = 0, 1, 2, \ldots \qquad (2.28)$$

where the function g is defined for all its arguments k, u_k, u_{k+1}, u_{k+2}, $\ldots, u_{k+n-1}$, and where the initial values (at $k = 0$) of u_k, u_{k+1}, u_{k+2}, $\ldots, u_{k+n-1}$ are given as $u_0, u_1, u_2, \ldots, u_{n-1}$, respectively. Then such a difference equation with its given initial values, i.e., the *initial value problem* associated with this difference equation (2.28) has a unique solution corresponding to each arbitrary set of the n initial values $u_1, u_2, \ldots, u_{n-1}$.

This theorem shall remind us of its parallel for the initial value problem associated with nth order differential equation. However the proof of the present theorem may be surprisingly simple. It starts by substituting the initial value $u_0, u_1, \ldots, u_{n-1}$ for $k = 0$ in the right hand side of (2.28), which uniquely determines u_n on the left hand side. Once this u_n is found, it can be used in the right hand side as $u_n \equiv u_{1+(n-1)}$, the last argument of g, to give u_{1+n} on the left hand side. Continuing this way by substituting $u_{n+1} \equiv u_{2+(n-1)}$ next as the last argument of g, we obtain $u_{2+n} \equiv u_{n+2}$. This *iterative* procedure can be continued to generate all $u_k, k \geq n$, i.e, $u_n, u_{n+1}, u_{n+2}, \ldots$, which constitutes the unique solution to the initial value problem of the difference equation (2.28) and its given n initial conditions $u_0, u_1, u_2, \ldots, u_{n-1}$.

Even though the most basic methods of solving difference equations will be the subject of Chapter 3 in addition to Chapters 4-7, it may be instructive now to have a glimpse of such methods, and how they compare with the methods of solving differential equations.

This we shall do in the following example with the simplest of the two type equations, i.e., linear, homogeneous and constant coefficient.

Example 2.1   Solution of Differential and Difference Equations

(a). Consider the differential equation

$$\frac{d^2u}{dx^2} - 6\frac{du}{dx} + 8u = 0, \qquad (E.1)$$

it is a linear, second order, homogeneous and with constant coefficients. Our basic method for solving it is to let $u = e^{mx}$ in (E.1),

$$m^2 e^{mx} - 6m e^{mx} + 8e^{mx} = 0,$$

$$e^{mx}(m^2 - 6m + 8) = 0,$$

and since $e^{mx} \neq 0$, we have the *characteristic equation* of (E.1),

$$m^2 - 6m + 8 = 0,$$

$$(m - 2)(m - 4) = 0, \qquad (E.2)$$

which yields two distinct real roots $m_1 = 2$ and $m_2 = 4$. So the two corresponding solutions for (E.1) are $u_1(x) = e^{2x}$ and $u_2(x) = e^{4x}$, which are linearly independent since $m_1 \neq m_2$. Hence the general solution to (E.1) is their linear combination

$$u(x) = c_1 e^{2x} + c_2 e^{4x}; \qquad (E.3)$$

where c_1 and c_2 are two arbitrary constants. In case we have two equal roots $m_1 = m_2$ for the characteristic equation, then it is known that we can modify the one solution $e^{m_1 x}$ by multiplying it by x to generate the second *linearly independent* solution $x e^{m_1 x}$ for the given second order differential equation.

To have a unique solution, these two constants must be determined, which is done through imposing two auxiliary conditions. For example, the following two initial conditions

$$u(0) = 2, \qquad (E.4)$$

and

$$u'(0) = 5 \qquad (E.5)$$

which make (E.1), (E.4) and (E.5) constitute an *initial value problem*. If we impose the initial conditions (E.4) and (E.5) on the solution in (E.3), we have

$$2 = c_1 + c_2, \quad c_2 = 2 - c_1,$$

$$5 = 2c_1 + 4c_2,$$

$$5 = 2c_1 + 8 - 4c_1, \quad 2c_1 = 3,$$

$$c_1 = \frac{3}{2}, \quad c_2 = \frac{1}{2}.$$

and the final solution to the initial value problem (E.1), (E.4) and (E.5) becomes

$$u(x) = \frac{3}{2}e^{2x} + \frac{1}{2}e^{4x}.$$

(b). Now we consider the difference equation in u_k,

$$u_{k+2} - 6u_{k+1} + 8u_k = 0, \qquad (E.6)$$

which is also linear, second order, homogeneous and with constant coefficients. The method of solution, again parallels that which we did for its analogous differential equation (E.1). The method starts by letting $u_k = \lambda^k$ (as a sequence) in (E.6) to also result in an *auxiliary equation* in parallel to the characteristic equation (E.2) of the differential equation (E.1),

$$u_{k+2} - 6u_{k+1} + 8u_k = \lambda^{k+2} - 6\lambda^{k+1} + 8\lambda^k = 0,$$

$$\lambda^k(\lambda^2 - 6\lambda + 8) = 0,$$

and since $\lambda^k \neq 0$, we have the auxiliary equation

$$\lambda^2 - 6\lambda + 8 = 0, \quad (\lambda - 2)(\lambda - 4) = 0, \qquad (E.7)$$

$$\lambda_1 = 2 \quad \text{and} \quad \lambda_2 = 4.$$

With these two *distinct* values $\lambda_1 = 2$ and $\lambda_2 = 4$, we have the two corresponding sequences solutions of the difference equation (E.6) as $u_k^{(1)} = \lambda_1^k = 2^k$ and $u_k^{(2)} = \lambda_2^k = 4^k$. Since the difference equation is linear, we can easily show, parallel to what we did for the differential equation, that the linear combination of these two solutions,

$$u_k = c_1 2^k + c_2 4^k \qquad (E.8)$$

is the general solution to the difference equation (E.6). We may note here that for the differential equation (E.1), we were after m in e^{mx} to obtain from its characteristic equation (E.2), while with the difference equation we are after the base λ in λ^k to obtain from its auxiliary equation (E.7). Indeed this can be reconciled when we write $e^{mx} = (e^m)^x = \lambda^x$, and hence determining m is equivalent to determining $\lambda = e^m$. What remains now is to determine the two arbitrary constants in the general sequence solution (E.8). This will require two auxiliary conditions, which we can take here as the given initial sequences

$$u_0 = 1, \qquad\qquad\qquad (E.9)$$

$$u_1 = 3. \qquad\qquad\qquad (E.10)$$

So, imposing (E.9) and (E.10) on (E.8), we obtain

$$u_0 = 1 = c_1 2^0 + c_2 4^0 = c_1 + c_2 = 1,$$

$$u_1 = 3 = c_1 2 + c_2 4 = 2c_1 + 4c_2 = 3,$$

$$c_2 = 1 - c_1, \quad 2c_1 + 4(1 - c_1) = 3,$$

$$2c_1 = 1, \quad c_1 = \frac{1}{2}; \quad c_2 = \frac{1}{2}.$$

The unique solution to the initial value problem (E.6), (E.9) and (E.10), associated with the difference equation (E.6), becomes

$$u_k = \frac{1}{2}2^k + \frac{1}{2}4^k = 2^{k-1} + 2^{(2k-1)}.$$

Exercises 2.1

1. Classify the following difference equations with regard to order, linearity, homogeneity, and variable or constant coefficient.

   (a) $u_{k+3} - 2u_{k+1} + 4u_k = k^2$

   (b) $u_{k+2} + ku_k = 1$

   (c) $u_{k+1} - u_k^2 = 0$

(d) $u_{k+3} - 5u_{k+2} + 6u_{k+1} + 8u_k = k + 3$

(e) $(2k+1)u_{k+2} - 3ku_{k+1} + 4u_k = 5k^2 - 3k$

(f) $(2k+1)u_{k+1}u_k - 3ku_{k+1} + 4u_k = 1$

(g) $u_{k+2} + u_{k+1} = k$

2. Write the *linear* difference equations of problem 1 with the aid of the forward shift operator E.

3. (a) Write the most general difference equation in (2.27) with the aid of the forward shift operator.

   (b) Write the *nonlinear* difference equations of problem 1 with the aid of the E operator.

4. For the following initial value problem, verify that the given sequence is a solution of the corresponding initial value problem.

   Hint: Make sure that the sequence satisfies the difference equation as well as its associated initial condition or conditions.

   (a) $\{u_k\} = \{k^2\}$,

   $$u_{k+2} - 4u_{k+1} + (3 + \frac{4}{k})u_k = 0, \quad k \geq 1,$$

   $$u_0 = 0, \quad u_1 = 1.$$

   (b) $\{u_k\} = \{\frac{1}{k}\}$,

   $$u_{k+1} + u_{k+1}u_k - u_k = 0, \quad k \geq 1,$$

   $$u_1 = 1.$$

   (c) $\{u_k\} = \sin(\alpha k)$,

   $$u_{k+1} - (2\cos\alpha)u_k - u_{k-1} = 0, \quad k \geq 1, \quad \alpha \in R,$$

   $$u_0 = 0, \quad u_1 = \sin\alpha.$$

(d) $\{u_k\} = \{k^2\}$,

$$\Delta u_k - \frac{2}{k} u_k = u_{k+1} - u_k - \frac{2}{k} u_k = 1, \quad k \geq 1,$$

$$u_0 = 0.$$

5. Verify that the Fibonacci sequence u_k in (2.25) satisfies the recurrence relation (difference equation) in (2.24).

6. Classify the difference equations in Problem 4.

7. Write the linear difference equation in Problem 6 (or 4) with the aid of the E operator.

8. Verify that the sequence

$$u_n = \frac{3}{5}(3)^n - \frac{3}{5}(-2)^n$$

is the solution to the initial value problem

$$u_{n+2} - u_{n+1} + 6u_n = 0, \quad n \geq 0,$$

$$u_0 = 0,$$

$$u_1 = 3.$$

9. Attempt to solve the initial value problem in Exercise 8 by starting with $u_n = \lambda^n$ as we did in part (b) of Example 2.1.

10. Attempt to solve the following *homogeneous* difference equation of second order

$$u_{k+2} - 3u_{k+1} + 2u_k = 0 \tag{E.1}$$

by assuming the form of the solution $u_k = \lambda^k$ as we did in part (b) of Example 2.1, where first you

(a) Find the characteristic equation, then find its root λ_1 and λ_2

(b) If $\lambda_1 \neq \lambda_2$ in the characteristic equation, then find the general solution $u_k = A\lambda_1^k + B\lambda_2^k$.

11. Do problem 10 for the difference equation

$$u_{k+2} + 3u_{k+1} + u_k = 0.$$

12. Consider the *homogeneous* difference equation of order 2

$$u_{k+2} + u_k = 0,$$

Show that the roots of its characteristic equation are complex numbers $\lambda_1 = i = \sqrt{-1}$ and $\lambda_2 = -i = -\sqrt{-1}$.

13. Consider the *homogeneous* difference equation

$$u_{k+2} - 2u_{k+1} + u_k = 0.$$

(a) Show that the two roots of its characteristic equation are equal; $\lambda_1 = \lambda_2 = 1$.

(b) In parallel to what is done and we mentioned in Example 2.1 about equal roots of the characteristic equation for a linear constant coefficient differential equation, how can you write a general sequence solution for this case?

14. Attempt to answer the same question in Problem 13 for the following difference equation

$$4u_{k+2} + 4u_{k+1} + u_k = 0.$$

15. Solve the difference equation,

$$u_{k+2} - 4u_{k+1} + 4u_k = 0.$$

Hint: See Example 2.1, and problems 9 to 14, particularly Problem 13.

16. Verify that

$$u_k = (A + Bk)2^k + 3k + \frac{1}{8}2^k \qquad (E.1)$$

is a (general) solution to the following *nonhomogeneous* difference equation (associated with the homogeneous equation of Problem 15)

$$u_{k+2} - 4u_{k+1} + 4u_k = 2^k + 3k. \qquad (E.2)$$

Hint: Just substitute the solution (E.1) in the left hand side.

2.2 Summation by Parts and The Fundamental Theorem of Sum Calculus

We will now present a few of the essential elements of the difference and summation calculus which show a striking correspondence to familiar properties of the differential and integral calculus. For example, given two sequences $\{u_k\}$ and $\{v_k\}$, the *difference of the product* sequence $\{u_k v_k\}$ may be computed as follows, which was done for (1.18).

$$
\begin{aligned}
\Delta(u_k v_k) &= u_{k+1}v_{k+1} - u_k v_k \\
&= u_{k+1}v_{k+1} - u_{k+1}v_k + u_{k+1}v_k - u_k v_k \\
&= u_{k+1}(v_{k+1} - v_k) + v_k(u_{k+1} - u_k) \\
&= u_{k+1}\Delta v_k + v_k \Delta u_k \quad\quad\quad (2.33)
\end{aligned}
$$

after adding and substituting, the term $u_{k+1}v_k$ in the second line, grouping, then using the definition of the difference operator in the third line as was done earlier for (1.19). This result may be compared to the differential of the product of two functions

$$ d(uv) = u\,dv + v\,du. $$

In a similar way, it is possible to find the *difference of the quotient* of two sequences, as we did for (1.18)

$$ \Delta\left(\frac{u_k}{v_k}\right) = \frac{v_k \Delta u_k - u_k \Delta v_k}{v_k v_{k+1}}. \quad\quad\quad (2.34) $$

The important operation of *summation by parts* may now be illustrated by returning to the formula for the difference of a product. Summing equation (2.33) between the indices $k = 0$ and $k = N - 1$, we have

$$ \sum_{k=0}^{N-1} \Delta(u_k v_k) = \sum_{k=0}^{N-1} (u_{k+1}\Delta v_k + v_k \Delta u_k). $$

Expanding the left hand side and summing term by term, we see that all of the interior terms of the sum cancel and only two "boundary terms" survive. This leaves

$$ u_N v_N - u_0 v_0 = \sum_{k=0}^{N-1} (u_{k+1}\Delta v_k + v_k \Delta u_k). $$

Rearranging this expression gives the *summation by parts formula*

$$\sum_{k=0}^{N-1} v_k \Delta u_k = u_N v_N - u_0 v_0 - \sum_{k=0}^{N-1} u_{k+1} \Delta v_k. \qquad (2.35)$$

To help illustrate this summation by parts formula (2.35), we need a few basic elementary results. The geometric series is indispensable in much of what follows, and it serves as a good starting point. If $a \neq 0$ and $a \neq 1$, is a real number and $N > 1$ is a positive integer, then the sum $\sum_{k=0}^{N-1} a^k$ may always be evaluated explicitly. However, it is usually far simpler to use the fact that

$$\sum_{k=0}^{N-1} a^k = \frac{1-a^N}{1-a}, \quad a \neq 1. \qquad (2.36)$$

The usual proof for this is simple if we write it as S_N,

$$S_N = 1 + a + a^2 + \ldots + a^{N-2} + a^{N-1} \qquad (2.37)$$

and multiply this by a, to have

$$a S_N = a + a^2 + \ldots + a^{N-2} + a^{N-1} + a^N. \qquad (2.38)$$

Then subtract S_n in (2.37) from $a S_N$ in (2.38), where all terms cancel out except the term 1 in (2.37) and the term a^N in (2.38), to have

$$a S_N - S_N = (a-1) S_N = a^N - 1,$$

$$S_N = \frac{a^N - 1}{a - 1} = \frac{1 - a^N}{1 - a}, \quad a \neq 1. \qquad (2.39)$$

This is for the general case of $a \neq 1$. The special case of $a = 1$ for the geometric series on the left hand side of (2.37) clearly gives

$$\sum_{k=0}^{N-1} 1 = \overbrace{1 + 1 + \ldots + 1}^{N} = N.$$

This result in (2.39) was also proved in (E.7) of Example 1.14 via the use of the inverse difference (sum) operator Δ^{-1} as a finite sum in (1.58).

Example 2.2   A Differencing Pair

Having just summed the sequence $\{a^k\} = \{1, a, a^2, \ldots\}$, let us look again at the forward difference operator applied to the sequence. The operation is simple:

$$\Delta(a^k) = a^{k+1} - a^k = a^k(a - 1) \qquad (1.28)$$

which was one of our first *differencing* pairs that we discussed in Section 1.2. This is not much of a surprise when we think of the parallel in integral and differential calculus, where differentiation is, in general, much simpler than integration.

Example 2.3   Summation by Parts - the First Illustration

We now have the results needed to apply the summation by parts expression (2.35). Again, let $a \neq 0$, $a \neq 1$ be a real number and $N > 1$ be a positive integer. Consider the problem of evaluating the sum $\sum_{k=0}^{N-1} ka^k$. To apply the summation by parts rule, we should identify the given sum with the left hand side of (2.35). Letting $v_k = k$ and $\Delta u_k = a^k$, we see that the right hand side of (2.35) requires Δv_k and u_{k+1}. The first of these, Δv_k, is easy to find:

$$\Delta v_k = \Delta(k) = (k + 1) - k = 1, \qquad (E.1)$$

or we can, simply, consult the difference Δ and anti-difference Δ^{-1} pairs that we had derived or listed in Sections 1.2 and 1.3, respectively. Finding u_k when $\Delta u_k = a^k$ is given, is more difficult as we saw for (1.58). In fact, what we are seeking is the "anti-difference" of the sequence $\{a^k\}$, i.e., $\Delta^{-1}(a^k)$, that we have already covered in Section 1.3. The analogy with finding the antiderivative of a function is very close. We must find a sequence $\{u_k\}$ that satisfies $\Delta u_k = a^k$. The preceding Example 2.2 in which we showed that $\Delta(a^k) = a^k(a - 1)$ holds the key for this very special case, since it implies that

$$\Delta(\frac{a^k}{a - 1}) = a^k, \qquad (E.2)$$

a simple result that was also derived in (E.6) of Example 1.14. Hence our sequence $\{u_k\}$ is $\frac{a^k}{a-1}$. Now we may apply the summation by parts rule. By (2.35) we have

$$\sum_{k=0}^{N-1} ka^k = \frac{a^N}{a-1} \cdot N - \frac{a^0}{a-1} \cdot 0 - \sum_{k=0}^{N-1} \frac{a^{k+1}}{a-1} \cdot 1$$

$$= \frac{Na^N}{a-1} - \frac{a}{a-1} \sum_{k=0}^{N-1} a^k$$

$$= \frac{Na^N}{a-1} - \frac{a}{a-1} \cdot \frac{1-a^N}{1-a}$$

$$= \frac{a}{(a-1)^2} \left[(N-1)a^N - Na^{N-1} + 1 \right].$$

(E.3)

As an illustration of this formula, we take

$$\sum_{k=0}^{4} k2^k = 2 + 8 + 24 + 64 = 98$$

by explicit calculation, which agrees with the expression derived above when $a = 2$ and $N = 5$ as $2[4 \cdot 2^5 - 5 \cdot 2^4 + 1] = 98$.

At this point it seems worthwhile to digress for a moment and explore an idea that we have already discussed in Section 1.3. This is the issue of finding the *anti-difference* of a given sequence. This concept will be used in the work ahead, since it completes the correspondence between the difference calculus and the infinitesimal calculus in an elegant way. Recall that the problem of finding the *anti-difference* of a sequence $\{u_k\}$ is that of finding a sequence $\{v_k\}$ that satisfies $\Delta v_k = u_k$. Since this operation is the inverse of finding the difference of $\{u_k\}$, we write the relationship $v_k = \Delta^{-1} u_k$. Any time we find the difference of a sequence, we have also found the *anti-difference* of another sequence, in exact parallel to differentiation and anti-differentiation (integration). For completeness, we repeat here a few of the pairs that we had in Sections 1.2 and 1.3.

1. Since $\Delta(k) = 1$,

$$\Delta^{-1}(1) = k$$

(2.40)

2. Since $\Delta(a^k) = (a-1)a^k$,

$$\Delta^{-1}(a^k) = \frac{1}{a-1}a^k, \quad a \neq 0, 1$$

(2.41)

3. Since $\Delta(k(k-1)) = 2k$,

$$\Delta^{-1}(2k) = k(k-1) \tag{2.42}$$

4. In general, since $\Delta(k^{(m)}) = mk^{(m-1)}$, we have that

$$\Delta^{-1}(k^{(m)}) = \frac{k^{(m+1)}}{m+1}, \ m \neq -1. \tag{2.43}$$

For these and other pairs, see (1.57) to (1.69). We may point out again that, as in the case of antiderivatives, the *anti-difference* of a given sequence is determined only up to an *additive constant* sequence. If $v_k = \Delta^{-1}(u_k)$, then $v_k + C = \Delta^{-1}(u_k)$ also where C represents any sequence, all of whose elements are identical. This follows since $\Delta(C) = 0$.

The analogy between the discrete calculus and the continuous (infinitesimal) calculus is completed when we recall the role which antiderivatives play in the *Fundamental Theorem of Calculus*. It is now tempting to ask if there is an analogous result involving *anti-differences*. Indeed there is, and it might be called the *Fundamental Theorem of Sum Calculus*. Just as the Fundamental Theorem of (Integral) Calculus is a rule for evaluating definite integrals, the Fundamental Theorem of Sum Calculus is a rule for evaluating definite sums or sums with limits. It has the following form:

Theorem 2.2   The Fundamental Theorem of Sum Calculus

For $v_k = \Delta^{-1}u_k$, or in other words $u_k = \Delta v_k$, we have

$$\sum_{k=m}^{m+N-1} u_k = \Delta^{-1}(u_k)\Big|_{k=m}^{k=m+N}$$

$$= v_k\Big|_{k=m}^{k=m+N} = v_{m+N} - v_m. \tag{2.44}$$

We will illustrate this very important tool of *difference* and *sum* calculus in the following two examples.

Example 2.4   Application of the Fundamental Theorem of Sum Calculus

In order to find the finite sum $\sum_{k=1}^{N} k$, we let $u_k = k$ in the left hand side of (2.44) with the limits $k = m = 1$ to $k = m + N - 1 = 1 + N - 1 = N$. For the right hand side of (2.44), we recall that the anti-difference of $u_k = k$ is $v_k = \frac{1}{2}k(k-1)$, i.e., $\Delta^{-1}(k) = \frac{1}{2}k(k-1)$. So, if we substitute this v_k in (2.44) with limits $k = 1$ to N, we have

$$\sum_{k=1}^{N} k = \frac{1}{2}k(k-1) \Big|_{k=1}^{k=N+1} = \frac{1}{2}(N+1)N - 0 = \frac{1}{2}(N+1)N.$$

In the above example we note that $\Delta(\frac{1}{2}k(k-1)) = \Delta\frac{1}{2}k^{(2)} = \frac{2}{2}k^{(1)} = k$, where $\Delta^{-1}k = \frac{1}{2}k(k-1)$ as the anti-difference, which corresponds to the antiderivative in integral calculus. However the finite sum of (2.44)

$$\sum_{k=m}^{m+N-1} u_k = \Delta^{-1}(u_k) \Big|_{k=m}^{N+m} \tag{2.44}$$

corresponds to the anti-difference at the upper limit $k = m + N$, minus its value at the lower limit $k = m$. This makes the definite sum in (2.44) as the parallel of the definite integral $\int_a^b f(x)\,dx$, while the anti-difference $\Delta^{-1}u_k$ without the limits is the parallel of indefinite integral $\int f(x)\,dx$.

We should also note that the result of the above example,

$$\sum_{k=1}^{N} k = \frac{1}{2}N(N+1)$$

is no more than the very familiar formula for the simplest algebraic series, where it can be proved in a very direct simple way. Here we note that the given sequence $c_k = k$ is a linear function of k, hence its Nth sum is N times its average value of $\frac{N+1}{2}$ between the two limits $k = 1$ and $k = N$, which gives the sum $\sum_{k=1}^{N} k$ as $\frac{1}{2}N(N+1)$. However, more difficult calculations of more involved sums may be simplified considerably by the use of the Fundamental Theorem of Sum Calculus. This is in the sense that just as we have various applications for the definite integral in calculus, here we can also use the Fundamental Theorem of Sum Calculus to evaluate complicated finite sums that are intractable by simple or direct means, unlike the above summing of the (linear) sequence $\{u_k\} = \{k\}$, which can be derived very easily using its average value $\frac{N+1}{2}$. The following example illustrates the application of the fundamental theorem of sum calculus in evaluating a complicated sum.

Example 2.5   Summing of Finite Series

Consider the following rather complicated sum

$$S = \sum_{k=0}^{50} k^{(-3)} = \sum_{k=0}^{50} \frac{1}{(k+1)(k+2)(k+3)} \qquad (E.1)$$

Since $\Delta^{-1}(k^{(-3)}) = \frac{k^{(-2)}}{-2}$ as in (1.51), we apply the Fundamental Theorem of Sum Calculus to the left hand side of (E.1),

$$S = -\frac{1}{2}\frac{1}{(k+1)(k+2)}\Big|_0^{51} = -\frac{1}{2}\left(\frac{1}{52 \cdot 53} - \frac{1}{1 \cdot 2}\right) = \frac{1377}{2756}.$$

Exercises 2.2

1. Show that the difference of the product of two sequences may also be expressed as

$$\Delta(u_k v_k) = v_{k+1}\Delta u_k + u_k \Delta v_k$$

$$\Delta(u_k v_k) = \Delta v_k A u_k + \Delta u_k A v_k$$

where A is the averaging operator of (1.7).

Hint: To involve the averaging operator A in $Au_k = \frac{1}{2}(u_k + u_{k+1})$, add and subtract $\frac{1}{2}u_k v_{k+1}$ and $\frac{1}{2}v_k u_{k+1}$ to the resulting expression of $\Delta(u_k v_k)$.

2. Use the geometric series and/or summation by parts (when necessary) to evaluate the following sums.

    (a) $\sum_{k=0}^{19} e^{-k}$

    (b) $\sum_{k=1}^{9} 3^k$

    (c) $\sum_{k=0}^{9} k2^{-k}$

(d) $\sum_{k=0}^{24} k^2 3^k$

3. Use the Fundamental Theorem of Sum Calculus Theorem 2.2 and/or summation by parts to evaluate the following sums.

   (a) $\sum_{k=1}^{n} k^2$

   (b) $\sum_{k=1}^{n} \frac{1}{(k+2)(k+1)}$

   (c) $\sum_{k=1}^{n} k^3 + k$

   (d) $\sum_{k=1}^{n} \frac{k}{(k+3)(k+2)(k+1)}$

4. Find the value of the following infinite series by determining the general nth partial sum $S_n = \sum_{k=0}^{n} a^k$ and then evaluating $\lim_{n \to \infty} S_n$.

   (a) $\sum_{k=0}^{\infty} \frac{1}{(k+1)(k+2)}$

   (b) $\sum_{k=1}^{\infty} \frac{1}{k(k+2)}$

5. Two sums which will be important in work that follows are

$$\sum_{k=0}^{N-1} \sin \frac{2\pi k}{N} \quad \text{and} \quad \sum_{k=0}^{N-1} \cos \frac{2\pi k}{N}.$$

Evaluate these sums in two ways:

   (a) Use the Fundamental Theorem of Sum Calculus.

   (b) Note that

$$\sin\left(\frac{2\pi k}{N}\right) = Im\left\{\exp\left(\frac{i2\pi k}{N}\right)\right\}$$

$$\cos\left(\frac{2\pi k}{N}\right) = Re\left\{\exp\left(\frac{i2\pi k}{N}\right)\right\},$$

since $e^{ix} = \cos x + i \sin x$, and use the geometric series to sum the complex exponential sequence in $\sum_{k=0}^{N-1} e^{\frac{i2\pi k}{N}}$.

2.3   Discrete Transforms and Review of Integral and Finite Transforms

In the previous section we have begun to see the striking parallels between certain operations of the infinitesimal calculus as they apply to functions, and analogous discrete operations as they apply to sequences or sampled functions. In this and the following section, these parallels will be pursued further as we investigate the idea of integral transforms and their discrete analogs. To clarify such parallels, we will briefly discuss the method of using integral transforms and finite transforms (Fourier series and coefficients) and their main property in transforming derivatives to (a simpler) algebraic expression in addition to involving the given auxiliary conditions. The method that is based on this idea is called "transform method", "operational calculus method," or the "Heaviside calculus." The first exposure of the reader to such a method may, most likely, be that of the Laplace transform. This brief introduction of the integral and finite transforms is to be followed by a parallel use of the discrete version of such transforms, and primarily, the discrete Fourier transforms for solving difference equations associated with boundary conditions. Such discrete Fourier transforms involve only a *finite sum* operator, as we shall see shortly in Section 2.4, applied to a difference equation in order to transform it to (the simpler) algebraic equation in addition to involving the boundary conditions at the outset. Thus, we shall term such a method as the "operational sum calculus" method. The z-transform, which is a discrete analog of the Laplace transform, is defined as an *infinite sum*, can also be embodied under the operational sum calculus method, but will be introduced on its own with the necessary details and some relevant applications in Chapter 6. As it is the case with its continuous analog, the z-transform is relevant for solving initial value problems, whereas the Fourier transforms are compatible with boundary value problems. In addition to the presentation in Section 2.4, more details of the discrete Fourier transforms with a few illustrative applications are found in Chapters 4 and 5. More applications are found in Chapter 7. This section, then, will start with a brief introduction of integral transforms in part A, followed by the familiar finite Fourier transforms in part B.

A. Integral Transforms - The Operational Calculus Method

For now, we shall use T to denote a general *integral transform*. The transform T operates on functions u which are suitably defined on a particular interval I. The result of applying T to a function G is a new function g which has its own independent variable t. This may be expressed as

$$g(t) \equiv T\{G\} = \int_I K(t,x)G(x)\,dx. \qquad (2.45)$$

We note that it may be customary to use capital letters for the transformed function as it is the case with the Laplace transform. Also, as we shall see shortly, the discrete exponential, sine and cosine Fourier transforms are the closest to their parallels of the finite Fourier transforms, where the latters are the Fourier coefficients c_n, a_n and b_n, respectively, and where the use of these lower case symbols, especially a_n and b_n, are well-entrenched in the writings of the Fourier trigonometric series analysis. So our choice of notation in (2.45) is in preparation for the notation used for the finite transforms, and most importantly, the discrete transforms. Another point, though not so forceful, is that we find many of the original practical discrete problems in the literature modeled as difference equations in sequences such as $\{C_n\}, \{G_n\}$, where a capital letter is used in the equation. So, for us to apply a discrete transform on the difference equation in G_n, we had to adopt the lower case g_k for its discrete transform. This is the same notation we adopted in our book on the integral and discrete transforms, when we introduced (in Section 4.1.5) what we termed then as "the Operational Difference Calculus Method" of the discrete Fourier transforms for solving difference equations. Besides, as we shall soon see in Section 2.4, the formulas for a discrete Fourier transform (2.93) and its inverse (2.94) are, for all practical purposes, *symmetric*, which is in contrast to both the Laplace transform (2.46) and its discrete analog, the z-transform of (6.3) with their rather complicated forms of inverses. The latters' inverses are considered beyond the mathematical preparation assumed for this book to suit as a text for the elementary course in difference equations.

The function $K(t,x)$ (of two variables) which appears in the integrand in (2.45) is called the *kernel (or nucleus) of the transform T*. The kernel, K, and the interval I are the two components which define the transform T. The *transform variable t* takes on values from an interval, often different than I, which is also characteristic of the particular transform. In summary, T in (2.45), transforms a function $G(x)$ by integrating it against a special function $K(t,x)$ to produce a new function $g(t)$, and it is often written in *operator* notation as $g = \mathcal{K}\{G\}$. Of course, there may exist an inverse to the transform T termed T^{-1}, which is defined to take g back to G via, in general, a different kernel $M(t,x)$, and we write $G = \mathcal{K}^{-1}\{g\} = \mathcal{M}\{g\}$.

Some examples of an integral transform are:

1. Fourier transform:

$$K(t,x) = e^{ixt}, \quad I = (-\infty, \infty).$$

2. Laplace transform:

$$K(t,x) = e^{-tx}, \quad I = (0, \infty).$$

In this book we shall limit our study to the Fourier and Laplace transforms, and most importantly in regards to their discrete analogs for their use in solving linear difference equations.

We note that in general the interval I of the above integral transform (2.45) is infinite.

The Laplace (Integral) Transform

We may point out again that the first exposure of the readers to integral transforms may very well be that of the Laplace transform of $f(t), t > 0$, where the typical notation used is

$$F(s) = \int_0^\infty e^{-st} f(t)\, dt \equiv \mathcal{L}\{f\} \tag{2.46}$$

where $F(s)$ is the Laplace transform of $f(t)$, $0 < t < \infty$, and most often the variable t represents *time* while s represents *frequency*. A simple example is the Laplace transform of $f(t) = e^{at}$, $t > 0$, which we can show to be

$$
\begin{aligned}
F(s) &= \frac{1}{s-a}, \quad s > a \\
\mathcal{L}\{e^{at}\} &= \int_0^\infty e^{-st}e^{at}\,dt = \int_0^\infty e^{-(s-a)t}\,dt \\
&= -\frac{1}{s-a}e^{-(s-a)t}\mid_{t=0}^\infty = -\frac{1}{s-a}(0-1), s > a, \\
\mathcal{L}\{e^{at}\} &= \frac{1}{s-a}, s > a
\end{aligned}
\tag{2.47}
$$

In the following example we will present the most important property of the Laplace transform in reducing derivatives defined on $(0,\infty)$ to an algebraic expression, i.e., its basic operational calculus property.

Example 2.6   Basic Operational Calculus Properties of Laplace Transform

As we intend to use the discrete transforms method in Section 2.4, Chapters 4, 5, and 6, besides the typical methods of solving difference equations in Chapter 3, it is instructive to show (or remind of) how the Laplace transform in (2.46) algebraizes the derivative $\frac{df}{dx}$, or derivatives, on $(0,\infty)$, hence it can transform a differential equation to a much simpler to solve algebraic equation,

$$
\mathcal{L}\left\{\frac{df}{dx}\right\} = sF(s) - f(0)
\tag{2.48}
$$

This is a simple example of what we have already termed the *"operational"* or *"Heaviside"* calculus. Here, we also note that, besides transforming the derivative $\frac{df}{dx}$ in the x-space to an algebraic expression $sF(s) - f(0)$ in the s-space, the Laplace transform also involves the given initial condition $f(0)$ in this process. The proof of (2.48) is simple, and it involves one integration by parts, which we shall present here for exercising the operational process, or for more of a review,

$$
\begin{aligned}
\mathcal{L}\left\{\frac{df}{dx}\right\} &= \int_0^\infty e^{-sx}\frac{df}{dx}dx \\
&= e^{-sx}f(x)\,|_{x=0}^\infty - \int_0^\infty -se^{-sx}f(x)dx \\
&= [0 - f(0)] + s\int_0^\infty e^{-sx}f(x)dx \qquad\qquad (2.48) \\
&= s\int_0^\infty e^{-sx}f(x)dx - f(0) = sF(s) - f(0)\ , \\
\mathcal{L}\left\{\frac{df}{dx}\right\} &= sF(s) - f(0)
\end{aligned}
$$

after letting $u(x) = e^{-sx}$ and $dv = df$ in the first integral, and assuming the condition on $f(x)$ such that $\lim_{x\to\infty} e^{-sx}f(x) = 0$ for the vanishing of the first part of the first term in the second line. The above algebraization applies to higher derivatives (primarily with constant coefficients), for example

$$
\mathcal{L}\left\{\frac{d^2 f}{dx^2}\right\} = s^2 F(s) - sf(0) - f'(0) \qquad\qquad (2.49)
$$

whose proof follows as that of (2.48) and we leave it as the subject of exercise 1.

We emphasize that the Laplace transformation in (2.49) has the added advantage of involving the *initial conditions* $f(0)$ and $f'(0)$ at the outset. This makes the Laplace transform compatible with differential equations associated with initial conditions, i.e., *initial value problems.*

With the Laplace transform pairs in (2.47) and (2.48), we will try in the following example, to complete our illustration of the operational calculus method of Laplace transform in algebraizing derivatives as in (2.48), (2.49), hence simplifying solving differential equations.

Example 2.7   The Operational Calculus Method of Laplace Transform

Consider the simple first order homogeneous differential equation in $u(t)$ and its initial condition $u(0) = 1$,

$$\frac{du}{dt} - au = 0 \qquad\qquad (E.1)$$

$$u(0) = 1 \qquad\qquad (E.2)$$

where we can easily find their solution by inspection as $u(t) = e^{at}$, that can be verified at a glance. We illustrate here the use of the Laplace transform in reducing the differential equation (E.1) in $u(t)$ to an algebraic equation in $U(s) = \mathcal{L}\{u(t)\}$, the Laplace transform of $u(t)$. Hence the latter algebraic equation should, in principle, be easier to solve than the differential equation (E.1). We have already seen that, in using the Laplace transform pair (2.48), the Laplace transformation of the differential equation in (E.1) will involve the given initial condition (E.2), hence there is no need for determining the arbitrary constant necessary for the usual methods of, basically, integrating the (solvable) differential equation.

If we let $U(s) = \mathcal{L}\{u(t)\}$, and Laplace transform both sides of the differential equation in (E.1), using the important operational property (2.48) (of the Laplace transform) on the derivative $\frac{du}{dt}$ in (E.1), we obtain

$$\begin{aligned} sU(s) - u(0) \quad &- \quad aU(s) = 0 \\ sU(s) \quad &- \quad aU(s) = (s - a)U(s) = u(0) = 1 \end{aligned} \qquad (E.3)$$

after using the initial condition $u(0) = 1$ from (E.2).

Now, we solve this simple *algebraic* equation in $U(s)$ to have

$$U(s) = \frac{1}{s - a}, \qquad s > a. \qquad\qquad (E.4)$$

What remains is to return to the original solution $u(t)$ of (E.1) - (E.2), or, in other words, find the inverse Laplace transform $u(t) = \mathcal{L}^{-1}\{U(s)\}$ of $U(s) = \frac{1}{s-a}, s > a$ in (E.4). For this, we most often have to depend on the, fortunately, extensive available Laplace transform tables. In our present case, we have already planned for this in deriving the Laplace transform pair,

$$\mathcal{L}\{e^{at}\} = \frac{1}{s - a}, \qquad s > a \qquad\qquad (2.47)$$

in (2.47). Hence, if we compare this pair with (E.4), we obtain $u(t) = e^{at}$ as the solution to the differential equation (E.1), and which also satisfies the initial condition (E.2), since the latter is always involved in the process of Laplace transforming the derivative $\frac{df}{dx}$, as seen in (2.48).

This is only an illustration of a very simple example for showing how the operational (or Heaviside) calculus method works in the case of the Laplace transform.

It is clear from this example, and especially the Laplace transform of $\frac{d^2 f}{dx^2}$ in (2.49) that the Laplace transform is compatible with initial value problems, since as seen in (2.49) it requires the initial value $f(0)$ and the initial derivative $f'(0)$ while it algebraizes the second derivative $\frac{d^2 f}{dx^2}$. The z-transform, as the discrete analog of the Laplace transform, is compatible with difference equations associated with *initial value problems*. This is the main reason behind relegating a special Chapter 6 to it, while Section 2.4, and Chapters 4 and 5 cover the discrete Fourier transforms, which are used, mainly, for solving difference equations associated with *boundary value problems*.

Next we present a brief introduction to the (complex) exponential Fourier transform and its (symmetric) inverse for the important purpose of drawing on their properties in establishing their discrete analog, i.e., the discrete Fourier transform and its inverse, which are of our concern for their operational method of solving difference equations.

The Exponential Fourier (Integral) Transform

We note here again the use of capital letter F in $F(s)$ for the transform, which is the convention for Laplace transform, and it is in a clear variance with what we wrote above in (2.45) for the transform as $g(t)$. We have started with the convention in (2.45) because it is the one that fits more with the common use of the transforms of our concern, namely, the Fourier transforms, and more importantly their discrete versions, i.e., the discrete Fourier transforms that we shall cover in detail in Section 2.4 and in Chapters 4, 5. Also in this book, where our aim is to explore the methods of solving difference equations, we will not be concerned much with the Laplace transform, but with another transform that plays its role for solving difference equations, namely, the *z-transform*, which we shall cover in Chapter 6. Another important point for the use of the notation of (2.45) is that the Fourier Transform $g(t)$ and its inverse $G(x)$ are, (aside from a minor scaling factor of $\frac{1}{2\pi}$ and a minus sign in the exponent of the kernel,) symmetric, namely,

$$g(t) = \int_{-\infty}^{\infty} e^{ixt} G(x) \, dx \equiv \mathcal{F}\{G\}, \qquad (2.50)$$

$$G(x) = \frac{1}{2\pi} \int_{-\infty}^{\infty} e^{-ixt} g(t) \, dt \equiv \mathcal{F}^{-1}\{g\}. \qquad (2.51)$$

where it does not make much difference which one of the above Fourier transforms, (2.50) or (2.51) has which symbol, as long as we are *consistent* in the use of the same notation in our writing or reporting.

Example 2.8   A Fourier Transform Pair

A simple example of a Fourier transform is that of the *gate function* $p_a(x)$, which as we shall show next, is

$$G(x) = p_a(x) = \begin{cases} 1, & |x| < a, \\ 0, & |x| > a, \end{cases} \qquad (E.1)$$

which, as we shall show next, is

$$\mathcal{F}\{p_a(x)\} = \frac{2 \sin at}{t} \qquad (E.2)$$

If we substitute $p_a(x)$ of (E.1) in (2.50) we have

$$\begin{aligned} g(t) &= \mathcal{F}\{p_a(x)\} = \int_{-\infty}^{\infty} p_a(x) e^{-ixt} dx \\ &= \int_{-\infty}^{-a} 0 e^{-ixt} dx + \int_{-a}^{a} 1 e^{-ixt} dx + \int_{a}^{\infty} 0 e^{-ixt} dx \\ &= \int_{-a}^{a} e^{-ixt} dx = \frac{1}{-it} e^{-ixt} \mid_{x=-a}^{a} \qquad (E.3) \\ &= -\frac{1}{it} [e^{-iat} - e^{iat}] = \frac{2}{2i} [e^{iat} - e^{-iat}] \\ &= \frac{2 \sin at}{t} \end{aligned}$$

after using the definition of $p_a(x)$ from (E.1) (with its three branches) in the integral, integrating, and involving the Euler identity $\sin\theta = \frac{1}{2i}[e^{i\theta} - e^{-i\theta}]$ in the last step.

As we shall see in Section 2.4 and Chapters 4 and 5, the discrete Fourier transforms are also symmetric, which is what concerns us most here as we use them to solve *boundary value problems* associated with difference equations. We have alluded to this application very briefly in Section 2.1, and in particular, for modeling and solving a *traffic flow problem* in (2.21)-(2.22), as we shall discuss it in Section 4.4.

Speaking of notations, the variable x in the Fourier transform and its inverse (2.50), (2.51) is in radians. However, for the applications in electrical engineering, the function $G(x)$ is usually taken in the frequency space, where the frequency f is given in cycles per second (Hertz), thus $x = 2\pi f$ radians. So in that field, and for the convenience, it is very natural for the Fourier transform and its inverse (2.50) and (2.51) to be written as follows:

$$g(t) = \int_{-\infty}^{\infty} e^{i2\pi ft} G(f)df \tag{2.52}$$

$$G(f) = \int_{-\infty}^{\infty} e^{-i2\pi ft} g(t)dt \tag{2.53}$$

where $j = \sqrt{-1}$ is used instead of i for the obvious reason of not to be confused with the symbol i, which is used for the current in circuit theory.

Indeed, since the Fourier transforms are used extensively in electrical engineering, much of the discrete transforms analysis is done with this notation of (2.52), (2.53). More importantly, the Fast Fourier Transform (FFT), which is the most efficient algorithm for computing the discrete Fourier transforms, was developed in association with the electrical engineering literature in 1965, and hence most forms of this valuable algorithm are found with the notation of (2.52) and (2.53). For such variation of notations, see problems 4 and 5 in the Exercises of Section 4.2.

In the same way we did for the Laplace transform in Example 2.6, we will show in the following example the basic operational calculus

property of the complex exponential Fourier transform (2.50) in re-
ducing the derivative $\frac{dG}{dx}$, defined on the whole real line $-\infty < x < \infty$
in the x-space, to a simple algebraic form $-itg(t)$ in the Fourier
transform t-space, and with the assumption that $G(x)$ vanishes as
$x \to \mp\infty$, i.e.,

$$\mathcal{F}\{\frac{dG}{dx}\} = itg(t) \tag{2.54}$$

Example 2.9   Basic Operational Calculus Property of the
                   Fourier Exponential Transform

Consider $\frac{dG}{dx}$ on $(-\infty, \infty)$ in the x-space, such that $G(x)$ vanishes as
$x \to \mp\infty$, its Fourier transform, according to (2.50) is

$$\mathcal{F}\{\frac{dG}{dx}\} = \int_{-\infty}^{\infty} \frac{dG}{dx} e^{ixt} dx$$

As we did for the case of the Laplace transform of $\frac{df}{dx}$ on $(0, \infty)$ in
(2.48) and Example 2.6, we will use one integration by parts letting
$u(x) = e^{ixt}$ and $dv = dG(x)$;

$$
\begin{aligned}
\mathcal{F}\{\frac{dG}{dx}\} &= \int_{-\infty}^{\infty} \frac{dG}{dx} e^{ixt} dx \\
&= G(x)e^{ixt} \big|_{-\infty}^{\infty} - \int_{-\infty}^{\infty} G(x) \cdot (it)e^{ixt} dx, \\
&= -it \int_{-\infty}^{\infty} G(x)e^{ixt} dx = -itg(t),
\end{aligned}
$$

$$\mathcal{F}\left\{\frac{dG}{dx}\right\} = -itg(t) \tag{2.54}$$

after invoking the condition that $G(x)$ vanishes as $x \to \mp\infty$ for the
first term in the first line to vanish, and recognizing the final integral
as the Fourier transform $g(t)$ of $G(x)$ as given in (2.50).

   In the same way, we can use two integration by parts to show
that

$$\mathcal{F}\left\{\frac{d^2G}{dx^2}\right\} = (-it)^2 g(t) = -t^2 g(t) \tag{2.55}$$

with the assumption that, here, $G(x)$ as well as its first derivative
$G'(x)$ vanish as $x \to \mp\infty$. We shall leave the details of proving (2.55)
for the exercises (see problem 3).

Fourier Sine and Cosine (Integral) Transforms

For completeness, and where we may need to compare them with their discrete analogs, the discrete Fourier sine (DST) and cosine transforms (DCT) in Section 2.4, and Chapter 5, we will present here the Fourier sine and cosine integral transforms for functions $G(x)$ on $(0, \infty)$ along with their (symmetric) inverses.

$$\mathcal{F}_s\{G\} = g_s(t) = \int_0^\infty G(x) \sin xt\, dx, \qquad (2.56)$$

$$\mathcal{F}_s^{-1}\{g_s\} = G(x) = \frac{2}{\pi} \int_0^\infty g_s(t) \sin xt\, dt \ , \qquad (2.57)$$

and

$$\mathcal{F}_c\{G\} = g_c(t) = \int_0^\infty G(x) \cos xt\, dx, \qquad (2.58)$$

$$\mathcal{F}_c^{-1}\{g_c\} = \frac{2}{\pi} \int_0^\infty g_c(t) \cos xt\, dt \qquad (2.59)$$

These Fourier sine and cosine transforms in (2.56) and (2.58) are considered as special cases of the Fourier (complex) exponential transform (2.50) of $G(x)$ on the symmetric interval $(-\infty, \infty)$, when $G(x)$ is odd or even, respectively. This observation is shown when we write $e^{ixt} = \cos xt + i \sin xt$ in the integral of (2.50), and note that for the *odd* function $G(x) \equiv G_o(x)$ for example, only the product $G(x) \sin xt$, which is an even function, contributes to the integration on the symmetric interval $(-\infty, \infty)$ to render it twice its integral on $(0, \infty)$. The product $G(x) \cos xt$, on the other hand, is an odd function, whose integration on the symmetric interval $(-\infty, \infty)$ vanishes. This says that for $G_o(x) \equiv G(x)$ odd, we have

$$
\begin{aligned}
\int_{-\infty}^\infty G_o(x) e^{ixt} dx &= \int_{-\infty}^\infty G_o(x) \cos xt\, dx \\
&\quad + i \int_{-\infty}^\infty G_o(x) \sin xt\, dx \\
&= 0 + 2i \int_0^\infty G(x) \sin xt\, dx
\end{aligned}
$$

In the same manner we can show that for an *even* function $G_e(x)$, its exponential Fourier transform (2.50) reduces to a Fourier cosine transform, i.e.,

$$\int_{-\infty}^{\infty} G_e(x)e^{ixt}dx = \int_{-\infty}^{\infty} G_e(x)\cos xt dx + i \int_{-\infty}^{\infty} G_e(x)\sin xt\, dx$$

$$= 2 \int_0^{\infty} G(x)\cos xt$$

since $G_e(x)\sin xt$ is an odd function, which makes the second integral vanish, while $G_e(x)\cos xt$ is an even function, which gave the final doubling of the integral on $(0, \infty)$. We conclude this section by listing the basic operational calculus properties of these Fourier sine and cosine transforms,

$$\mathcal{F}_s\left\{\frac{d^2G}{dx^2}\right\} = tg(0) - t^2g_s(t) \qquad (2.60)$$

and

$$\mathcal{F}_c\left\{\frac{d^2G}{dx^2}\right\} = -g'(0) - t^2g_s(t) \qquad (2.61)$$

assuming for both (2.60) and (2.61) that the limits of $G(x)$ as well as its derivative $\frac{dG}{dx}$ vanish as $x \to \mp\infty$. The derivation of (2.60)-(2.61) involves two integrations by parts, just as in the case of the Laplace transform in (2.49). We shall leave this derivation, for exercise 4(a), which is supported by the necessary hints.

For now, we may just observe that for algebraizing second order derivatives, while the Laplace transforms requires the value of the function as well as its derivative, i.e., $f(0)$ and $f'(0)$ at $x = 0$ in (2.49), the Fourier sine transform in (2.60) requires only the value of the function $g(0)$ at $x = 0$, while the Fourier cosine transform in (2.61) requires the derivative of the function $g'(0)$ at $x = 0$. In physical terms, the Laplace transform of $\frac{d^2f}{dt^2}$ requires the initial displacement $f(0)$ as well as the initial velocity $f'(0)$ for the displacement $f(t)$ as a function of time $t \epsilon (0, \infty)$. On the other hand, the Fourier sine and cosine transforms require, respectively, the say, temperature $G(0)$ and its gradient $G'(0)$ at the end of a semi-infinite bar where $x \epsilon (0, \infty)$. More details on these and other observations, of the sine and cosine transforms, including the (apparent) limitations of handling only *even order derivatives* compared to the Laplace and (exponential) Fourier transforms, are the subject of exercise 4(b).

Double Fourier Transforms for Partial Differential Equations

Here we would like only to remark that while Fourier and other transforms are used to solve ordinary differential equations in $G(x)$ by transforming them to the simpler algebraic equations, a *double* Fourier transform, for example, of functions of two variables $G(x, y)$, i.e., $g(t, s) = \mathcal{F}_{(2)}\{G(x, y)\}$ as defined in (2.164)-(2.165), is used to algebraize *partial differential equations* in $G(x, y)$. The method is a rather simple extension of the results developed above, and it is done for integral as well as the following finite transforms, where the basic details and applications are found in the author's book on the subject among other references. The point we would like to make here is that while we shall develop the similar "operational sum calculus method" of using discrete Fourier transforms to solve ordinary difference equations with boundary conditions in the sequence G_n, an extension can be made of defining double discrete Fourier transforms, for example, to be used for transforming partial difference equations in the sequence $G(n, m) \equiv G_{n,m}$ to the simpler algebraic equation in the transformed sequence $g_{k,l} = \tilde{\mathcal{F}}_{(2)}\{G_{n,m}\}$, where we used the symbol $\tilde{\mathcal{F}}_{(2)}$ here to indicate a *double* discrete Fourier transformation. The treatment in this book, of using the discrete Fourier transforms, will be limited to solving ordinary difference equations with boundary conditions. However, most of the well established methods of solving partial difference equations are discussed and illustrated in Section 3.5. We shall limit our acquaintance with the double discrete Fourier transform $\tilde{\mathcal{F}}_{(2)}$ to its definition as well as its inverse, which will be given at the end of Section 2.4.

Next we present a brief introduction to the finite Fourier transforms (or Fourier series and its coefficients), which should stand as sort of a bridge between the above Fourier integral transforms and the discrete Fourier transforms of Section 2.4, and Chapters 4 and 5.

B. Finite Transforms (Fourier Series Coefficients) and the Orthogonal Expansions

Now there are situations in which the *transform variable* t takes on only certain discrete values which we may denote $\{t_k\}$. In this case, it is desirable to work with a kernel $K(t,x)$ which possesses the property of *orthogonality* on the interval I' (which is generally different from I of (2.45), and often finite). A kernel $K(t_k, x)$ is *orthogonal* if

$$\int_{I'} K(t_k, x)\overline{K}(t_l, x)\, dx = \begin{cases} h(k) & \text{if } k = l \\ 0 & \text{if } k \neq l, \end{cases} \tag{2.62}$$

$$h(k) = \int_{I'} |K(t_k, x)|^2\, dx.$$

The notation $\overline{K}$ stands for the *complex conjugate* of K for cases in which K is complex valued, which simply means that every imaginary number $i = \sqrt{-1}$ in the kernel K must be replaced by $-i$. For example, in the case of the Fourier transform $K(t, x) = e^{ixt}$, while its complex conjugate $\overline{K}(t, x) = e^{-ixt}$. An orthogonal kernel $\phi(t_k, x)$ is called *orthonormal* if

$$\int_{I'} \phi(t_k, x)\overline{\phi}(t_l, x)\, dx = \begin{cases} 1 & \text{if } k = l \\ 0 & \text{if } k \neq l. \end{cases} \tag{2.63}$$

This follows easily when we normalize the kernel $K(t_k, x)$ of (2.62) and let

$$\phi(t_k, x) = \frac{K(t_k, x)}{\sqrt{h(k)}},$$

$$\begin{aligned} \int_{I'} \phi(t_k, x)\overline{\phi}(t_l, x)\, dx &= \int_{I'} \frac{K(t_k, x)}{\sqrt{h(k)}} \cdot \frac{\overline{K}(t_l, x)}{\sqrt{h(k)}}\, dx \\ &= \frac{1}{h(k)} \int_{I'} K(t_k, x)\overline{K}(t_l, x)\, dx \\ &= \begin{cases} \frac{h(k)}{h(k)} = 1 & \text{if } k = l \\ \frac{0}{h(k)} = 0 & \text{if } k \neq l \end{cases} \equiv \delta_{k,l}. \end{aligned}$$

where $\delta_{k,l}$ is the *Kronecker delta function*.

When the transform variable t is *discrete*, we may write the transform T, which we now call a *finite transform*, as

$$g_{I'}(t_k) = T\{G\} = \int_{I'} \phi(t_k, x) G(x)\, dx. \tag{2.64}$$

This defines a set of transformed values $\{g(t_k)\}$ which, because of the generally finite interval I', would be different from the samples $g(t_k)$ of the transform $g(t)$ in (2.50). For this reason we should use a different symbol for the sequence of the finite transform of (2.64), for example $g_{I'}(t_k)$. However, since such finite transforms (2.64) are often called the *Fourier coefficients* of the function $G(x)$, it is sufficient to label them g_k to distinguish them from the samples $g(t_k)$ of the Fourier transform in (2.50).

We may add that because of the orthonormal property of the kernel $\phi(t_k, x)$ it is possible to express the function $G(x)$ on the interval I' in terms of the following series with coefficients $g_{I'}(t_k) \equiv c_k$.

$$G(x) = \sum_k g_{I'} \bar{\phi}(t_k, x) = \sum_k c_k \bar{\phi}(t_k, x). \tag{2.65}$$

This representation takes the form of a series of *orthogonal expansion or Fourier series* where the Fourier coefficients $c_k = g_{I'}(t_k)$ are defined in (2.64) as the general *finite transform*. This form of c_k in (2.65) can be established if we multiply both sides of (2.65) by $\phi(t_l, x)$ then integrate over the generally finite interval I'

$$
\begin{aligned}
\int_{I'} G(x)\phi(t_l, x)dx &= \int_{I'} \left[\sum_k c_k \phi(t_l, x)\bar{\phi}(t_k, x) \right] dx \\
&= \sum_k c_k \int_{I'} \phi(t_l, x)\bar{\phi}(t_k, x)dx \\
&= c_k \delta_{k,l} = c_l = g_{I'}(t_l)
\end{aligned}
\tag{2.66}
$$

after allowing the integral on the right hand side of the first line to enter inside the, generally, infinite sum, then using the orthonormality property of the set $\{\phi(t_k, x)\}$ on the interval I'.

In the following example, we shall illustrate the concept of orthonormality for the complex exponential set of functions $\{e^{\frac{in\pi x}{a}}\}_{n=-\infty}^{\infty}$ on the symmetric finite interval $(-a, a)$, then follow it by writing the

Fourier series expansion of $G(x)$ in (2.65) on the symmetric interval $I' \equiv (-a, a)$ in terms of the same orthonormal set of complex exponential functions $\{\frac{1}{\sqrt{2a}} e^{\frac{in\pi x}{a}}\}_{n=-\infty}^{\infty}$.

Example 2.10   Orthonormality, Finite Transforms and Fourier
                   Series

First we show that the kernel

$$\{K(x, t_n)\} = \left\{ \frac{1}{\sqrt{2a}} e^{\frac{i\pi n x}{a}} \right\}_{n=-\infty}^{\infty}$$

is orthonormal on the symmetric interval $I' = (-a, a)$. From the definition of *orthonormality* in (2.61) we must show that

$$\int_{-a}^{a} \frac{1}{\sqrt{2a}} e^{\frac{i\pi k x}{a}} \cdot \frac{1}{\sqrt{2a}} e^{-\frac{i\pi l x}{a}} \, dx = \begin{cases} 1, & k = l \\ 0, & k \neq l \end{cases} \qquad (E.1)$$

The first part of $k = l$ is very easy to prove, since

$$\frac{1}{2a} \int_{-a}^{a} e^{\frac{i\pi k x}{a}} \cdot e^{-\frac{i\pi k x}{a}} \, dx = \frac{1}{2a} \int_{-a}^{a} dx = \frac{1}{2a} x \Big|_{x=-a}^{x=a}$$
$$= \frac{2a}{2a} = 1 \qquad (E.2)$$

The second part of $k \neq l$ in (E.1) is also easy after we perform the integration, then employ the *Euler identity*, $\sin\theta = \frac{1}{2i}(e^{i\theta} - e^{-i\theta})$

$$\int_{-a}^{a} \frac{1}{\sqrt{2a}} e^{\frac{ik\pi x}{a}} \cdot \frac{1}{\sqrt{2a}} e^{-\frac{il\pi x}{a}} \, dx = \frac{1}{2a} \int_{-a}^{a} e^{\frac{i\pi(k-l)}{a}} \, dx$$
$$= \frac{1}{2a} \frac{e^{\frac{i\pi(k-l)}{a}}}{\frac{i\pi(k-l)}{a}} \Big|_{x=-a}^{x=a}$$
$$= \frac{1}{2\pi i(k-l)} \left[e^{i\pi(k-l)} - e^{-i\pi(k-l)} \right] \qquad (E.3)$$
$$= \frac{1}{\pi(k-l)} \cdot \sin\pi(k-l) = 0, \quad k \neq l.$$

since $\sin\pi(k-l) = 0$ for the integer $k - l$.

Next we write the exponential Fourier series expansion of a function $G(x)$ on the symmetric interval $I' = (-a, a)$, which is a special case of the Fourier series (or general orthogonal expansion) as presented in (2.65) with its coefficients c_k as the finite Fourier transform in (2.64). Here we have $\phi(t_k, x) = \frac{1}{\sqrt{2a}} e^{\frac{ik\pi x}{a}}$ with $t_k = \frac{k\pi}{a}$, so its complex conjugate is $\overline{\phi}(t_k, x) = \frac{1}{\sqrt{2a}} e^{-\frac{ik\pi x}{a}}$, which is needed for (2.65).

$$G(x) = \sum_{k=-\infty}^{\infty} c_k \frac{1}{\sqrt{2a}} e^{-\frac{ik\pi x}{a}}, \quad -a < x < a, \qquad (E.4)$$

where we note that we have an infinite set for

$$\overline{\phi}(t_k, x) = \left\{ \frac{1}{\sqrt{2a}} e^{-\frac{ik\pi x}{a}} \right\}_{k=-\infty}^{k=\infty},$$

which we had to sum the series over. The Fourier coefficient c_k is the *finite* exponential Fourier transform, which is the following special case of (2.64),

$$c_k = g_a(t_k) = g_a\left(\frac{k\pi}{a}\right) = \int_{-a}^{a} G(x) \frac{1}{\sqrt{2a}} e^{\frac{ik\pi x}{a}} \, dx. \qquad (E.5)$$

Another more familiar form for the *complex exponential series* of $G(x)$ in (E.4) and its Fourier coefficient c_k in (E.5) is in the following Fourier series,

$$G(x) = \sum_{k=-\infty}^{\infty} d_k e^{-\frac{ik\pi x}{a}}, \qquad (E.6)$$

$$d_k = \frac{1}{2a} \int_{-a}^{a} G(x) e^{\frac{ik\pi x}{a}} \, dx \qquad (E.7)$$

after letting $d_k = \frac{1}{\sqrt{2a}} c_k$ in (E.4), (E.5).

The Finite Fourier Sine Transform

Another very familiar example of a finite Fourier transform is the following finite *Fourier sine transform* of $G(x)$ on the finite interval $(0, a)$,

$$f_s \{G\} = g_s\left(\frac{n\pi}{a}\right) \equiv g_n = \int_0^a f(x) \sin \frac{n\pi x}{a} \, dx \qquad (2.67)$$

with its kernel $K(t_k, x) = \sin \frac{n\pi}{a} x$, where $t_n = \frac{n\pi}{a}$. The *inverse* of this *finite Fourier sine transform* in the following *Fourier sine series* of $G(x)$ on $(0, a)$.

$$f_s^{-1}\{g_s(\frac{k\pi}{a})\} = G(x) = \frac{2}{a} \sum_{n=1}^{\infty} g_s(\frac{n\pi}{a}) \sin \frac{n\pi x}{a},$$

$$G(x) = \sum_{n=1}^{\infty} b_n \sin \frac{n\pi x}{a}$$

(2.68)

where $b_n = \frac{2}{a}g_s(\frac{n\pi}{a}) \equiv \frac{2}{a}g_n$ is called the *Fourier sine coefficient* of the Fourier sine series expansion (2.65) of $G(x)$ on $(0, a)$. We may remind again that it is such customary use of the lower case $b_n = \frac{2}{a}g_n$, the Fourier sine series coefficients as a finite Fourier sine transforms, for example, and its close parallel to the discrete Fourier sine transform, that lead us, somewhat, to adopt lower case symbols for the discrete transforms in this book starting with equation (2.45), which is in variance to what is found in most books on integral transforms including our book on the subject. The verification that $G(x)$ of (2.68) is indeed the inverse of the finite sine transform (2.67) is rather straight forward, and we leave it as the subject of exercise 5(a) with clear leading hint that includes the very important step of using the orthogonality property of the set $\{\sin \frac{n\pi x}{a}\}$ on $(0, a)$, the latter orthogonality property is proved in exercise 5(b).

The Finite Fourier Cosine Transform

Another finite Fourier transform is the following *finite Fourier cosine transform* of $G(x)$ on $(0, a)$ and its *inverse* $G(x)$ as the Fourier cosine series

$$f_c\{G\} = g_c(\frac{n\pi}{a}) = \int_0^a G(x) \cos \frac{n\pi x}{a} dx,$$

(2.69)

$$f_c^{-1}\{g_c(\frac{n\pi}{a})\} = G(x) = \frac{1}{a}g_c(0) + \frac{2}{a}\sum_{n=1}^{\infty} g_c(\frac{n\pi}{a}) \cos \frac{n\pi x}{a}$$

$$= a_0 + \sum_{n=1}^{\infty} a_n \cos \frac{n\pi x}{a}$$

(2.70)

where $a_0 = \frac{1}{a}g_c(0)$ and $a_n = \frac{2}{a}g_c(\frac{n\pi}{a}), n = 1, 2, \ldots$

As mentioned above, for the finite Fourier sine transform (2.67) and its inverse (2.68), the proof that $G(x)$ in (2.70) is the inverse transform of $g_c(\frac{n\pi}{a})$ in (2.69) is left for exercise 5(e) where the important orthogonality property of the cosine set of functions $\left\{1, \cos\frac{k\pi x}{a}\right\}_{k=1}^{\infty}$ on $(0, a)$ is proved in exercise 4(b). In this verification (with using the orthogonality property of the cosine functions), it will become clear why we have a different form for the Fourier cosine coefficient $a_0 = \frac{1}{a}g_c(0)$ from all the rest of the coefficients $a_n = \frac{2}{a}g_c(\frac{n\pi}{a}), n = 1, 2, \ldots$ (see problem 5(c), (d)).

The Finite (Complex) Exponential Fourier Transform

The more general example of a finite Fourier transform is the following *(complex) exponential Fourier transform* of $G(x)$ on the symmetric interval $(-a, a)$, and its inverse $G(x)$ as the *(complex) exponential Fourier series* of $G(x)$, on $(-a, a)$, which we have already illustrated in Example (2.10)

$$f_e\{G\} = g_e(\frac{n\pi}{a}) = \int_{-a}^{a} G(x)e^{\frac{-in\pi}{a}x}dx, \qquad (2.71)$$

$$
\begin{aligned}
f_e^{-1}\{g_e(\frac{n\pi}{a})\} &= G(x) = \frac{1}{2a}\sum_{n=-\infty}^{\infty} g_e(\frac{n\pi}{a})e^{\frac{in\pi x}{a}} \\
&= \sum_{n=-\infty}^{\infty} c_n e^{\frac{in\pi x}{a}}, \qquad -a < x < a
\end{aligned}
\qquad (2.72)
$$

where $c_n \equiv \frac{1}{2a}g_e\left(\frac{n\pi}{a}\right)$.

The verification that $G(x)$ in (2.27) is the inverse of $g_e(n)$ in (2.71) parallels exactly what was done in exercise 5(e) for the finite Fourier sine and cosine transforms of (2.67) - (2.68) and (2.69 - (2.70), respectively. For that, the needed orthogonality property of $\{e^{\frac{in\pi x}{a}}\}_{n=-\infty}^{\infty}$ on $(-a, a)$ was proved already in Example 2.10.

The Basic Operational Properties of the Finite Fourier Transforms

We may now discuss the three basic finite Fourier transforms, namely, the (complex) exponential, the sine, and the cosine Fourier transforms with emphasis on their respective important properties of algebraizing derivatives on a *finite* interval, and also involving the given auxiliary (boundary) conditions at the boundary or boundaries of the finite interval. As in the case of the integral transforms of part A, we do this for the finite Fourier transforms because their operational calculus properties parallel even closer that of the discrete Fourier transforms. We are, of course, after such properties of the latter discrete Fourier transforms to use for the direct discrete transforms method of solving difference equations, with, generally nonhomogeneous boundary conditions, or periodic boundary conditions with jump discontinuity at the boundary points, which is a relatively new feature for this book. The second reason is that the finite transforms, and the Fourier ones in particular stand as a bridging point between the integral transforms of part A and the discrete transforms of the next Section 2.4, which are of a main concern in this book. This is in the sense that the finite transform $g_{I'}(t_k)$ is already defined for infinite but discrete values t_k of its argument as in (2.64) compared to the continuous argument t of the integral transform $g(t)$ in (2.45). In addition, the inverse of the finite transform (or the Fourier series expansion of $G(x)$ in (2.65), though an infinite sum, is still a sum of discrete sequence $u_k \equiv c_k \overline{\phi}(t_k, x)$, which is in contrast to, in general, an integral of continuous function for the inverse of the integral transform, for example, the inverse Fourier transform $G(x)$ in (2.51). As we mentioned earlier, and which we shall introduce soon, all the discrete Fourier transforms, as well as their inverses, are finite sums. This means that there is a long way in approximating a Fourier integral transform by a discrete one, however, this distance may be a bit closer for the finite transforms (or Fourier series). Fortunately, for the subject of this book, we are dealing here with a set of generally finite samples for the difference equations, where operating on the difference equation by a discrete Fourier transform involves a finite sum over these sample points. This emphasis makes (linear) difference equations the natural setting for the use of the discrete Fourier transforms, where they transform these equations to algebraic equations, as well as involve the given auxiliary (boundary) conditions up front, hence simplifying the solutions of discrete problems modeled by difference equations as their most suitable mathematical representation. For simplicity, we shall use the symmetric interval $(-\pi, \pi)$ for the finite (complex) exponential Fourier transform of (2.71)-(2.72) and $(0, \pi)$ for the finite sine and cosine transforms of (2.67)-(2.68) and (2.69)-(2.70), respectively.

1. THE FINITE EXPONENTIAL FOURIER TRANSFORM.

As was introduced in (2.71)-(2.72), the finite (complex) exponential transform of $G(x)$, defined on the symmetric interval $(-\pi, \pi)$ becomes

$$f_e\{G\} = g_e(n) = \int_{-\pi}^{\pi} G(x)e^{-inx}dx \qquad (2.73)$$

with its inverse as the following Fourier series expansion of $G(x)$ on $(-\pi, \pi)$,

$$f_e^{-1}\{g_e(n)\} = G(x) = \frac{1}{2\pi}\sum_{n=-\infty}^{\infty} g_e(n)e^{inx} \qquad (2.74)$$

An extremely important property of this Fourier series representation (2.74) of $G(x)$ on the interval $(-\pi, \pi)$ is that $G(x)$ here is *periodic* with period 2π. The proof is simple, which stems from the periodic property of the orthogonal function e^{inx} with period 2π, since $e^{in(x+2\pi)} = e^{inx}e^{in2\pi} = e^{inx}\cos 2\pi n = e^{inx}$.

So with this in mind, we can show that for $G(x)$ of the Fourier series (2.74), $G(x + 2\pi) = G(x)$, since

$$G(x + 2\pi) = \frac{1}{2\pi}\sum_{n=-\infty}^{\infty} g_e(n)e^{in(x+2\pi)}$$

$$= \frac{1}{2\pi}\sum_{n=-\infty}^{\infty} g_e(n)e^{inx}e^{in2\pi}$$

$$= \frac{1}{2\pi}\sum_{n=-\infty}^{\infty} g_e(n)e^{inx}$$

The *basic operational calculus property* of this finite Fourier transform is again that of algebraizing a derivative $\frac{dG}{dx}$ or $(-\pi, \pi)$, and involving the given boundary conditions at the ends of the interval $x = \mp\pi$ as follows

$$f_e\{\frac{dG}{dx}\} = (-1)^n[G(\pi) - G(-\pi)] + ing_e(n) \qquad (2.75)$$

As we did for the case of the Laplace transform in (2.48) and the Fourier transform in (2.54), as proved in Example 2.9, this result (2.75) can be obtained with one integration by parts. In the same manner, two integration by parts give the following result for transforming a second order derivative $\frac{d^2G}{dx^2}$ on $(-\pi, \pi)$,

$$
\begin{aligned}
f_e\{\frac{d^2G}{dx^2}\} = (-1)^n[G'(\pi) - G'(-\pi)] \\
+ in\left[(-1)^n\{G(\pi) - G(-\pi)\}\right] + ing_e(n)]
\end{aligned}
\tag{2.76}
$$

The proofs of (2.75) and (2.76) are left for exercise 6 of this section. We note how in (2.75) the finite Fourier transformation reduced the first derivative $\frac{dG}{dx}$ in the x- space to an algebraic expression in $g_e(n)$ in the transform n-space, plus that it involved the difference (or jump) of the values of the function $G(\pi) - G(-\pi)$ between the two end points $x = \mp\pi$ of the interval $(-\pi, \pi)$. In (2.76) the second derivative is also algebraized to become $(in)^2g_e(n)$ in the transform space, plus a term involving the jump in the value of the function $G(\pi) - G(-\pi)$ and another one involving the jump $G'(\pi) - G'(-\pi)$ in the derivative of the function $G(x)$ between the two end points $\mp\pi$. We soon will see a parallel to these results in using the discrete Fourier transforms for algebraizing difference equations and involving the jumps in *the boundary values*.

We may note that in (2.75), if there is no jump between the values at the end points, i.e., for the *periodic boundary condition* $G(\pi) = G(-\pi)$, then the derivative $\frac{dG}{dx}$ is transformed to the simple one term $ing_e(n)$,

$$
f_e\{\frac{dG}{dx}\} = ing_e(n)
\tag{2.77}
$$

The same thing happens for (2.76), where in the case that there are no jumps in the value of the function $G(x)$ and its derivative $G'(x)$ at the end points, i.e., $G(\pi) = G(-\pi)$ and $G'(\pi) = G'(-\pi)$, then $\frac{d^2G}{dx^2}$ is transformed to $(in)^2g_e(n)$

$$
f_e\{\frac{d^2G}{dx^2}\} = (in)^2g_e(n) = -n^2g_e(n)
\tag{2.78}
$$

Indeed such simple results (2.77), (2.78), are usually obtained when we take the derivative of both sides of (2.74) and, formally allow ourselves the differentiation inside the infinite series,

$$\frac{dG}{dx} = \frac{d}{dx}\frac{1}{\pi}\sum_{n=-\infty}^{\infty} g_e(n)e^{inx} = \frac{1}{2\pi}\sum_{n=-\infty}^{\infty} g_e(n)\frac{de^{inx}}{dx}$$

$$= \frac{1}{2\pi}\sum_{n=-\infty}^{\infty} ing_e(n)e^{inx}$$

(2.79)

This *formal* result shows that $ing_e(n)$ is the finite Fourier transform of $\frac{dg}{dx}$ according to (2.79) where we can let $H(x) = \frac{dG}{dx}$, and see that its finite transform is $h(n) = ing_e(n)$. An operation like this, obviously, does not seem to bring about the jump term $(-1)^n[G(\pi) - G(-\pi)]$ of (2.75) that we obtained via the operation of the integration by parts. The variance in these two results (2.75) and (2.79) may be attributed intuitively to the goodness of the integration operation compared to the price we have to pay for the differentiation operation in (2.79). It is indeed more explicit than that, where this variance can be traced back to allowing the differentiation operation in the infinite series of (2.79), an operation that we have to pay a price for in terms of some limitations on $G(x)$, not the least of which is the uniform convergence of the Fourier series of $G(x)$ in (2.74), which disallows any jump discontinuity, such as $G(\pi) - G(-\pi)$ of (2.76), i.e., it should satisfy the typical periodic boundary condition $G(\pi) = G(-\pi)$. We shall leave the more detailed reasoning and discussion of this observation for (2.79) versus (2.75) and also (2.78) versus (2.76), to exercise 7.

2. THE FINITE FOURIER SINE TRANSFORM.

For $G(x)$ defined on $(0, \pi)$, its finite Fourier sine transform as a special case of that in (2.67) with $a = \pi$ is

$$f_s\{G(x)\} = g_s(n) = \int_0^\pi G(x)\sin nx\, dx \qquad (2.80)$$

which, as we explained in the case of the Fourier sine transform of (2.56), can be shown to result from the (complex) exponential Fourier transform (2.73) of an *odd* function $G(x)$ on the symmetric interval $(-\pi, \pi)$. From (2.68), the inverse of this finite Fourier sine transform is the following Fourier sine series of $G(x)$ on $(0, \pi)$,

$$G(x) = \frac{2}{\pi} \sum_{n=1}^{\infty} g_s(n) \sin nx = \sum_{n=1}^{\infty} b_n \sin nx \qquad (2.81)$$

where $b_n = \frac{2}{\pi} g_s(n)$ is the Fourier sine coefficient.

The basic operational calculus property of the finite Fourier sine transform (2.70) in algebraizing, practically, only *even order* derivatives, as we explained for the case of the Fourier sine integral transform in (2.56) (see exercise 4(b)), is

$$f_s\{\frac{d^2G}{dx^2}\} = n[G(0) - (-1)^n G(\pi)] - n^2 g_s(n) \qquad (2.82)$$

We note here that this finite sine transform involves the values of the function $f(0)$ and $f(\pi)$ at the two end points $x = 0$, π. Hence, as expected, this finite sine transform is compatible with boundary value problems that involve the values of the function at the boundary points, which are termed *Dirichlet*–type boundary conditions. The derivation of (2.82) involves two integrations by parts as was done for the Fourier sine transform of (2.56), which we leave for exercise 6(b).

3. THE FINITE FOURIER COSINE TRANSFORM.

In the same manner we introduce the finite Fourier cosine transform of $G(x)$, on $(0, \pi)$ as a special case of that in (2.69) with $a = \pi$.

$$f_c\{G(x)\} = g_c(n) = \int_0^{\pi} G(x) \cos nx\, dx \qquad (2.83)$$

with its inverse as the following Fourier cosine series expansion of $G(x)$ on $(0, \pi)$, which is a special case of (2.70) for $a = \pi$.

$$\begin{aligned} G(x) &= \frac{1}{\pi} g_c(0) + \frac{2}{\pi} \sum_{n=1}^{\infty} g_c(n) \cos nx \\ &= a_0 + \sum_{n=1}^{\infty} a_n \cos nx \end{aligned} \qquad (2.84)$$

where $a_0 = \frac{1}{\pi} g_c(0)$, while $a_n = \frac{2}{\pi} g_c(n)$ for $n = 1, 2, \ldots$

The important operational calculus property of the finite cosine transform (2.83) in algebraizing, practically, only *even order derivative*, is

$$f_c\{\frac{d^2G}{dx^2}\} = (-1)^n G'(\pi) - G'(0) - n^2 g_c(n) \qquad (2.85)$$

whose derivation involves two integration by parts, and is left for exercise 6(c). We note here that this finite Fourier cosine transform involves the first order derivatives of the function $G'(0)$ and $G'(\pi)$ at the end points $x = 0, \pi$, which contrasts that of the *values* of the function $G(0)$ and $G(\pi)$ at the two end points $x = 0, \pi$ (*Dirichlet conditions*) for the finite Fourier sine transform of $\frac{d^2G}{dx^2}$ in (2.82). The conditions of involving the slope of $G(x)$ at the two end points as in (2.85) are termed *Neumann*–type boundary conditions.

Other modification to the finite Fourier transforms besides those in (2.73), (2.80) and (2.83) are available, and often are constructed in advance so that they can involve the particular given auxiliary conditions on the solution of a differential equation. For example, the following "modified" finite sine transform of $G(x)$ on $(0, \pi)$

$$f\{G\} = g(n) = \int_0^\pi G(x) sin(n - \frac{1}{2})x dx \qquad (2.86)$$

can be shown to algebraize the second order derivative $\frac{d^2G}{dx^2}$ on $(0, \pi)$ in the following way

$$f\{\frac{d^2G}{dx^2}\} = -(n - \frac{1}{2})g(n) + (n - \frac{1}{2})G(0) - (-1)^n G'(\pi) \qquad (2.87)$$

where we note that it involves the value of the function at the first end $x = 0$, and the value of its derivative $G'(\pi)$ at the second end point $x = \pi$. Such transform complements the finite Fourier sine transform which involves the values of the function $G(0)$ and $G(\pi)$ at the two end points as shown in (2.82), and the finite cosine transform which involves the derivatives $G'(0), G'(\pi)$ at the two end points as in (2.85). From an application point of view, this makes such a modified finite sine transform compatible with, for example, a temperature distribution problem where the temperature $G(0)$ is given at the first end $x = 0$, and the gradient $G'(\pi)$ is given at the other end $x = \pi$.

It is hoped that after introducing the involvement of the discrete Fourier transform, in Section 2.4 and Chapters 4 and 5, for solving difference equations in this book, that a variation or "modification" of, the discrete exponential (DFT), the discrete sine (DST) and the discrete cosine (DCT) transforms can be derived to accommodate the different auxiliary conditions imposed on the two ends of a finite sequence as the sought solution of the given difference equation associated with such boundary conditions. Of course, the auxiliary conditions here are, primarily, boundary conditions as one would expect from the discrete Fourier transforms in parallel to their corresponding finite Fourier transforms, where, as we saw above, they involve variations of boundary conditions. This is in contrast with the Laplace transform and its discrete analog, the z-transform that involve initial conditions, hence their compatibility with initial value problems.

We shall leave the proof of (2.87) for an exercise, where, as we expect, we need two integration by parts. The subject of constructing compatible transforms, including the one in (2.86), is discussed in some detail in the author's book on "Integral and Discrete Transforms," in particular Section 1.6 of Chapter 1.

A general remark that may stand to relate the Fourier integral transform representations of part A or the finite transforms in part B, and the infinite sum (Fourier series) of (discrete values) in the inverse of the finite transforms, is that of the Shannon sampling theorem. It simply stands to use the *discrete* values of $f_e(n)$ of the (complex) exponential Fourier transform in an infinite sum to generate its continuous value $f_e(t)$. The explicit result and its derivation are left for exercise 8 with detailed leading hints.

All what we have presented of integral and finite transforms in parts A and B of this section, respectively, was mainly to illustrate the idea of their role in algebraizing derivatives as well as involving the given boundary conditions at the outset, and draw upon their parallel for the discrete transforms in solving difference equations associated with *boundary values* as we shall introduce in the next Section 2.4. Our main emphasis will be on the discrete Fourier transforms, which will also be discussed in detail in Chapters 4 and 5. The z-transform, as the discrete analog of the Laplace transform is discussed in Chapter 6 as another discrete transform that is useful in solving difference equations associated with *initial* values.

Exercises 2.3

1. (a) Prove the following basic operational calculus property (2.49) of the Laplace transform:

$$\mathcal{L}\{\frac{d^2 f}{dt^2}\} = s^2 F(s) - sf(0) - f'(0) . \qquad (2.49)$$

Hint: Apply two integrations by parts twice, starting with $u(x) = e^{-sx}$, $dv = \frac{d^2 f}{dx^2} dx = \frac{df'}{dx} dx = df'(x)$, then continue the second integration by parts with $u(x) = e^{-sx}$ and $dv = df(x)$. See also the derivation of (2.48).

   (b) Use the result (2.48) only to derive (2.49).

Hint: Let $h(t) = \frac{df}{dt}$, and apply (2.48) on $\frac{dh}{dt} = \frac{d^2 f}{dt^2}$, remembering that $\mathcal{L}\{h(t)\} = H(s) = \mathcal{L}\{\frac{df}{dt}\} = sF(s) - f(0)$.

2. Show that if $F(s) = \mathcal{L}\{f(t)\}$, then $\mathcal{L}\{e^{bt} f(t)\} = F(s - b)$, which is what is called the *shifting* property of the Laplace transform.

3. (a) For a similar way to the Laplace transform operational property (2.49) of problem 1, prove the basic operational calculus property (2.55) of the (inverse) exponential Fourier transform (2.50), assuming that $\lim_{x \to \mp\infty} G(x) = 0$ and $\lim_{x \to \mp\infty} G'(x) = 0$.

Hint: See the hint to problem 1 and the derivation of (2.54), where in the present case we need two integrations by parts starting with $u(x) = e^{ixt}$, $dv = \frac{d^2 G}{dx^2} dx = \frac{dG'}{dx} dx = df$, where $f(x) = G'(x)$, then follow this by a second integration by parts with $u(x) = e^{ixt}$ and $dv = \frac{dG}{dx} dx = dG$.

   (b) Attempt to use the method of problem 1(b) in employing the result in (2.54) to derive (2.55). Note that you still need $\lim_{x \to \mp\infty} G'(x) = 0$ besides $\lim_{x \to \mp\infty} G(x) = 0$.

4. Consider the (inverse) Fourier sine and cosine transforms in (2.57) and (2.59), respectively.

(a) Prove their basic operational calculus properties in (2.60) and (2.61), respectively, with the assumptions that $\lim_{x \to \mp\infty} G(x) = 0$ and $\lim_{x \to \mp\infty} G'(x) = 0$.

   Hint: For both (2.60) and (2.61), use integration by parts twice with, for example in (2.60), $u(x) = \sin(xt)$ and $dv = \frac{d^2G}{dx^2}dx = \frac{dG'}{dx}dx = df$, where $f(x) = G'(x)$ for the first integration by parts, then continue the second integration by parts.

(b) From (2.48) and (2.49) of the Laplace transform and (2.54) and (2.55) of the exponential Fourier transform, we note that the Laplace transform in (2.48)–(2.49) as well as the Fourier transform in (2.54)–(2.55) algebraize the first and the second order derivatives, i.e., we can show that they cover *all order derivatives*. In comparison, what do you observe about the operational calculus properties of the Fourier sine and cosine transforms in (2.60)-(2.61).

   Hint: Watch for the integration by parts since, for example, letting $u(x) = \sin(xt)$ needs two integrations by parts to return back to an integral transform of the original function with the same kernel $\sin(xt)$ in order that the sine transform of $\frac{d^2G}{dx^2}$ ends up in terms of the same Fourier sine transform of $G(x)$.

5. (a) Prove that the set $\{\sin\frac{n\pi x}{a}\}_{n=1}^{\infty}$ is orthogonal on the interval $(0, a)$, and show the following relation:

$$\int_0^a \sin\frac{k\pi x}{a}\sin\frac{l\pi x}{a}dx = \begin{cases} 0, & k \neq l \\ \dfrac{a}{2}, & k = l \end{cases} \qquad (E.1)$$

   Hint: Consult the definition of orthogonality in (2.63) with $K(t_k, x) = \sin\frac{k\pi x}{a}$, and for the integral of the product in (E.1) consult the trigonometric identity

$$\sin A \sin B = \frac{1}{2}[\cos(A - B) - \cos(A + B)]. \qquad (E.2)$$

(b) Verify that the Fourier sine series (2.68) does indeed have its Fourier sine coefficient as $b_n = \frac{2}{a}g_s(\frac{n\pi}{a})$, where $g_s(\frac{n\pi}{a})$ is given by (2.67).

Hint: Multiply both sides of (2.68) by $\sin\frac{n\pi x}{a}$, then integrate from $x = 0$ to a, using the orthogonality property of $\{\sin\frac{n\pi x}{a}\}_{n=1}^{\infty}$ on $(0, a)$, and the result (E.1) in part (a).

(c) Similar to part (a), show that the set $\{\cos\frac{k\pi x}{a}\}_{k=1}^{\infty}$ is orthogonal on $(0, a)$, i.e., prove the following relation:

$$\int_0^a \cos\frac{k\pi x}{a}\cos\frac{l\pi x}{a}dx = \begin{cases} 0, & k \neq l \\ \frac{a}{2}, & k = l \neq 0 \\ a, & k = l = 0 \end{cases} \qquad (E.3)$$

Hint: See the hint to part (a), and for the integral use the trigonometric identity

$$\cos A \cos B = \frac{1}{2}[\cos(A - B) + \cos(A + B)]. \qquad (E.4)$$

(d) In part (c), we showed two values for the integral in (E.3) (of a for $k = l = 0$, and $\frac{a}{2}$ for $k = l \neq 0$). Explain why we have different formulas in (2.70) for the Fourier cosine coefficient $a_0 = \frac{1}{a}g_c(0)$ for $n = 0$, and $a_n = \frac{2}{a}g_c(n)$ for the rest of $n = 1, 2, \ldots$?

Hint: Note that in verifying that such a_0 and a_n, $n = 1, 2, \ldots$, are indeed the Fourier cosine coefficients, we need a special integral for a_0, namely the one with value a in (E.3), while all other a_n's, $n = 1, 2, \ldots$ have different integral with value $\frac{a}{2}$ in (E.3).

(e) As in part (b), verify the form of the Fourier cosine coefficients in (2.70).

Hint: See the hint in part (b), use the orthogonality of the cosine functions $\{1, \cos\frac{n\pi x}{a}\}_{n=1}^{\infty}$ on $(0, a)$ and its result in (E.2) of part (c).

(f) Prove the following orthogonality property of $\{\sin\frac{n\pi x}{a}, \cos\frac{n\pi x}{a}, \sin\frac{2\pi x}{a}, \cos\frac{2\pi x}{a}, \ldots\}$ on $(0, a)$:

$$\int_0^a \sin\frac{k\pi x}{a}\cos\frac{l\pi x}{a}dx = 0, \qquad k \neq l \qquad (E.5)$$

Hint: See the hint for part (a), and for the integral use the trigonometric identity

$$\sin A \cos B = \frac{1}{2}[\sin(A - B) + \sin(A + B)] . \quad (E.6)$$

6. (a) Prove the results in (2.75) and (2.76).

    Hint: Use one and two integration by parts for (2.75) and (2.76), respectively, and watch carefully for the boundary terms evaluated at $x = \pi$ and $x = -\pi$.

    (b) Derive the result in (2.82).

    Hint: See the hint in part (a).

    (c) Derive the result in (2.85).

7. Explain the reason behind the difference between the results in (2.75), (2.76), with their jumps, and their corresponding ones (2.77), (2.78) without the jump.

    Hint: Since allowing the differentiation operation inside the infinite series in the first line of (2.71) requires, aside from $G(x)$ and its derivative $G'(x)$ being continuous on $(-\pi, \pi)$, that $G(-\pi) = G(\pi)$, i.e., the result (2.69) is valid provided that $G(-\pi) - G(\pi) = 0$, a result that is a special case of (2.67). The same reasoning can be given to the difference between (2.70) and (2.68), where for differentiating twice inside the Fourier series (2.66) for $\frac{d^2 G}{dx^2}$, one of the necessary conditions is that $G(\pi) - G(-\pi) = 0$ and $G'(\pi) - G'(-\pi) = 0$.

8. Prove the following result of the Shannon sampling theorem. For

    $$f_e(t) = \int_{-a}^{a} e^{\frac{-in\pi}{a}x} G(x)dx, \quad (E.1)$$

    show that $f_e(t)$ can be expressed in terms of its samples $\{f_e(\frac{n\pi}{a})\}_{n=-\infty}^{\infty}$ as the following sampling series,

    $$f_e(t) = \sum_{n=-\infty}^{\infty} f_e(\frac{n\pi}{a}) \frac{\sin(at - n\pi)}{(at - n\pi)} \quad (E.2)$$

Hint: Write the Fourier exponential series for $G(x)$ as in (2.72), multiply both sides by $e^{-\frac{in\pi x}{a}}$ then integrate from $-a$ to a and allow the term by term integration. Use $f_e(t)$ in (E.1) to recognize the expression $f_e(\frac{n\pi}{a})$ for its samples.

2.4 The Discrete Fourier Transforms and Their Sum Calculus Method

In parallel to the last section, where the operational calculus method of the integral and finite Fourier transforms was used to solve differential equations associated with, generally, nonhomogeneous boundary conditions, we introduce here the discrete Fourier transforms to show their similar advantage in solving difference equations with, generally, nonhomogeneous boundary conditions. This is in the sense that the discrete Fourier transforms also algebraize the difference operator of the, mainly, constant coefficient difference equations, and at the same time, involve the given boundary conditions.

We began this general discussion with the integral transforms case in which both the original independent variable, x, and the transform variable, t, are *continuous* (taking on *all* values in certain intervals) as in (2.45). Next we considered a discrete transform variable $\{t_k\}$ (taking on *discrete* values in an interval) for the finite transform in (2.64). Now what happens if the function $G(x)$ is sampled on a discrete set of grid points $\{x_n\}$? That is, we now consider both variables t and x to be discrete. The integrals of (2.62) and (2.64) now become sums over the discrete set $\{x_n\}$. In particular, if we label the grid points $\{x_0, x_1, \cdots, x_{N-1}\}$, then the *discrete orthogonality property* becomes

$$\sum_{n=0}^{N-1} K(t_k, x_n)\overline{K}(t_l, x_n) = \begin{cases} h(N) & \text{if } k = l \\ 0 & \text{if } k \neq l. \end{cases} \quad (2.88)$$

The *discrete transform* itself may now be written as

$$\tilde{g}(t_k) = \tilde{T}\{\tilde{G}\} = \sum_{n=0}^{N-1} \phi(t_k, x_n)\tilde{G}(x_n) \text{ for } k = 0, 1, \cdots, N-1, \quad (2.89)$$

$$\phi(t_k, x_n) = \frac{K(t_k, x_n)}{h(N)} \tag{2.90}$$

where the new notation $\tilde{g}$, $\tilde{G}$, and $\tilde{T}$ have been introduced to indicate that these are the discrete analogs of the continuous transforms g, G, and T of (2.45). Clearly the samples $g(t_k)$ of the transform (2.45) with its infinite interval I, the finite transform $g_k = g_{I'}(t_k)$ in (2.64) and the discrete transform $\tilde{g}(t_k)$ in (2.89) are different. The analysis for the relations between them is very important, but is not in the main stream of this book. However, very often this is the case where the discrete transform $\tilde{g}(t_k)$, and especially the Fourier series, are used to *approximate* the integral transform $g(t)$. This obviously, makes a difference or *error* $\epsilon \equiv g(t_k) - \tilde{g}(t_k)$, which manifests itself in a number of very well known errors such as the *truncation* and *aliasing* or *sampling* (*discretizing*) errors. This vital subject is covered, on a very accessible level to the readers, in the author's recent book on this same subject entitled "Integral and Discrete Transforms with Applications and Error Analysis," Marcel Dekker Inc., 1992, and in particular its Chapter 4 on the discrete transforms with its very detailed and illustrated treatment of, principally, the discrete Fourier transforms and the varied errors associated with their use in practice.

Since our topic here is difference equations, our aim will be the development and use of the discrete transforms for solving difference equations. In this case, as we shall see in Chapters 4 and 5, we will use the simple sequence notation g_k and G_n to stand for our main transforms, the discrete transforms $\tilde{g}(t_k)$ and $\tilde{G}(x_n)$ of (2.89). We will present next an illustration of the discrete Fourier Transforms (DFT) and their discrete orthogonality, as special cases of (2.89) and (2.88), respectively, which will also be discussed in Chapters 4 and 5 with much more details.

The Discrete Fourier Transform (DFT)

As we mentioned earlier, very often, the discrete Fourier transforms are expressed in the notation of (2.52), (2.53) as opposed to (2.50), (2.51). Consider the sequence $\tilde{G}(f_n)$, $n = 0, 1, 2, \ldots, N-1$, as a function of the *discrete frequency* f_n in cycles per second (Hertz). The discrete Fourier transform (DFT) of the sequence is the following sequence $\tilde{g}(t_k)$, $k = 0, 1, 2, \cdots, N - 1$ with its N elements,

$$\tilde{g}(t_k) = \frac{1}{N} \sum_{n=0}^{N-1} \tilde{G}(f_n) e^{i2\pi t_k f_n}, \quad k = 0, 1, 2, \ldots, N - 1. \qquad (2.91)$$

with its (symmetric–modulo a constant factor of $\frac{1}{N}$ and i replaced by $-i$) inverse (IFDT) as

$$\tilde{G}(f_n) = \sum_{k=0}^{N-1} \tilde{g}(t_k) e^{-i2\pi t_k f_n}, \quad n = 0, 1, 2, \ldots, N - 1. \qquad (2.92)$$

With T as the *samples spacing* in time, i.e., $\Delta t_k = t_{k+1} - t_k = T$, we have $t_k = kT$, and for reasons that will become very clear in Chapter 4, we have the frequency spacing as $\Delta f_n = f_{n+1} - f_n = \frac{1}{NT}$, thus $f_n = \frac{n}{NT}$. With these values given to t_k and f_n in (2.91), (2.92), we have one of the most common forms of the discrete Fourier transform and its inverse (see (4.5)-(4.6)),

$$\tilde{g}(kT) = \frac{1}{N} \sum_{n=0}^{N-1} \tilde{G}(\frac{n}{NT}) e^{\frac{i2\pi nk}{N}}, \quad k = 0, 1, 2, \ldots, N - 1, \qquad (2.93)$$

$$\tilde{G}(\frac{n}{NT}) = \sum_{k=0}^{N-1} \tilde{g}(kT) e^{-\frac{i2\pi nk}{N}}, \quad n = 0, 1, 2, \ldots, N - 1. \qquad (2.94)$$

There are about half a dozen variations of equivalent forms for the DFT and its inverse, which will be discussed in problems 4 and 5 of the exercises.

The discrete kernel here is

$$\begin{aligned} K(t_k, f_n) &= K(kT, \frac{n}{NT}) = e^{i2\pi t_k f_n} \\ &= e^{i2\pi kT \cdot \frac{n}{NT}} = e^{\frac{i2\pi nk}{N}} = e^{(\frac{i2\pi}{N})^{nk}} \equiv \omega^{nk} \end{aligned} \qquad (2.95)$$

where $\omega = e^{\frac{i2\pi}{N}}$ is the Nth *root of unity*, since $\omega^N = e^{\frac{i2\pi N}{N}} = e^{i2\pi} = \cos 2\pi + i \sin 2\pi = 1$. We note that $W = \omega^n, m = 0, 1, 2, \ldots, N - 1$ are, by definition, all Nth roots of unity, since $W^N = \omega^{mN} = e^{i2\pi m} = \cos 2\pi m + i \sin 2\pi m = 1$. Note that in some books the capital letter W is used instead of ω for the Nth root of unity.

To illustrate the discrete orthogonality in (2.88), we can easily show that this discrete kernel is orthogonal. If we use $K(t_k, f_n)$ from (2.95) in (2.88), remembering the complex conjugation for $\overline{K}(t_l, f_n)$ and the use of $2\pi f_n$ for x_n, we have

$$
\begin{aligned}
\sum_{n=0}^{N-1} K(t_k, 2\pi f_n)\overline{K}(t_l, 2\pi f_n) &= \sum_{n=0}^{N-1} e^{\frac{i2\pi nk}{N}} \cdot e^{\frac{-i2\pi ln}{N}} \\
&= \sum_{n=0}^{N-1} e^{\frac{i2\pi n(k-l)}{N}}
\end{aligned} \tag{2.96}
$$

We will start producing the first part of (2.88) for the case $k = l$, which is very simple, where each term in the sum becomes 1, and this adds to N as a special case of the constant $h(N)$ in the first part of (2.88). For the second part of $k \neq l$, we recognize the sum in (2.96) as a simple *geometric series*,

$$
\begin{aligned}
\sum_{k=0}^{N-1} e^{\frac{i2\pi n(k-l)}{N}} &= \sum_{k=0}^{N-1} \left[e^{\frac{i2\pi(k-l)}{N}} \right]^n = \sum_{k=0}^{N-1} W^n \\
&= \frac{1 - W^N}{1 - W} = 0, \quad W \neq 1, \quad k \neq l.
\end{aligned} \tag{2.97}
$$

after letting $W \equiv e^{\frac{i2\pi(k-l)}{N}}$, and using the geometric series formula (2.39), and noting that $W \neq 1$, since

$$
W = e^{\frac{i2\pi(k-l)}{N}} = \cos\frac{2\pi(k-l)}{N} + i\sin\frac{2\pi(k-l)}{N}, \tag{2.98}
$$

and this is never equal to 1 unless $k = l$, which we are not considering here, or that $\frac{k-l}{N}$ is a multiple of N, which cannot be, as k and l run through $k, l = 0, 1, 2, \ldots, N$. However the numerator $1 - W^N = 0$, since

$$
\begin{aligned}
W^N = e^{i2\pi(l-k)} &= \cos 2\pi(l-k) + i\sin 2\pi(l-k) \\
&= 1 + 0 = 1.
\end{aligned} \tag{2.99}
$$

In parallel to what we presented of the basic operational (integral) calculus properties for the integral and finite Fourier transforms in parts A, B of the last section, we will present here two basic "Operational Sum Calculus (or Algebra)" properties of the discrete Fourier transforms. This is followed by the application of this method to solving a difference equation with boundary conditions.

The Basic Operational Sum Calculus Properties of the Discrete Fourier Transform (DFT)

As we shall see in Chapter 4, it is customary to simplify the notation of the discrete Fourier transforms of (2.93)-(2.95) by writing g_k for $\tilde{g}(kt)$ in (2.93), and G_n for $\tilde{G}(\frac{n}{NT})$ in (2.95), and letting $\omega \equiv e^{\frac{i2\pi}{N}}$ as the Nth root of unity, where (2.93)-(2.95) are written now as

$$\tilde{\mathcal{F}}\{G_n\} = g_k = \frac{1}{N} \sum_{n=0}^{N-1} G_n \omega^{nk}, 0 \le k \le N-1 \qquad (2.100)$$

$$\tilde{\mathcal{F}}^{-1}\{g_k\} = G_n = \sum_{k=0}^{N-1} g_k \omega^{-nk}, \quad \omega \equiv e^{\frac{i2\pi}{N}}, \quad 0 \le n \le N-1 \ (2.101)$$

as seen in (4.12) and (4.11), respectively, where we also write $\mathcal{F}$ for the above operators $\tilde{\mathcal{F}}$ and $\tilde{\mathcal{F}}^{-1}$ for simplicity as we shall do in Chapter 4, and we caution not to confuse them with the integral operators of the exponential Fourier integral transform and its inverse in (2.50)-(2.51). We remind again that in some books the capital letter W is used instead of ω for the Nth root of unity.

Our primary aim here is to use the discrete Fourier transforms to algebraize the action of the (forward shift) E-operator, or a combination of, in a difference equation, compared to algebraizing derivatives with the integral and finite transforms of the last two parts of this section. This means we should first try to transform $EG_n = G_{n+1}$, to an algebraic expression in terms of g_k, the discrete Fourier transform of G_n, i.e., to remove the action of E. We can then follow this by algebraizing the higher order operation $E^l G_n = G_{n+l}$, or the combinations of, in a difference equation that are possible for the particular transform. The other important question is that, in the process of the transformation we must involve the given auxiliary conditions for the method to be called an operational calculus method. This idea was observed clearly in the examples of the various integral and finite transforms that we have already discussed in the preceding parts A, B of the previous section.

As it is the case for the Fourier series in (2.74), which is periodic with period 2π, we can show here that both g_k and G_n are periodic with period N. Thus their sum in (2.100)-(2.101) can have the limits, for example in (2.100), from $k = -m$ to $k = N - 1 - m$ as long as the difference between the two limits is the period of N samples $((N-1)$ increments$)$. To prove the periodicity of G_n with its period N, we will show that for r integer, $G_{n+rN} = G_n$, which as we shall see is a consequence of the periodicity, with period N, of the kernel $\omega^{-nk} = e^{\frac{-2\pi i n k}{N}}$ in (2.101). If we use (2.100) with n replaced by $n + rN$, we obtain

$$
\begin{aligned}
G_{n+rN} &= \sum_{k=0}^{N-1} g_k \omega^{-k(n+rN)} = \sum_{k=0}^{N-1} g_k \omega^{-rNk} \omega^{-nk} \\
&= \sum_{k=0}^{N-1} g_k \omega^{-nk} = G_n, \\
G_{n+rN} &= G_n
\end{aligned}
\tag{2.102}
$$

after recognizing that $\omega^{-rNk} = e^{-2\pi i r k} = \cos 2\pi r k = 1$.

Next, we use this periodic property of the discrete Fourier transform to show that

$$
\tilde{\mathcal{F}}\{EG_n\} = \tilde{\mathcal{F}}\{G_{n+1}\} = \omega^{-k} g_k
\tag{2.103}
$$

which gives us the first result of how the discrete Fourier transform algebraizes the forward shift operation EG_n on G_n. To prove (2.103), we substitute G_{n+1} for G_n in (2.100) to have

$$
\tilde{\mathcal{F}}\{G_{n+1}\} = \frac{1}{N} \sum_{n=0}^{N-1} G_{n+1} \omega^{nk}
\tag{2.104}
$$

Then to bring the sum on the right hand side to the form of (2.101), by involving G_n instead of G_{n+1} inside the sum, we change variables by letting $m = n + 1$, where the sum in (2.104), with due attention to the change in the limits of the summation from $m = 1$ to $m = N - 1 + 1$ because of now $m = n + 1$, becomes

$$
\begin{aligned}
\frac{1}{N} \sum_{n=0}^{N-1} G_{n+1} \omega^{nk} &= \frac{1}{N} \sum_{m=1}^{N-1+1} G_n \omega^{k(m-1)} \\
&= \frac{1}{N} \sum_{m=1}^{N-1+1} G_m \omega^{mk} \omega^{-k} \\
&= \omega^{-k} \frac{1}{N} \sum_{m=1}^{N-1+1} G_m \omega^{mk} = \omega^{-k} \frac{1}{N} \sum_{m=o}^{N-1} G_m \omega^{mk} \\
&= \omega^{-k} g_k
\end{aligned}
$$

(2.105)

where in the last step we used the periodicity property of $G_m \omega^{mk}$ with its period N, hence the sum over m can be taken from $m = 0$ to $N - 1$ instead of $m = 1$ to $N - 1 + 1 = N$. In the same way we can show that

$$
\tilde{\mathcal{F}} \left\{ E^l G_n \right\} = \tilde{\mathcal{F}} \left\{ G_{n+l} \right\} = \omega^{-kl} g_k \qquad (2.106)
$$

In (2.105) and (2.106) we note that although the forward shift operator was algebraized, the, generally, nonhomogeneous boundary conditions with values G_0 and G_N were not involved. This is in contrast to what we had in (2.75) and (2.76) for the finite Fourier transform, where the boundary values were involved, which was due to and using the integration by parts process in deriving the latter results. This situation, for the discrete Fourier transform, is remedied when we use the parallel process of *summation by parts* very shortly to obtain the parallel in (2.108) for the DFT, and later in (2.140) and (2.143) for the DST and the DCT, respectively.

The result in (2.103) should stand as the parallel, of how the discrete Fourier transform algebraizes the forward shifting of the E operator, to that of how the finite exponential Fourier transform algebraizes the first order derivative as seen in (2.73),

$$
f_e \left\{ \frac{dG}{dx} \right\} = (-1)^n \left[G(\pi) - G(-\pi) \right] + i n g_e(n) \qquad (2.73)
$$

and in particular, its following special case (2.75) when $G(x)$ has no jump discontinuity between the two end points $x = -\pi$ and $x = \pi$, i.e., it has the typical periodic boundary condition $G(\pi)-G(-\pi)=0$, or $G(-\pi)=G(\pi)$,

$$f_e\left\{\frac{dG}{dx}\right\} = ing_e(n) \qquad (2.75)$$

As we discussed following (2.73) and having (2.75), the appearance of the jump $G(\pi)-G(-\pi)$ between the two boundary values $G(\pi)$ and $G(-\pi)$ at $x=\pi$ and $x=-\pi$, was the result of using the integration by parts on the integral

$$f_e\left\{\frac{dG}{dx}\right\} = \int_0^\pi \frac{dG}{dx}e^{inx}dx.$$

The other method that involves differentiating $G(x)$ of the Fourier series expansion in (2.74), gives the special result of (2.75). As we explained then, the reason for this result of (2.75) instead of (2.73) is that the differentiation operation $\frac{d}{dx}$ is not allowed to enter inside the infinite Fourier series (2.74) of $G(x)$, except under some rather strict conditions not the least of which that $G(\pi)=G(-\pi)$.

We should clarify an important point here regarding the presence of the *jump* $J=G(\pi)-G(-\pi)$ between the boundary values $G(\pi)$ and $G(-\pi)$, and the *periodicity* of the Fourier series representation of $G(x)$ on $(-\pi,\pi)$ in (2.74). It is clear from (2.74) that the Fourier series of $G(x)$ on $(-\pi,\pi)$,

$$G(x) = \frac{1}{2\pi}\sum_{n=-\infty}^{\infty} g_e(n)e^{inx}$$

gives its own *periodic extension* to $G(x)$ with period 2π, and, in particular, it gives equal (periodic) values at the two end points $x=-\pi$ and $x=\pi$. For the special case of continuous $G(x)$ on $(-\pi,\pi)$, this, of course, coincides with the periodic extension of the continuous $G(x)$ for the special case of $G(-\pi)=G(\pi)$, i.e., when $G(x)$ does not have a jump between the two ends $x=-\pi,\pi$ of its principal interval $(-\pi,\pi)$. In the presence of a jump, i.e., $G(\pi)-G(-\pi)\neq 0$, the above two periodic extensions of $G(x)$ (with its jumps at the boundary points) and the Fourier series (with its assigned values at the middle of the jump for the end points) differ at the end points. The first has the values $G(-\pi_+)$ and $G(\pi_-)$ at the end points $-\pi$ and

π, respectively, while the Fourier series converges to the same value $\frac{1}{2}[G(-\pi_+) + G(\pi_-)]$ at the two end points $x = -\pi$ and $x = \pi$. The latter sounds like preserving the typical description of periodicity of the Fourier series, where equal values are expected at $x = -\pi$ and $x = \pi$. In summary, the Fourier series repeats the function with its jump discontinuities periodically with period 2π, but when it comes to the Fourier series' own value at such end points with jumps (or any points with jumps in the interior of $(-\pi, \pi)$) it assigns values at the middle of such jumps.

In parallel to what we had in (2.73) for the finite Fourier transform, and in order to involve the boundary values G_0 and G_N at the two ends $n = 0$ and $n = N$ of the domain of the sequence $\{G_n\}_{n=0}^{N-1}$, in addition to algebraizing the operation $EG_n = G_{n+1}$, we look to the discrete analog of integration by parts, namely the *summation by parts* that was introduced and illustrated in Section 2.2, whose statement is

$$\sum_{k=0}^{N-1} v_k \Delta u_k = u_N v_N - u_0 v_0 - \sum_{k=0}^{N-1} u_{k+1} \Delta v_k \qquad (2.35)$$

Using this summation by parts formula once, we will show that

$$\tilde{\mathcal{F}}\{G_{n+1}\} = \frac{\omega^{-k}}{N}[G_N - G_0] + \omega^{-k} g_k \qquad (2.107)$$

as the parallel to (2.73) of the finite exponential Fourier transform, and where the jump $G_N - G_0$ between the values of the transformed sequence g_k at the two end points of the sequence $n = 0$ and $n = N$ is abundantly clear. So we can take (2.105) to accommodate the auxiliary periodic condition with period N on the sequence G_n, while (2.107) has the advantage of accommodating more general boundary conditions of the sequence being given at the two end points $n = 0, N$, or just the difference of the two, i.e., the jump $G_N - G_0$. The latter may accommodate the situation of a thermocouple between two ends $n = 0$ and $n = N$, where the reading of the meter supplies information related to the difference $G_n - G_0$, and not to G_0 and G_n individually. To prove (2.106), we substitute G_{n+1} for G_n in (2.100) to have

$$\tilde{\mathcal{F}}\{G_{n+1}\} = \frac{1}{N}\sum_{n=0}^{N-1}\omega^{nk}G_{n+1} = \frac{1}{N}\sum_{n=0}^{N-1}\omega^{nk}\left[G_{n+1} - G_n + G_n\right]$$

$$= \frac{1}{N}\sum_{n=0}^{N-1}\omega^{nk}G_{n+1} = \frac{1}{N}\sum_{n=0}^{N-1}\omega^{nk}\Delta G_n + \frac{1}{N}\sum_{n=0}^{N-1}\omega^{nk}G_n$$

$$= \frac{1}{N}\sum_{n=0}^{N-1}\omega^{nk}\Delta G_n + g_k.$$

Now the first term on the right hand side is ready for the summation by parts of (2.35) above, where we let $v_n = \omega^{nk}$ and $u_n = G_n$, and note then that $v_N = \omega^{Nk} = v_0 = \omega^0 = 1$ by the definition of ω being the Nth root of unity,

$$\sum_{n=0}^{N-1}\omega^{nk}G_{n+1} = G_N v_N - G_0 v_0 - \sum_{n=0}^{N-1}G_{n+1}\Delta\omega^{nk} + N g_k$$

$$= G_N - G_0 - \sum_{n=0}^{N-1}G_{n+1}\left[\omega^{k(n+1)} - \omega^{nk}\right] + N g_k$$

$$= G_N - G_0 - \sum_{n=0}^{N-1}[\omega^k - 1]\omega^{nk}G_{n+1} + N g_k ,$$

$$\sum_{n=0}^{N-1}\left[1 - \omega^k + 1\right]\omega^{nk}G_{n+1} = G_N - G_0 + N g_k ,$$

$$\omega^k\sum_{n=0}^{N-1}\omega^{nk}G_{n+1} = G_N - G_0 + N g_k , \qquad (2.107)$$

$$\frac{1}{N}\sum_{n=0}^{N-1}\omega^{nk}G_{n+1} = \frac{\omega^{-k}}{N}[G_N - G_0] + \omega^{-k}g_k$$

The same method of summation by parts can be used twice to algebraize $E^2 G_n = G_{n+2}$, where the jump $G_N - G_0$ between the two ends $n = 0$ and $n = N$ as well as the jump in their forward shifted value by 1, i.e., $G_{N+1} - G_1 \equiv E(G_N - G_0)$ are now involved in the process of the discrete Fourier transformation of $E^2 G_n$

$$\tilde{\mathcal{F}}\{E^2 G_n\} = \tilde{\mathcal{F}}\{G_{n+2}\} = \frac{\omega^{-k}}{N}[G_{N+1} - G_1]$$
$$+ \frac{\omega^{-2k}}{N}[G_N - G_0] \qquad (2.108)$$
$$+ \omega^{-2k} g_k$$

Of course, this result can be established by employing the same summation by parts, used for (2.107), but repeat it twice, which we leave as exercise 4. The derivation of the general case of (2.108) for $\tilde{\mathcal{F}}\{G_{n+l}\}$,

$$\tilde{\mathcal{F}}\{G_{n+l}\} = \frac{1}{N} \sum_{j=1}^{l} [G_{N+l+j} - G_{l-j}] \omega^{-kj} + \omega^{-kl} g_k \qquad (2.109)$$

follows the same method of deriving (2.107), (2.108), or the following shorter method of deriving (2.108). Such shorter method that uses only (2.107) and a simple change of variable, which is along the lines we suggested for the Laplace transform of second order derivatives in exercise 1(b)of Section 2.3, is to apply (2.107) to $H_n = G_{n+1}$ instead of G_n to have

$$\frac{1}{N} \sum_{n=0}^{N-1} \omega^{nk} G_{n+2} = \frac{1}{N} \sum_{n=0}^{N-1} \omega^{nk} H_{n+1} = \frac{\omega^{-k}}{N}[H_N - H_0] + \frac{\omega^{-k}}{N} h_k$$
$$(2.110)$$

but, according to (2.107),

$$h_k = \tilde{\mathcal{F}}\{H_n\} = \tilde{\mathcal{F}}\{G_{n+1}\} = \frac{\omega^{-k}}{N}[G_n - G_0] + \omega^{-k} h_k$$

So, if we substitute this h_k in (2.110), we obtain (2.108),

$$\frac{1}{N} \sum_{n=0}^{N-1} \frac{\omega^{-nk}}{N} G_{n+2} = \frac{\omega^{-k}}{N}[G_{N+1} - G_1] + \omega^{-k}\{\frac{\omega^{-k}}{N}[G_N - G_0]$$

$$+ \omega^{-k} g_k\} = \frac{\omega^{-k}}{N}[G_{N+1} - G_1] + \frac{\omega^{-2k}}{N}[G_N - G_0]$$

$$+ \omega^{-2k} g_k \qquad (2.108)$$

We have already seen how the discrete Fourier transforms in (2.100) and its inverse in (2.101), are essentially symmetric, except for the minor minus sign in ω^{-nk} of (2.101) and the scaling factor $\frac{1}{N}$ in (2.100). So, it should be easy to see that we expect the same results (2.107)-(2.108) from the inverse discrete Fourier transformation (2.101) for $Eg_k = g_{k+1}$ and $E^2 g_k = g_{k+2}$, respectively, which are

$$\tilde{\mathcal{F}}^{-1}\{Eg_k\} = \tilde{\mathcal{F}}^{-1}\{g_{k+1}\} = \omega^n[g_N - g_0] + \omega^n G_n \qquad (2.111)$$

and

$$\tilde{\mathcal{F}}^{-1}\{E^2 g_k\} = \tilde{\mathcal{F}}^{-1}\{g_{k+2}\} = \omega^n[g_{N+1} - g_1] + \omega^{2n}[g_N - g_0] + \omega^{2n} G_n \qquad (2.112)$$

These results (2.107)-(2.108) and (2.111)-(2.112) can be generated for the discrete Fourier transform of $E^{-1}G_n = G_{n-1}, E^{-2}G_n = G_{n-2}, E^{-1}g_k = g_{k-1}$ and $E^{-2}g_k = g_{k-2}$ as follows

$$\tilde{\mathcal{F}}\{E^{-1}G_n\} = \mathcal{F}\{G_{n-1}\} = \frac{\omega^k}{N}[G_N - G_0] + \omega^k g_k \qquad (2.113)$$

$$\tilde{\mathcal{F}}\{E^{-2}G_n\} = \mathcal{F}\{G_{n-2}\} = \frac{\omega^k}{N}[G_{N+1} - G_1] + \omega^{2k}[G_N - G_0]$$

$$+ \omega^{2k} g_k \qquad (2.114)$$

$$\tilde{\mathcal{F}}^{-1}\{E^{-1}g_k\} = \tilde{\mathcal{F}}^{-1}\{g_{k-1}\} = \omega^{-n}[g_N - g_0] + \omega^{-n} G_n \qquad (2.115)$$

$$\tilde{\mathcal{F}}^{-1}\{E^{-2}g_k\} = \tilde{\mathcal{F}}\{g_{k-2}\} = \omega^{-n}[g_{N+1} - g_1] + \omega^{-2n}[g_N - g_0] + \omega^{-2n} G_n \qquad (2.116)$$

For the derivation of (2.115), for example, we need to show that

$$\tilde{\mathcal{F}}^{-1}\{g_{k-1}\} = \sum_{k=0}^{N-1} g_{k-1}\omega^{-nk} = \sum_{m=-2}^{N-1-2} g_{m+1}\omega^{-n(m+2)}$$

$$= \sum_{k=0}^{N-1} g_{k-1}\omega^{-nk} = \omega^{-2n}\sum_{m=0}^{n-1} g_{m+1}\omega^{-nk} \qquad (2.117)$$

after making the change of variable $m = k - 2$, and using the periodicity with period N of the product $g_{m+1}\omega^{mk}$. With this identity

(2.117), we can then appeal to (2.111) for the sum on the right hand side to establish (2.115),

$$\sum_{k=1}^{N-1} g_{k-1}\omega^{-nk} = \omega^{-2n}[\omega^n\{g_N - g_0\} + \omega^n G_n] \tag{2.115}$$
$$= \omega^{-n}[g_N - g_0] + \omega^{-n}G_n$$

The above important pairs of the operational sum calculus that are used to algebraize the forward and backward shift operators and also involve the boundary conditions, may now be compared to the pairs (2.73)-(2.76) of the finite (complex) exponential transform,

$$f_e\{\frac{dG}{dx}\} = (-1)^n[G(\pi) - G(-\pi)] + ing_e(n) \tag{2.75}$$

$$f_e\{\frac{d^2G}{dx^2}\} = (-1)^n[G'(\pi) - G'(-\pi)] + \\ in[(-1^n)\{G(\pi) - G(-\pi)\}] + (in)^2 g_e(n) \tag{2.76}$$

where we can see a very close parallel that we have long aimed for, which was one of the main reasons behind presenting all the integral and finite transforms of Section 2.3.

Now we are prepared to illustrate how the operational sum calculus method algebraizes difference equations and involves the given boundary conditions.

Example 2.11   The Operational Sum Calculus Method for Difference Equations.

Consider the following difference equation (E.1) in G_n with its periodic boundary condition (E.2),

$$G_n - G_{n-1} + F_n = 0, \qquad 0 \le n \le N - 1 \tag{E.1}$$

$$G_0 = G_N, \tag{E.2}$$

which describes the steady state of a traffic flow in a loop of N intersections and N streets, where we will have its detailed discussion in Section 4.4, and in particular a complete solution in Example 4.8 (see Figs. 4.7, 4.8). The sequence $\{G_n\}$ represents the number of vehicles in the nth loop, while F_n represents the traffic flow into the nth loop from the nth entrance.

In this example, we will concentrate on showing how the present (exponential) discrete Fourier transform (DFT) algebraizes the difference equation (E.1) and at the same time involves the (periodic) boundary condition (E.2). This is in the same manner that the well known operational calculus method of Laplace and Fourier transforms, for example, does for the differential equations and their auxiliary conditions. We will leave the rest of the details for finding the final answer for Example 4.8 of Section 4.4, where by that time, in Chapter 4, we would have developed most of the basic properties and enough pairs of the discrete Fourier transforms, which will make it possible to complete the problem to its final solution.

If we apply the discrete Fourier transform (2.100) to both sides of the difference equation, recognizing, of course, the linear property of this transform as a finite sum, we have

$$\tilde{\mathcal{F}}\{G_n - G_{n-1}\} + F_n\} = \tilde{\mathcal{F}}\{G_n\} - \tilde{\mathcal{F}}\{G_{n-1}\} + \tilde{\mathcal{F}}\{F_n\} = 0$$
$$= g_k - \omega^k[G_N - G_0] - \frac{\omega^k}{N}g_k + f_k = 0 \qquad (E.3)$$

where $f_k = \tilde{\mathcal{F}}\{F_n\}$, and that we used the vital DFT pair (2.113) for the discrete Fourier transform of G_{n-1}. If we use the given boundary condition $G_N - G_0 = 0$ of (E.2) in (E.3), we have the final algebraic equation in the transformed sequence $\{g_k\}$,

$$g_k - \omega^k g_k = -f_k, \qquad (E.4)$$

whose (simple) solution is

$$g_k = \frac{f_k}{1 - \omega^k}, \quad k = 1, 2, \ldots, \quad N - 1 \qquad (E.5)$$

and, of course $1 - \omega^k \neq 0$, since $\omega^k \neq 1$, for $k = 1, 2, \ldots, N - 1$. We note that the *excluded* value g_0 can be obtained directly from the definition of the DFT in (2.100) with $k = 0$

$$g_0 = \frac{1}{N} \sum_{n=0}^{N-1} G_n \qquad (E.6)$$

i.e., as the simple *average* of the transformed sequence $\{G_n\}_{n=0}^{N-1}$. However, in case we want to obtain g_0 from (E.4) or (E.5), we note

here that $1 - \omega^0 = 0$ in (E.4), and g_0 cannot be determined, i.e.,
no solution g_0 unless $f_0 = 0$. In the latter case, when g_0 exists, it is
arbitrary and the solution g_k, $k = 0, 1, 2, \ldots, N - 1$ of (E.5) exists,
but it is not unique due to the arbitrary value of g_0. So, to remedy
this we may have to rely on our intuition to assign a specific g_0
from our knowledge, or a good guess of the average value of the •
sought sequence $\{G_n\}_{n=0}^{N-1}$. Such an initial guess may be improved
by an iterative procedure on the first resulting approximation of the
solution G_n.

As we mentioned earlier, at this stage we leave finding the original solution G_n,(of the difference equation (E.1) and its periodic
boundary condition (E.2)), via the inverse discrete Fourier transform (2.111) for Example 4.8 of Section 4.4. In Example 4.8, we will
assign a particular, but realistic function F_n in (E.1), where its discrete Fourier transform f_k is known, then we use the available tools
developed up until Section 4.4 to find the inverse discrete transform
of g_k in (E.4) as the explicit solution G_n for (E.1)-(E.2).

We may point out again to the advantage of the operational sum
calculus pair (2.113), obtained via the integration by parts, for the
not necessarily typical periodic condition at the two end points, i.e.,
boundary conditions with jump discontinuity $J = G_N - G_0 \neq 0$. So
in case we have the periodic boundary condition with a jump at the
end points $J = G_N - G_0$, then, with the help of the pair (2.113), the
discrete transform of the difference equation (E.1) along with the
new boundary condition becomes

$$g_k - \frac{\omega^k}{N}[B - A] - \omega^k g_k + f_k = 0,$$

$$g_k = \frac{\frac{(B-A)}{N}\omega^k - f_k}{1 - \omega^k}, \qquad k = 1, 2, \ldots, N - 1$$

**The Discrete Fourier Sine (DST) and Cosine (DCT) Trans-
forms**

The discrete sine and cosine transforms are the subjects of Chapter 5, but we introduce them here to discuss their basic operational
sum calculus properties, since they represent two important cases of

the discrete Fourier transforms. They, of course, are derived from
the discrete exponential Fourier transform, (DFT) but as we saw
in the case of the Fourier sine and cosine integral transforms and
their finite versions in Section 2.3, they have their special roles to
play, especially with regard to involving different types of *boundary
conditions* at the end points.

1. The Discrete Fourier Sine Transform (DST)

As we shall develop it in Chapter 5, the discrete Fourier sine
transform of the sequence $\{G_n\}_{n=1}^{M-1}$ is defined as

$$b_k = \mathcal{S}\{G_n\} = \frac{2}{M} \sum_{n=1}^{M-1} G_n \sin \frac{\pi nk}{M}, \qquad 1 \le k \le M-1 \qquad (2.118)$$

with its inverse as

$$G_n = \mathcal{S}^{-1}\{b_k\} = \sum_{n=1}^{M-1} b_k \sin \frac{\pi nk}{M}, \quad 1 \le n \le M-1, \qquad (2.119)$$

which are also found in (5.4) and (5.3), respectively. We note here
that the discrete sine transform of (2.118) satisfies the boundary condition $b_0 = b_M = 0$, and the same is for its inverse where $G_0 = G_M =
0$. These boundary conditions of assigning values of the sequence at
the two end points are termed *"Dirichlet"* conditions. So, the DST
of (2.118) is clearly compatible with vanishing boundary values at
the end points $n = 0$ and $n = M$, i.e., when $G_0 = G_N = 0$. An example is the fixed zero temperature at the two ends of a bar. This is
to be compared with the compatibility of the (exponential) discrete
Fourier transform (2.100) and its inverse (2.101) with the typical
periodic condition $G_0 = G_N$, or with the more general periodic condition with a jump $J = G_N - G_0 \ne 0$ between the two boundary
points $N, 0$. Of course, these comparisons stem from using the direct
definitions of the DST and the DFT, and before involving the summation by parts method that will bring about the involvement of the
nonhomogeneous Dirichlet boundary conditions $G_0 = A, G_N = B$ as
we shall show in (2.140) in the same way we did for the DFT in
(2.108), for example, where a more general periodic condition with

jumps $G_N - G_0$, and $G_{N+1} - G_1$ at or near the two end points (of the transformed sequence $\{G_n\}$) were brought about.

2. THE DISCRETE FOURIER COSINE TRANSFORM (DCT)

The Discrete Fourier cosine transform of the sequence $\{G_n\}_{n=0}^{M}$ is defined as

$$a_k = \mathcal{C}\{G_n\} = \frac{1}{M}[G_0 + 2 \sum_{n=1}^{M-1} G_n \cos \frac{\pi n k}{M} + (-1)^k G_M], \quad (2.120)$$

$$0 \le k \le M$$

and its inverse

$$G_n = \mathcal{C}^{-1}\{a_k\} = \frac{a_0}{2} + \sum_{k=1}^{M-1} a_k \cos \frac{\pi n k}{M} + (-1)^n \frac{a_M}{2}, \quad 0 \le n \le M$$

$$(2.121)$$

which are found in (5.2) and (5.1) respectively.

In comparison with the discrete Fourier sine transform, the discrete Fourier cosine transform have the "evenness" property $G_{-n} = G_n$ about the end point $n = 0$, i.e., $G_{-1} = G_1$ and $G_{M-n} = G_{M+n}$ about the other end point $n = M$, i. e., $G_{M-1} = G_{M+1}$. The first property $G_{-n} = G_n$ can be shown easily when we substitute $-n$ for n in (2.121) to have

$$
\begin{aligned}
G_{-n} &= \frac{a_0}{2} + 2 \sum_{n=1}^{M-1} a_k \cos(\frac{-\pi n k}{M}) + (-1)^{-n} \frac{a_M}{2} \\
&= \frac{a_0}{2} + 2 \sum_{n=1}^{M-1} a_k \cos \frac{\pi n k}{M} + (-1)^n \frac{a_M}{2}
\end{aligned}
$$

$$(2.122)$$

Showing $G_{M-n} = G_{M+n}$ follows in the same way.

From this evenness property we can deduce the conditions at the boundary points $n = 0$ and $n = M$ that this DCT transform is compatible with. The evenness around a point reminds us of an "approximation" to a zero slope, hence we can use it to model, for example, an insulated end at that point, where the temperature gradient vanishes. So, such boundary condition can be modeled by the evenness of the discrete cosine transform with $G_1 - G_{-1} = 0$ at the end point $n = 0$, and $G_{M+1} - G_{M-1} = 0$ at the other end point $n = M$. Thus, we say that the discrete cosine transform is compatible with the evenness conditions which are termed as *Neumann boundary conditions*. This is in contrast to the *vanishing* conditions at the two end points $G_0 = G_M = 0$ of the discrete Fourier sine transform (DST) or *Dirichelet conditions*, and the *periodic* condition $G_0 = G_N$ of the discrete exponential Fourier transform (DFT).

We may remind again that, as we saw in the two pairs (2.107) and (2.103) for the discrete Fourier transform of $EG_n = G_{n+1}$, the first pair (2.107) was derived with the help of the summation by parts, which showed the advantage of involving the periodic "jump" $B-A$ at the boundaries instead of only a periodic boundary condition $G_N = G_0$ that the pair in (2.103) allows. For the present discrete sine and cosine transforms, as we have already remarked concerning the DST following (2.119), we will also attempt both methods, and in particular, the summation by parts one, which will allow us involving nonhomogeneous boundary conditions at the end points, as we shall see next.

The Basic Operational Sum Calculus of the Discrete Fourier Sine and Cosine Transforms.

For the cases of the Fourier sine and cosine integral transforms (2.60), (2.61) as well as their finite versions (2.82), (2.85), we saw how such transforms are compatible with (basically) only *even order* derivatives. The reason was that we needed to do an even number of integrations by parts on such integrals that involve a sine kernel, for example, to return to an integral with the same sine kernel. This was because we wanted the whole problem to be transformed in terms of the Fourier sine transform. Otherwise, with odd order derivatives, we will end up with Fourier sine as well as Fourier cosine transforms. This, of course, was not the case for the exponential Fourier transforms in (2.54)-(2.55), (2.75)-(2.76) or the Laplace transform in (2.48)-(2.49), since their exponential kernel can spring back after any number of integrations by parts, hence they can handle even as well as odd order derivatives.

So, for the present discrete versions of the Fourier sine and cosine transforms, we cannot expect $\mathcal{S}\{G_{n+1}\}$, for example, to result in an algebraic expression that involves $\mathcal{S}\{G_n\}$ only, and we may have to search for combinations of "shifted" G_n's, as compared to a "differentiated" $f(x)$ as $\frac{d^2 f}{dx^2}$, that may accomplish what we are after. In that case, such a combination would represent the closest parallel to the second order derivative that is algebraized by the Fourier sine and cosine integral transforms and their finite versions, which we have already discussed, respectively, in parts A and B of Section 2.3.

A clear example that the discrete Fourier sine transform in (2.118) does not transform the shifted G_{n+1} to a nonshifted sine transform $b_k = \mathcal{S}\{G_n\}$ in the transform space, can be shown when we write G_{n+1}, according to (2.119),

$$
\begin{aligned}
EG_n \;=\; G_{n+1} &= \frac{2}{M} \sum_{k=1}^{M-1} b_k \sin \frac{\pi k(n+1)}{M} \\
&= \frac{2}{M} \sum_{k=1}^{M-1} b_k \sin \frac{\pi k n}{M} \cos \frac{\pi k}{M} + \frac{2}{M} \sum_{k=1}^{M-1} b_k \cos \frac{\pi k n}{M} \sin \frac{\pi k}{M}
\end{aligned}
$$

$$(2.123)$$

where we recognize the first sum from (2.119) as $\mathcal{S}^{-1}\{b_k \cos \frac{\pi k}{M}\}$, which is what we would like to have if it was not for the second sum, that involves an inverse discrete *cosine* transform of $b_k \sin \frac{\pi k}{M}$, i.e., $\mathcal{C}^{-1}\{b_k \sin \frac{\pi k}{M}\}$ as shown in (2.121) for this special situation with $a_0 = b_0 \sin 0 = 0$ and $a_M = b_M \sin \pi k = 0$. This situation causes some mixing between the discrete sine and cosine transforms when we look at the two sums in (2.123). Of course, we prefer to see the whole right hand side of (2.123) expressed as $\mathcal{S}^{-1}\{h(k) \cdot b_k\}$, where $h(k)$ is just a (known) algebraic factor, so that we can jump at the conclusion of $\mathcal{S}\{G_{n+1}\} = \mathcal{S}\mathcal{S}^{-1}\{h(k)b_k\} = h(k)b_k$, and say that $\mathcal{S}$ algebraized G_{n+1} in transforming it in terms of b_k, a discrete sine transform without a shift. However, this is not the reality of (2.123), and it was not for the Fourier sine and cosine integrals (as well as their finite versions) of first order derivatives, where in the these cases we had to limit ourselves to transforming only *even* order derivatives as seen in (2.60)-(2.61) and (2.82),(2.85).

We mention in passing that for all these transforms, we may accept the above mixing of having a discrete sine transform of G_{n+1} in terms of the discrete sine and cosine transforms of G_n, if we are

willing to accept the idea of applying both the sine and cosine transforms, and hence receive the result of the transformation in terms
of both the sine and cosine transforms of the original sequence G_n.
This, indeed, is done sometimes for transforming systems with first
order (or mixed orders) derivatives. In our present situation, we are
not far away from a combination of shifted G_n's whose sine transform is $h(k)b_k$, i.e., it is algebraic in $b_k = \mathcal{S}\{G_n\}$, the non-shifted
discrete sine transform of G_n. The result (2.123) can be written in
the discrete sine-cosine operation forms as

$$G_{n+1} = \mathcal{S}^{-1}\{(\cos \frac{\pi k}{M}) \cdot b_k\} + \mathcal{C}^{-1}\{(\sin \frac{\pi k}{M}) \cdot b_k\} \qquad (2.124)$$

In the same way we established the expression for G_{n+1} in (2.123),
we can use (2.119) again for G_{n-1} to have

$$E^{-1}G_n = G_{n-1} = \frac{2}{M} \sum_{k=1}^{M-1} b_k \sin \frac{\pi k(n-1)}{M}$$

$$= \frac{2}{M} \sum_{k=1}^{M-1} b_k \sin \frac{\pi kn}{M} \cos \frac{\pi k}{M}$$

$$\qquad \qquad (2.125)$$

$$- \frac{2}{M} \sum_{k=1}^{M-1} b_k \cos \frac{\pi kn}{M} \sin \frac{\pi k}{M}$$

$$= \mathcal{S}^{-1}\{(\cos \frac{\pi k}{M} \cdot b_k\} - \mathcal{C}^{-1}\{(\sin \frac{\pi k}{M})b_k\}$$

which is exactly the wanted result that, when added to (2.124) will
remove the unwanted inverse discrete cosine transform term,

$$G_{n+1} + G_{n-1} = \mathcal{S}^{-1}\{2(\cos \frac{\pi k}{M})b_k\} \qquad (2.126)$$

or, written in terms of the direct (DST) transform notation as

$$\mathcal{S}\{G_{n+1} + G_{n-1}\} = \left(2\cos \frac{\pi k}{M}\right) b_k = \left(2\cos \frac{\pi k}{M}\right) \mathcal{S}\{G_n\} \quad (2.127)$$

This says that $G_{n+1} + G_{n-1}$ is the desired combination of shifts in
G_n that is compatible with the discrete sine transform, i.e., the latter

algebraizes $G_{n+1}+G_{n-1}$ in parallel to the algebraization of the second derivative $\frac{d^2 f}{dx^2}$ in the case of the Fourier sine and cosine integral and finite transforms in (2.60), (2.61) and (2.82), (2.85).

The result (2.127) does algebraize $(E + E^{-1})G_n = G_{n+1} + G_{n-1}$, $= (\mathcal{S}^2 - 2)G_n$, but, as was mentioned earlier, such a result involves, or satisfies implicitly, only the *homogeneous* Dirichlet boundary conditions $G_0 = 0$ and $G_M = 0$ at the two end points $n = 0$ and $n = M$, as it was clear from the definition of the inverse DST in (2.119). Here we present an improvement on such approach with the use of the operational sum method that will bring about the involvement of *nonhomogeneous Dirichlet* boundary conditions, namely, $G_0 = A$ and $G_M = B$. As we discussed in the case of the discrete exponential Fourier transform of G_{n+1}, the result (2.106) involved only the usual periodic boundary condition $G_0 = G_N$, however when we tried the method of summation by parts, a nonhomogeneous boundary term involving a jump $J = G_N - G_0$ was recovered at the end points $n = 0, N$ as seen in (2.107). Such recovery of nonhomogeneous terms at a boundary point is not new to the operational calculus methods, since, for example, the integration by parts did bring about such terms as seen in (2.60) for the Fourier sine transform, (2.61) for the Fourier cosine transform, (2.75)-(2.76) for the finite (complex) exponential Fourier transform, (2.82) for the finite sine transform, and in (2.85) for the finite cosine transform. So, instead of seeing the result (2.127) via the indirect way of using the inverse transform pair in (2.127), we may apply the discrete sine transform directly, and use this time *the summation by parts* method to bring about the nonhomogeneous boundary terms just as we did for the discrete (complex) exponential transform of G_{n+1} to have (2.107) with its nonhomogeneous boundary terms, as the jump $J = G_N - G_0$.

If we apply the discrete sine transform (2.118) directly on $(G_{n+1} + G_{n-1})$ we have

$$\mathcal{S}\{G_{n+1} + G_{n-1}\} = \frac{2}{M} \sum_{n=1}^{M-1} (G_{n+1} + G_{n-1}) \sin \frac{\pi n k}{M}$$

$$= \frac{2}{M} \sum_{n=1}^{M-1} G_{n+1} \sin \frac{\pi n k}{M} +$$

$$+\frac{2}{M}\sum_{n=1}^{M-1} G_{n-1}\sin\frac{\pi nk}{M} \quad (2.128)$$

We will compute the first sum, the second sum will follow in a very similar way. To simplify the computations, we will use $e^{i\pi nk}$ instead of $\sin \pi nk$, then when the computations are all done, we take the *imaginary* part, (Im: for brevity) of the final result, for example, we can start with

$$\begin{aligned}
\frac{2}{M}\sum_{n=1}^{m-1} G_{n+1}\sin\frac{\pi nk}{M} &= Im\left(\frac{2}{M}\sum_{n=1}^{M-1} G_{n+1}e^{\frac{i\pi nk}{M}}\right), \\
&= Im\left(\frac{2}{M}\sum_{n=1}^{M-1} G_{n+1}\lambda^{nk}\right),
\end{aligned} \quad (2.129)$$

where we let $\lambda \equiv e^{\frac{i\pi}{M}}$ to simplify writing the formulas. Of course, we are using the basic fact that

$$\sin\frac{\pi nk}{M} = Im[e^{\frac{i\pi nk}{M}}] = Im[\cos\frac{\pi nk}{M} + i\sin\frac{\pi nk}{M}]. \quad (2.130)$$

The same is done for the DST transform of G_{n-1}.

Since our aim is to use the summation by parts formula (2.35), we will in the following prepare writing $G_{n+1} = EG_n$ in terms of ΔG_n as $G_{n+1} = G_{n+1} - G_n + G_n = \Delta G_n + G_n$. So we consider

$$\frac{2}{M}\sum_{n=1}^{M-1} G_{n+1}\lambda^{nk} = \frac{2}{M}\sum_{n=1}^{M-1} \Delta G_n\lambda^{nk} + \frac{2}{M}\sum_{n=1}^{M-1} G_n\lambda^{nk} \quad (2.131)$$

after adding and subtracting $G_n\lambda^{nk}$ inside the first sum on the left hand side, as we did for the discrete (complex) exponential Fourier transform in the very first step that lead to (2.107). Now we apply the *the summation by parts* formula to the first sum on the right hand side of (2.131)

$$\sum_{n=1}^{M-1} v_n\Delta u_n = u_M v_M - u_1 v_1 - \sum_{k=1}^{M-1} u_{n+1}\Delta v_n, \quad (2.132)$$

letting $u_n = G_n$ and $v_n = \lambda^{nk}$, to have

$$\sum_{n=1}^{M-1} \Delta G_n \lambda^{Mk} = G_M \lambda^{Mk} - G_1 \lambda^k - \sum_{n=1}^{M-1} G_{n+1}[\lambda^{k(n+1)} - \lambda^{kn}]$$

$$= G_M(-1)^k - G_1 \lambda^k - (\lambda^k - 1) \sum_{n=1}^{M-1} G_{n+1} \lambda^{nk}$$

$$(2.133)$$

where we used $\lambda^{Mk} = e^{i\pi \frac{Mk}{M}} = e^{i\pi k} = \cos \pi k + i \sin \pi k = (-1)^k$.

If we substitute the result (2.133) in (2.131), we obtain

$$\frac{2}{M} \sum_{n=1}^{M-1} G_{n+1} \lambda^{nk} = \frac{2}{M}[(-1)^k G_M - \lambda^k G_1$$

$$-(\lambda^k - 1) \sum_{n=1}^{M-1} G_{n+1} \lambda^{nk} + \sum_{n=1}^{M-1} G_n \lambda^{nk}],$$

$$\frac{2}{M}(1 + \lambda^k - 1) \sum_{n=1}^{M-1} G_{n+1} \lambda^{nk} = \frac{2}{M}\left[\lambda^{Mk} G_M - \lambda^k G_1 + \sum_{n=1}^{M-1} G_n \lambda^{nk} \right],$$

$$\frac{2}{M} \sum_{n=1}^{M-1} G_{n+1} \lambda^{nk} = \frac{2}{M}\left[(-1)^k \lambda^{-k} G_M - G_1 + \lambda^{-k} \sum_{n=1}^{M-1} G_n \lambda^{nk} \right]$$

$$(2.134)$$

Now we will find the same transformation for G_{n-1}, then we add the two results and take the imaginary part to have the discrete sine transform of $(G_{n+1} + G_{n-1})$ in terms of the "nonshifted" discrete transform g_k of G_n as we started to work for it in (2.128). The steps will parallel exactly what we did in (2.131) to (2.134), except that in preparing for the use of the summation by parts in the step (2.131), we have to make ΔG_{n-1} out of G_{n-1} in the present case, where

$$G_{n-1} = G_{n-1} - G_n + G_n = -(G_n - G_{n-1}) + G_n$$
$$= -\Delta G_{n-1} + G_n \qquad (2.135)$$

So, for this case, we write

$$\sum_{n=1}^{M-1} G_{n-1} \lambda^{nk} = - \sum_{n=1}^{M-1} \Delta G_{n-1} \lambda^{nk} + \sum_{n=1}^{M-1} G_n \lambda^{nk} \qquad (2.136)$$

For the first sum on the right hand side we use summation by parts to have

$$\sum_{n=1}^{M-1} \Delta G_{n-1}\lambda^{nk} = G_{M-1}\lambda^{Mk} - G_0\lambda^k - \sum_{n=1}^{M-1} G_n[\lambda^{k(n+1)} - \lambda^{kn}]$$

$$= (-1)^k G_{M-1} - G_0\lambda^k - (\lambda^k - 1)\sum_{n=1}^{M-1} G_n\lambda^{nk}$$

$$(2.137)$$

Now we substitute this result in (2.136) to have

$$\sum_{n=1}^{M-1} G_{n-1}\lambda^{nk} = (-1)^{k+1}G_{M-1} + G_0\lambda^k + (\lambda^k - 1)\sum_{n=1}^{M-1} G_n\lambda^{nk},$$

$$\frac{2}{M}\sum_{n=1}^{M-1} G_{n-1}\lambda^{nk} = \frac{2}{M}\left[(-1)^{k+1}G_{M-1} + G_0\lambda^k + \sum_{n=1}^{M-1} G_n\lambda^{nk}\right.$$

$$\left. + \lambda^k \sum_{n=1}^{M-1} G_n\lambda^{nk}\right]$$

$$(2.138)$$

If we add these two results of (2.134) and (2.138), we have the main part of the answer to (2.128), and all that is left is to take the imaginary part of such sum,

$$\frac{2}{M}\sum_{n=1}^{M-1}(G_{n+1} + G_{n-1})\lambda^{nk} = \frac{2}{M}[(-1)^k\lambda^{-k}G_M - G_1$$

$$+\lambda^{-k}\sum_{n=1}^{M-1} G_n\lambda^{nk} + (-1)^{k+1}G_{M-1} + G_0\lambda^k + \lambda^k\sum_{n=1}^{M-1} G_n\lambda^{nk}]$$

$$= \frac{2}{M}[G_0\lambda^k - G_1 + (-1)^{k+1}G_{M-1} + (-1)^k\lambda^{-k}G_M$$

$$+(\lambda^k + \lambda^{-k})\sum_{n=1}^{M-1} G_n\lambda^{nk}]$$

$$= \frac{2}{M}[G_0(\cos\frac{\pi k}{M} + i\sin\frac{\pi k}{M}) - G_1 + (-1)^{k+1}G_{M-1}$$

$$+(-1)^k(\cos\frac{\pi k}{M} - i\sin\frac{\pi k}{M})G_M$$

$$+2\cos\frac{\pi k}{M}\sum_{k=1}^{M-1}(\cos\frac{\pi nk}{M} + i\sin\frac{\pi nk}{M})G_n]$$

$$(2.139)$$

Now we take the *imaginary part* of both sides of (2.139) to arrive at the desired result of $\mathcal{S}\{G_{n+1} + G_{n-1}\}$ as we had sought it starting with (2.128)

$$\mathcal{S}\{G_{n+1} + G_{n-1}\} = \frac{2}{M} \sum_{n=1}^{M-1} (G_{n+1} + G_{n-1}) \sin \frac{\pi n k}{M}$$

$$= \frac{2}{M} [G_0 \sin \frac{\pi k}{M} + (-1)^{k+1} \sin \frac{\pi k}{M} G_M]$$

$$+ \frac{2}{M} \cos \frac{\pi k}{M} \sum_{n=1}^{M-1} G_n \sin \frac{\pi n k}{M}$$

$$\mathcal{S}\{G_{n+1} + G_{n-1}\} = \frac{2}{M} [G_0 + (-1)^{k+1} G_M] \sin \frac{\pi k}{M}$$

$$+ (2 \cos \frac{\pi k}{M}) \cdot \frac{2}{M} \sum_{n=1}^{M-1} G_n \sin \frac{\pi n k}{M},$$

$$= \frac{2}{M} [G_0 + (-1)^{k+1} G_M] \sin \frac{\pi k}{M} + (2 \cos \frac{\pi k}{M}) \mathcal{S}\{G_n\}$$

$$(2.140)$$

Now we can see clearly that the discrete sine transform algebraizes the operator $E + E^{-1}$ in $E G_n + E^{-1} G n = (G_{n+1} + G_{n-1})$ to transform it to b_k, the transform of G_n without any shift. In addition, and in contrast to the result in (2.127), this summation by parts approach brought about involving the (nonhomogeneous) boundary values, G_0 and G_M as seen clearly in (2.140).

This method of computing the discrete sine transform as the imaginary part of an exponential discrete transform, has the added advantage that avails to us the result of the discrete Fourier cosine transform of $G_{n+1} + G_{n-1}$ by essentially taking, instead, the *real part* of the result of the above mentioned discrete exponential transform in (2.139). So for the discrete cosine transform of (2.120) we write

$$
\mathcal{C}\{G_{n+1} + G_{n-1}\} = \frac{1}{M}[G_1 + G_{-1} + 2\sum_{n=1}^{M-1}(G_{n+1} + G_{n-1})\cos\frac{\pi n k}{M}
$$
$$
+ (-1^k(G_{M+1} + G_{M-1})]
$$
$$
= \frac{1}{M}[G_1 + G_{-1} + 2\mathrm{Re}\sum_{n=1}^{M-1}(G_{n+1} + G_{n-1})e^{i\frac{\pi n k}{M}}
$$
$$
+ (-1)^k(G_{M+1} + G_{M-1})]
$$

(2.141)

where the sum has already been computed for (2.139) with its detailed steps done in (2.134) to (2.139). So, if we take the real part of the result of this sum in (2.139) we obtain the desired result of the important middle term in (2.141),

$$
\mathrm{Re}\sum_{k=1}^{M-1}(G_{n+1} + G_{n-1})e^{\frac{i\pi n k}{M}} = (\cos\frac{\pi k}{M})G_0 - G_1
$$
$$
+ (-1)^k[(\cos\frac{\pi k}{M})G_M - G_{M-1}]
$$
$$
+ 2(\cos\frac{\pi k}{M})\sum_{n=1}^{M-1}G_n\cos\frac{\pi n k}{M}
$$

(2.142)

$$
= (\cos\frac{\pi k}{M})G_0 - G_1 + (-1)^k[(\cos\frac{\pi k}{M})G_M - G_{M-1}]
$$
$$
+ (\cos\frac{\pi k}{M})[Ma_k - G_0 - (-1)^k G_M]
$$

after using the definition of the discrete cosine transform $a_k = \mathcal{C}\{G_n\}$ as in (2.120) for the last sum. Now we will substitute (2.142) for the real part of the sum in (2.141) to, finally have the discrete cosine transform of $\mathcal{C}\{G_{n+1} + G_{n-1}\}$ in terms of the nonshifted discrete Fourier cosine transform $a_k = \mathcal{C}\{G_n\}$, along with the expected differences $G_1 - G_{M-1}$ and $G_{M+1} - G_{M-1}$ around the end points $n = 0$

and $n = M$.

$$\mathcal{C}\{G_{n+1} + G_{n-1}\} = \frac{1}{M}[G_1 + G_{-1} + (2\cos\frac{\pi k}{M})G_0 - 2G_1$$
$$+ 2(-1)^k[(\cos\frac{\pi k}{M})G_M - G_{M-1}]$$
$$+ 2(\cos\frac{\pi k}{M})(Ma_k - G_0 - (-1)^k G_M)$$
$$+ (-1)^k(G_{M+1} + G_{M-1})],$$

$$\mathcal{C}\{G_{n+1} + G_{n-1}\} = \frac{1}{M}[(G_1 - G_{-1}) + (-1)^k(G_{M+1} - G_{M-1})]$$
$$+ (2\cos\frac{\pi k}{M})a_k$$

$$\mathcal{C}\{G_{n+1} + G_{n-1}\} = \frac{1}{M}[(G_1 - G_{-1}) + (-1)^k(G_{M+1} - G_{M-1})]$$
$$+ (2\cos\frac{\pi k}{M})\mathcal{C}\{G_n\}$$

$$(2.143)$$

In comparison to the discrete sine transform of $(G_{n+1} + G_{n-1})$ in (2.140), which involves *the values* of the original sequence G_0 and G_M at the end points $n = 0$ and $n = M$, respectively, as the Dirichlet conditions, the above discrete cosine transform of $(G_{n+1} + G_{n-1})$ involves *"the differences"* of the values of the sequence G_1, G_0 around the end point $n = 0$, and G_{M+1}, G_{M-1} around the end point $n = M$, as Neumann conditions, which as expected, relate to a sense of "a slope" of the sequence G_n at the two end points. This is not surprising when we know that the *finite sine transform* of a $\frac{d^2 G}{dx^2}$ on $(0, \pi)$ involves the *values* of the function $G(0)$ and $G(\pi)$ at the end points $x = 0$ and $x = \pi$ as in (2.82), while the *finite cosine* transform of the $\frac{d^2 G}{dx^2}$ involves the *derivatives* $G'(0)$ and $G'(\pi)$ at the end points $x = 0, \pi$, as seen in (2.85).

For the discrete sine transform of $(G_{n+1} + G_{n-1})$, we presented two methods, one that resulted in (2.127), where $E + E^{-1}$ was algebraized, but the (nonhomogeneous) boundary values G_0 and G_M were not involved, and the second that used the summation by parts method, which brought about such involvement as seen in (2.140). Here we used the second method for the discrete Fourier cosine transform to result in (2.143), where as expected, the (nonhomogeneous)

Neumann boundary conditions are involved. We may add here that we can also use the first method for the discrete cosine transform to give the following result, which parallels (2.127) of the discrete sine transform,

$$\mathcal{C}\{G_{n+1} + G_{n-1}\} = (2\cos\frac{\pi k}{M})a_k \qquad (2.144)$$

The derivation of (2.144) follows the same steps (2.123)-(2.127) used to derive (2.127) for the discrete sine transform, with due attention to the terms $\frac{a_0}{2}$ and $\frac{a_M}{2}$ outside the sum in the definition (2.120) of the inverse discrete cosine transform G_n of a_k. We start by using the inverse discrete cosine transform (2.120) for G_{n+1},

$$G_{n+1} = \frac{a_0}{2} + \sum_{k=1}^{M-1} a_k \cos\frac{\pi k(n+1)}{M} + (-1)^{n+1}\frac{a_M}{2}$$

$$= \frac{a_0}{2} + \sum_{k=1}^{M-1} (a_k \cos\frac{\pi k}{M})\cos\frac{\pi kn}{M}$$

$$- \sum_{k=1}^{M-1} a_k \sin\frac{\pi k}{M}\sin\frac{\pi kn}{M} + (-1)^{n+1}\frac{a_M}{2}$$

$$= \frac{a_0}{2} + \sum_{k=1}^{M-1} (a_k \cos\frac{\pi k}{M})\cos\frac{\pi kn}{M} + (-1)^n\frac{a_M}{2}\cos\frac{\pi M}{M}$$

$$- \sum_{k=1}^{M-1} \left(a_k \sin\frac{\pi k}{M}\right)\sin\frac{\pi kn}{M}$$

$$(2.145)$$

after setting $(-1)^{n+1}\frac{a_M}{2} = (-1)^n\frac{a_M}{2}\cos\frac{\pi M}{M}$ to prepare the first three terms in (2.145) for the sought inverse discrete cosine transform of $(a_k \cos\frac{\pi k}{M})$ inside the first sum,

$$G_{n+1} = \mathcal{C}^{-1}\{a_k \cos\frac{\pi k}{M}\} - \sum_{k=1}^{M-1} \left(a_k \sin\frac{\pi k}{M}\right)\sin\frac{\pi kn}{M}$$

This can also be written, after consulting the inverse discrete sine transform (2.119) for the last sum, as

$$G_{n+1} = \mathcal{C}^{-1}\{a_k \cos\frac{\pi k}{M}\} - \mathcal{S}^{-1}\{a_k \sin\frac{\pi k}{M}\} \qquad (2.146)$$

Following the same steps, leading to (2.146) above, we can write

$$G_{n-1} = C^{-1}\{a_k \cos \frac{\pi k}{M}\} + S^{-1}\{a_k \sin \frac{\pi k}{M}\}, \tag{2.147}$$

and when adding these two results (2.146) and (2.147), we obtain (2.144),

$$G_{n+1} + G_{n-1} = 2C^{-1}\{a_k \cos \frac{\pi k}{M}\} = C^{-1}\{2a_k \cos \frac{\pi k}{M}\}, \tag{2.148}$$

This result, of course, means that

$$C\{G_{n+1} + G_{n-1}\} = (2 \cos \frac{\pi k}{M})a_k, \tag{2.144}$$

We may remark that in difference operators notation we can write

$$G_{n+1} + G_{n-1} = EG_n + E^{-1}G_n = (E + E^{-1})G_n,$$

but this operation is the closest to a *second order central difference operator* δ^2, where

$$G_{n+1} + G_{n-1} - 2G_n = (E + E^{-1} - 2)G_n = \delta^2 G_n, \tag{2.149}$$

since

$$\delta G_n = G_{n+\frac{1}{2}} - G_{n-\frac{1}{2}}, \tag{2.150}$$

$$\delta^2 G_n = \delta(\delta G_n) = [G_{n+\frac{1}{2}+\frac{1}{2}} - G_{n-\frac{1}{2}+\frac{1}{2}}] - [G_{n+\frac{1}{2}-\frac{1}{2}} - G_{n-\frac{1}{2}-\frac{1}{2}}]$$

$$= G_{n+1} - G_n - G_n + G_{n-1},$$

$$\delta^2 G_n = G_{n+1} - 2G_n - G_{n-1} = (E + E^{-1} - 2)G_n$$

$$\tag{2.149}$$

The summary of the results of the discrete sine and cosine transforms (and their inverses) (2.118)-(2.119) and (2.120)-(2.121), respectively, in transforming difference equations into algebraic equations in the transform space, is that they both can algebraize a difference equation that has the operator $(E+E^{-1})$. This means that the difference equation must be *symmetric* in the E and E^{-1}, for example

$$\alpha G_{n+1} + \beta G_n + \alpha G_{n-1} = F_n$$

$$= \alpha(G_{n+1} + G_{n-1}) + \beta G_n = 2(E + E^{-1} + \beta)G_n = F_n \qquad (2.151)$$

is compatible with either of the discrete sine transform as shown in (2.140) or the discrete cosine transform as shown in (2.143) depending on the boundary conditions given at or around the end points $n = 0$ and $n = M$. In case the difference equation is *not symmetric* in E and E^{-1}, for example,

$$\alpha G_{n+1} + \beta G_n + \gamma G_{n-1} = F_n \qquad (2.152)$$

then, fortunately, a simple scaling of the amplitude of the unknown sequence G_n can be performed with $G_n = (\frac{\gamma}{\alpha})^{\frac{n}{2}} H_n$ to reduce (2.152) to a symmetric equation in H_n,

$$\sqrt{\alpha\gamma} H_{n+1} + \beta H_n + \sqrt{\alpha\gamma} H_{n-1} = \left(\frac{\gamma}{\alpha}\right)^{-\frac{n}{2}} F_n \qquad (2.153)$$

This can be verified by a simple substitution of $G_n = (\frac{\gamma}{\alpha})^{\frac{n}{2}} H_n$ in (2.152). The only price paid here is that the nonhomogeneous term, on the right hand side of the new (symmetric) equation (2.153), is a bit more complicated than F_n of the original (nonsymmetric) equation (2.152). For example, consider the difference equation

$$4G_{n+1} - 5G_n + 9G_{n-1} = n, \ 1 \leq n \leq N - 1 \qquad (2.154)$$

where we chose $\alpha = 4$ and $\gamma = 9$ to simplify the computations by removing the involved radicals in (2.153). According to (2.153), the substitution $G_n = (\frac{9}{4})^{\frac{n}{2}} H_n = (\frac{3}{2})^n H_n$ reduces the nonsymmetric equation (2.154) in $G_{n+1} = EG_n$ and $G_{n-1} = E^{-1}G_n$ to the following symmetric one in $H_{n+1} = EH_n$ and $H_{n-1} = E^{-1}H_n$,,

$$\sqrt{36} H_{n+1} - 5H_n + \sqrt{36} H_{n-1} = (\tfrac{9}{4})^{-\frac{3}{2}} F_n,$$
$$6H_{n+1} - 5H_n + 6H_{n-1} = (\tfrac{3}{2})^{-n} F_n \qquad (2.155)$$
$$= 6(EH_n + E^{-1}H_n) - 5H_n = (\tfrac{2}{3})^n F_n$$

We will conclude this part of the section, for introducing the operational sum calculus of the discrete Fourier sine and cosine transforms, with a few examples. The aim here is to show the advantages of the *algebraization* in the transform space, and also *the involvement of the (not necessarily homogeneous) boundary conditions.* The complete details of seeing the particular problem to its final solution, which, of course, involves using the inverse discrete transform to recover the solution of the difference equation, is left for Chapter 5, in particular Section 5.2.

Example 2.12   The Operational Sum Calculus of the DST for Solving a Difference Equation with Dirichlet Boundary Conditions

Consider the following difference equation in G_n,

$$\alpha G_{n+1} + \beta G_n + \alpha G_{n-1} = F_n, \quad 1 \leq n \leq N-1 \qquad (E.1)$$

with its (vanishing) boundary conditions at $n = 0$ and $n = N$,

$$G_0 = 0 \qquad (E.2)$$

$$G_N = 0 \qquad (E.3)$$

We recognize in the difference equation (E.1) that it is symmetric in the operators E and E^{-1},

$$\alpha G_{n+1} + \beta G_n + \alpha G_{n-1} \equiv [\alpha(E + E^{-1}) + \beta]G_n = F_n, \qquad (E.4)$$

hence we can use either the discrete sine $\mathcal{S}$ or cosine $\mathcal{C}$ transform to algebraize the operator $(E + E^{-1})$ as shown in (2.127), (2.140), (2.144), and (2.143) with $M = N$ for the discrete sine and cosine transforms, respectively. However, the boundary conditions, (E.2)-(E.3), involve the values of the transformed sequence G_0 and G_N at the end points, hence we must choose the discrete sine transform. But since these are vanishing boundary conditions $G_0 = 0$ and $G_N = 0$, it is natural to use the discrete sine transform pair (2.127). Of course, the more general pair (2.40) can also be used with $G_0 = 0$ and $G_N = 0$, which gives (2.127). So, if we take $M = N$ in (2.127) and apply the $\mathcal{S}$ transform to (E.1), letting $b_k = \mathcal{S}\{G_n\}$, we have

$$\mathcal{S}\{\alpha(G_{n+1} + G_{n-1}) + \beta G_n = F_n\}$$

$$= \mathcal{S}\{\alpha(G_{n+1}) + G_{n-1}) + \mathcal{S}\{\beta G_n\} = \mathcal{S}\{F_n\}$$

$$= \alpha(\cos\frac{\pi k}{N} b_k + \beta b_k = f_k, \quad 1 \leq k \leq N-1$$

$$= \left[\alpha \cos \frac{\pi k}{N} + \beta \right] b_k = f_k, \quad 1 \le k \le N - 1$$

$$(E.5)$$

where $f_k = \mathcal{S}\{F_n\}$.

In (E.5), we have our desired result of an algebraic equation in the discrete sine transform b_k, and where we have already incorporated the homogeneous boundary conditions (E.2)-(E.3) by our decision of selecting the discrete sine transform as our operational tool. This equation is easy to solve for b_k,

$$b_k = \frac{f_k}{2\alpha \cos \frac{\pi k}{N} + \beta}, 1 \le k \le N - 1 \qquad (E.6)$$

provided that $2\alpha \cos \frac{\pi k}{N} + \beta$ does not vanish, which can be assured if we insist that $|2\alpha| < |\beta|$. In case that for a particular k_1, for example, $2\alpha \cos \frac{\pi k}{N} + \beta$ vanishes at the same time, which renders b_{k_1} arbitrary in the resulting $0 b_{k_1} = 0$ in (E.5). In this case, the solution to the boundary value problem (E.1)–(E.3), as the inverse discrete Fourier sine transform of b_k in (E.5), is determined from (2.119) within the term that has an arbitrary amplitude b_{k_1},

$$G_n = b_{k_1} \sin \frac{\pi n k_1}{N} + \sum_{k=1, k \ne k_1}^{N-1} b_k \sin \frac{\pi n k}{N},$$

This also means that the boundary value problems (E.1)–(E.3), with $2\alpha \cos \frac{2\pi k_1}{M} + \beta = 0$, and even the unlikely coincidence that $f_{k_1} = 0$, does not have a unique solution because of the term $b_{k_1} \sin \frac{n \pi k_1}{M}$ with its arbitrary amplitude that makes infinite possibilities of solutions. Of course, if $2\alpha \cos \frac{\pi n k}{M} + \beta = 0$ and $f_k \ne 0$, then there is no solution to the boundary value problem (E.1)–(E.3).

What remains is to find the inverse of this explicit transform to have the final solution G_n of the boundary value problem (E.1)-(E.3),

$$G_n = \mathcal{S}^{-1}\{\frac{f_k}{2\alpha \cos \frac{\pi k}{N} + \beta}\}, \quad 1 \le n \le N - 1 \qquad (E.7)$$

$$= \sum_{k=1}^{N-1} \frac{f_k}{2\alpha \cos \frac{\pi k}{N} + \beta} \cdot \sin \frac{\pi n k}{N} \qquad (E.8)$$

The computation is left for Chapter 5, where more of the essential tools of the discrete sine and cosine transforms are developed, which are needed to handle this result (E.7) analytically. However, fortunately, this inverse as the finite sum in (E.8) can be computed very easily via the very fast algorithm, i.e., the fast Fourier transform (FFT), which will be discussed in Section 4.6, and which is available in hardware on (almost) all computers.

Example 2.13 Difference Equations with Nonhomogeneous Boundary Conditions and the Operational Sum Calculus Method

Here we consider the same problem of Example 2.12, except that the boundary conditions are nonhomogeneous. For the present case, the two ends $n = 0$ and $n = N$ are set at a non–vanishing constant value $G_0 = A$ and $G_N = B$,

$$\alpha G_{n+1} + \beta G_n + \alpha G_{n-1} = F_n \quad 1 \leq n \leq N - 1 \qquad (E.1)$$

$$G_0 = A \qquad (E.2)$$

$$G_N = B \qquad (E.3)$$

(a) The Series Substitution Method of the DST for Homogeneous Boundary Conditions

The usual method of doing this problem is by reducing it to another one with homogeneous boundary conditions, where the method of Example 2.12 can be applied to the latter. This meaning that we find a function Θ_n that subtracts the constant values at the end points. Such a function is the straight line between the two points $(0, G_0) = (0, A)$ and $(N, G_N) = (N, B)$, whose equations is

$$\Theta_n = A + \frac{(B - A)}{N} n \qquad (E.4)$$

With this, we let $H_n = G_n - \Theta_n = G_n - (A + \frac{B-A}{N}n)$ where the now homogeneous boundary conditions,

$$H_0 = G_0 - \Theta_0 = A - A = 0 \qquad (E.5)$$

$$H_N = G_N - \Theta_N = B - A - \frac{B-A}{N}N$$
$$= B - A - B + A = 0, \qquad (E.6)$$
$$H_N = 0$$

However, while the difference equation (E.1) reduces to a similar one in H_n, its nonhomogeneous term F_n is changed to a bit more complicated nonhomogeneous term, $V_n = F_n - (2\alpha + \beta)\Theta_n$,

$$\alpha(G_{n+1} + \beta G_n + \alpha G_{n-1}) = F_n$$

$$= \alpha[H_{n+1} + \Theta_{n+1} + \beta H_n + \beta\Theta_n + \alpha H_{n-1} + \alpha H_{n-1} + \alpha\Theta_{n-1}] = F_n,$$

$$\alpha H_{n+1} + \beta H_n + \alpha H_{n-1} = F_n - [\alpha\Theta_{n+1} + \beta\Theta_n + \alpha\Theta_{n-1}]$$

$$= F_n - \left[\alpha(A + \frac{B-A}{N}(n+1)) + \beta(A + \frac{(B-A)}{N}n)\right.$$

$$\left. + \alpha(A + \alpha\frac{B-A}{N}(n-1))\right]$$

$$\alpha H_{n+1} + \beta H_n + \alpha H_{n-1} = F_n - (2\alpha + \beta)\Theta_n \equiv V_n$$

$$(E.7)$$

after a simple grouping and some cancellation in the brackets on the right hand side.

So, the new problem (E.7) in H_n with its homogeneous boundary conditions (E.5), (E.6) can be solved first in $h_k = \mathcal{S}\{H_n\}$ as we did in Example 2.12 with F_n in (2.12) replaced by V_n of (E.7) above. So according to (E.6) of Example 2.12, we have

$$h_k = \frac{v_k}{2\alpha\cos\frac{\pi k}{N} + \beta} = \frac{f_k - (2\alpha + \beta)\theta_k}{2\alpha\cos\frac{\pi k}{N} + \beta} \qquad (E.8)$$

where $v_k = \mathcal{S}\{V_n\} = f_k - (2\alpha + \beta)\theta_k$. Of course, having h_k requires that we know θ_k, the discrete sine transform of Θ_n in (E.4), before we are able to take the inverse discrete sine transform of h_k in (E.8) to have $H - n$, then the final solution $G_n = H_n + \Theta_n$ of our original problem (E.1)-(E.3). This requires, as seen from Θ_n in (E.4), that we should have the discrete sine transform of 1 as well as that of n for the first and second terms of Θ_n in (E.4), respectively.

(b) The Direct Operational Sum Calculus Method of the DST for Nonhomogeneous boundary conditions

These intermediate computations, as well as the whole change of variables process of letting $G_n = H_n - \Theta_n$, can be avoided if we appeal to the operational difference-sum calculus of the discrete sine transform in its fullest, i.e., to consult the discrete sine transform pair of $(G_{n+1} + G_{n-1})$ that we derived in (2.140), and which would involve (if not asks for) the nonhomogeneous boundary conditions at the outset and in our first step of the operation.

So, if we use (2.140) on (E.1), letting $b_k = \mathcal{S}\{G_n\}$ and $f_k = \mathcal{S}\{F_n\}$, we have

$$\mathcal{S}\{\alpha G_{n+1} + \beta G_n + \alpha G_{n-1}\} = \mathcal{S}\{F_n\}$$

$$= (2\alpha \cos \frac{\pi k}{N})b_k + \frac{2}{N}[G_0 + (-1)^{k+1}G_N]\sin \frac{\pi k}{N} + \beta b_k = f_k$$

$$(2\alpha \cos \frac{\pi k}{N} + \beta)b_k = f_k - \frac{2\alpha}{N}\left[A + (-1)^{k+1}B\right],$$

$$b_k = \frac{f_k - \frac{2\alpha}{N}\left[A + (-1)^{k+1}B\right]\sin \frac{\pi k}{N}}{2\alpha \cos \frac{\pi k}{N} + \beta}$$

$$(E.9)$$

provided that $2\alpha \cos \frac{\pi k}{N} + \beta \neq 0$, which can be assured when $|2\alpha| < |\beta|$. Now we can apply the inverse discrete sine transform on this b_k to have the final solution

$$G_n = \mathcal{S}^{-1}\{b_k\} = \mathcal{S}^{-1}\{\frac{f_k - \frac{2\alpha}{N}\left[A + (-1)^{k+1}B\right]\sin \frac{\pi k}{N}}{2\alpha \cos \frac{\pi k}{N} + \beta}\} \qquad (E.10)$$

$$G_n = \sum_{h=1}^{N-1} \frac{f_k - \frac{2\alpha}{N}\left[A + (-1)^{k+1}B\right]\sin \frac{\pi k}{N}}{2\alpha \cos \frac{\pi k}{N} + \beta} \sin \frac{\pi n k}{N} \qquad (E.11)$$

without having to compute the inverse of Θ_n (E.4). This is a very clear advantage of the present operational sum calculus, which, of course, stems such strength from its parallel to the already proven operational (integral) calculus methods of solving differential equations with the use of Fourier and other integral and finite transforms.

An interesting illustration, that we shall cover in the next part (c) of this example, is to show the equivalence of b_k in (E.9) with

$b_k = g_k = h_k + \theta_k$ of (E.8), after θ_k, the discrete sine transform of Θ_n, have been computed. The transform pairs needed to find θ_k are, i.e., $\mathcal{S}\{1\}$ and $\mathcal{S}\{n\}$, as we mentioned earlier, are established in Examples 5.1 and 5.2 respectively. We may remind that the results of these examples for the discrete sine transform of simple sequences like $\{G_n\}_{n=1}^{N-1} = \{1\}_{n=1}^{N-1}$ and $\{G_n\}_{n=1}^{N-1} = \{n\}_{n=1}^{N-1}$ are "somewhat" cumbersome, the latter, in particular, as can be seen in (E.8) of Example 5.1 and (E.4) of Example 5.2, respectively. This may, again add to the advantage of using the operational operational sum calculus pair (2.140) that involves the nonhomogeneous boundary conditions for the discrete sine transform, and which gives the result (E.9) that does not need the evaluation of $\theta_k = \mathcal{S}\{\Theta\}$ in (E.8).

(c) The Typical Series Substitution Versus the Operational Sum Calculus Methods of the DST

With the rather long expression of $\mathcal{S}\{n\}$ as in (E.4) of Example 5.2, we limit ourselves here to dealing with the rather modest expression of $\mathcal{S}$ in (E.8) of Example 5.1,

$$\mathcal{S}\{1\} = \frac{2}{N}\left[\frac{(1-(-1)^k)\sin\frac{\pi k}{N}}{2(1-\cos\frac{\pi k}{N})}\right] \qquad (E.12)$$

to illustrate the equivalence of the answer (E.9) of the direct operational sum method of the discrete sine transform method of part (b) with that of the usual method that involves h_k with its θ_k in (E.8). With such limitation to the above DST pair (E.12), we consider the special case of the nonhomogeneous boundary conditions $G_0 = A$ and $G_N = A$, where θ_k in (E.8) is

$$\theta_k = \mathcal{S}\{A\} = \frac{2A}{N}\frac{(1-(-1)^k)\sin\frac{\pi k}{N}}{2(1-\cos\frac{\pi k}{N})} \qquad (E.13)$$

Now we use θ_k for $b_k = g_k = \mathcal{S}\{G_n\} = g_k = h_k + \theta_k$ with h_k as in (E.8) to show that we get the same answer as the one we have obtained by the direct DST operational sum method in (E.9),

$$b_k = g_k = h_k + \theta_k = \frac{f_k - (2\alpha + \beta)\theta_k}{2\alpha \cos \frac{\pi k}{N} + \beta}$$

$$= \frac{f_k - (2\alpha + \beta)\theta_k - (2\alpha \cos \frac{\pi k}{N} \theta_k + \beta\theta_k)}{2\alpha \cos \frac{\pi k}{N} + \beta}$$

$$= \frac{f_k - 2\alpha(1 - \cos \frac{\pi k}{N})\theta_k}{2\alpha \cos \frac{\pi k}{N} + \beta} \tag{E.14}$$

$$= \frac{f_k - 2\alpha(1 - \cos \frac{\pi k}{N}) \cdot \frac{2A(1-(-1)^k)\sin \frac{\pi k}{N}}{2N(1-\cos \frac{\pi k}{N})}}{2\alpha \cos \frac{\pi k}{N} + \beta}$$

$$b_k = g_k = \frac{f_k - \frac{2\alpha A}{N}\left[1 + (-1)^{k+1}\right]\sin \frac{\pi k}{N}}{2\alpha \cos \frac{\pi k}{N} + \beta},$$

which is the same result of (E.9) when we take $B = A$.

An interesting problem that can be solved by the present method of the DST is that of Example 7.4 in Chapter 7. It models a discrete system of coupled springs and masses.

Our avoidance of the linear term n for computing the result (E.8) for the typical method of part (a) in the above example, may be exemplified when we deal with the nonhomogeneous *Neumann boundary conditions*,

$$\delta G_0 = G_1 - G_{-1} = A, \tag{2.156}$$

$$\delta G_N = G_{N+1} - G_{N-1} = B \tag{2.157}$$

In this case, the typical method of part (a) would require $G_n = H_n + \Phi_n$, where

$$\Phi_n = \frac{nA}{2} + \frac{B - A}{4N}n^2 \tag{2.158}$$

to give

$$\delta H_0 = \delta G_0 - \delta\Phi_0 = 0,$$

$$\delta H_n = \delta G_N - \delta\Phi_N = 0,$$

since

$$\delta\Phi_0 = \Phi_1 - \Phi_{-1} = \frac{A}{2} + \frac{B-A}{4N} - \left[-\frac{A}{2} + \frac{B-A}{4N} \right] = A, \tag{2.159}$$

$$\delta\Phi_0 = 0$$

$$\delta\Phi_N = \Phi_{N+1} - \Phi_{N-1} = \left[\frac{A(N+1)}{2} + \frac{(B-A)(N+1)^2}{4N} \right]$$

$$- \left[\frac{A(N-1)}{2} + \frac{(B-A)(N-1)^2}{4N} \right] \tag{2.160}$$

$$= \frac{A}{2} + \frac{A}{2} + \frac{(B-A)}{4N}(2N+2N) = A + B - A = B,$$

$$\delta\Phi_N = 0$$

In this case, of the Neumann boundary conditions, of course, we will use the discrete Fourier cosine transform as we shall illustrate in the next Example 2.14, where the typical method would require knowing $\phi_k = \mathcal{C}\{\Phi_n\}$, which in turn, as seen in (2.158), requires knowing the discrete cosine transforms of n as well as n^2. In contrast, and parallel to what we saw in (E.9)-(E.11) in part (b) of the above Example 2.13, and what we have already developed for the important DCT operational sum pair in (2.143), the DCT operational sum method would involve the nonhomogeneous Neumann boundary conditions *without* having to find the "possibly involved" $\phi_k = \mathcal{C}^{-1}\{\Phi_k\}$ in the intermediate step of the typical method that starts with (2.158). We can clearly see how the nonhomogeneous Neumann boundary conditions terms $G_1 - G_{-1} = A$ and $G_{N+1} - G_{N-1} = B$, with $M = N$ in the "rather new" DCT operational transform of $(G_{n+1} + G_{n-1})$ in (2.143).

Next we present a simple example of using the discrete Fourier cosine transform for solving the same difference equation of Example 2.12, but with Neumann boundary conditions instead of the Dirichelet boundary conditions of Example 2.12.

Example 2.14   The Compatibility of the Discrete Cosine Transform with the Neumann Boundary Conditions

For this example we choose the same difference equation of Example 2.12,

$$\alpha G_{n+1} + \beta G_n + \alpha G_{n-1} = F_n,$$
$$= [\alpha(E + E^{-1}) + \beta]G_n = F_n, \quad 1 \leq n \leq N - 1 \tag{E.1}$$

with the follwing Neumann–type boundary conditions around the end points $n = 0$, $n = M$,

$$\delta G_0 = G_1 - G_{-1} = 0 \tag{E.2}$$

and

$$\delta G_N = G_{N+1} - G_{N-1} = 0. \tag{E.3}$$

These conditions can be considered as approximations to the vanishing of the gradient around $n = 0$ and $n = N$, respectively. Also, as we have discussed them earlier, they represent the *"evenness"* about the end points $n = 0$ and $n = N$. Hence, the discrete cosine transform and its pair in (2.144) with $M = N$ is very compatible with this problem, since it transforms the difference equation to an algebraic equation in $a_n = \mathcal{C}\{G_n\}$, and it also automatically satisfies the Neumann boundary conditions (E.2), (E.3) of the evenness at the boundary points $n = 0$ and $n = N$, where we take $M = N$ in (2.143)). Of course, in (E.1) we have a symmetric equation in E and E^{-1}, and for non-symmetric case such as

$$\alpha G_{n+1} + \beta G + \gamma G_{n-1} = F_n \tag{E.4}$$

we can employ a scaling with $G_n = (\frac{\gamma}{\alpha})^{\frac{n}{2}} H_n$ that reduces (E.1) to another equation in H_n that is symmetric in E and E^{-1} as we have already discussed in (2.152)–(2.153), and illustrated in (2.154)–(2.155). So, if we take the discrete cosine transform of (E.1), using (2.144) with $M = N$, letting $a_k = \mathcal{C}\{G_n\}$, we have

$$\mathcal{C}\{\alpha(G_{n+1} + G_{n-1}) + \beta G_n\} = \mathcal{C}\{F_n\},$$
$$= \mathcal{C}\{\alpha(G_{n+1} + G_{n-1})\} + \mathcal{C}\{\beta G_n\} = \mathcal{C}\{F_n\} \tag{E.5}$$
$$= \alpha(2\cos\frac{\pi k}{N})a_k + \beta a_k = f_k$$

where $f_k = \mathcal{C}\{F_n\}$. The algebraic equation (E.5) in the discrete cosine transform a_k, can be solved easily to give

$$a_k = \frac{f_k}{2\alpha\cos\frac{\pi k}{N} + \beta}. \tag{E.6}$$

provided that $2\alpha \cos \frac{\pi k}{N} + \beta \neq 0$, which can be assured if $|2\alpha| < |\beta|$. All what remains is to find the inverse discrete cosine transform $G_n = \mathcal{C}^{-1}\{a_k\}$, as defined in (2.121), of the above a_k to have the final solution of the boundary value problem (E.1)–(E.3),

$$G_n = \frac{a_0}{2} + \sum_{k=1}^{N-1} a_k \cos \frac{\pi n k}{N} + \frac{a_N}{2}(-1)^n , \quad 0 \leq n \leq N \qquad (E.7)$$

$$= \frac{f_0}{2(2\alpha + \beta)} + \sum_{k=1}^{N-1} \frac{f_k}{2\alpha \cos \frac{\pi k}{N} + \beta} \cos \frac{\pi n k}{N} + (-1)^n \frac{f_N}{2(\beta - 2\alpha)} . \quad (E.8)$$

The case of the nonhomogeneous Neumann boundary conditions of (2.156)-(2.157) can be covered in parallel to what was done for the DST in Example 2.13, and with the help of the important DCT operational pair in (2.143).

A very interesting application for the diffusion process in the altruistic neighborhood model is covered in Example 7.7 of Section 7.2 in Chapter 7. There we use the DCT for solving its resulting boundary value problem.

The Operational Sum Method of the Double DFT's for Partial Difference Equations

At the end of part A in Section 2.3, we remarked very briefly on how the Fourier transforms can be extended to two (or higher) dimensions and used in a similar way to solve partial differential equations in the function of two variables $G(x, y)$. This was done in order to bring about its parallel for the present case of the operational sum calculus method. It is in the sense that a double discrete Fourier transform is easily defined, and its operational properties should be developed to solve partial difference equations associated with boundary conditions. Of course, a triple (and higher dimensional) Fourier integral or discrete transform can also be defined and utilized in a similar fashion.

The Double Fourier (Integral) Transform and its Inverse.

As an extension of the (single) Fourier transform and its inverse in (2.50)-(2.51), the *double* Fourier transform of the function of two variables $G(x, y), -\infty < x < \infty; -\infty < y < \infty$, is defined as

$$g(t, s) = \mathcal{F}_{(2)}\{G(x, y)\} = \int_{-\infty}^{\infty} \int_{-\infty}^{\infty} e^{i(xt+ys)} G(x, y) dx dy \quad (2.161)$$

with its inverse as

$$G(x, y) = \mathcal{F}_{(2)}^{-1}\{g(t, s)\} = \frac{1}{4\pi^2} \int_{-\infty}^{\infty} \int_{-\infty}^{\infty} e^{-i(xt+ys)} g(t, s) dt ds$$
$$(2.162)$$

Just for the sake of showing how this double Fourier transform algebraizes partial derivatives of $G(x, y)$, and hence facilitate solving partial difference equations in $G(x, y)$, we present the following important operational pair,

$$\mathcal{F}_{(2)}\{\frac{\partial^2 G}{\partial x^2} + \frac{\partial^2 G}{\partial y^2}\} = -(t^2 + s^2) g(t, s) \quad (2.163)$$

which should stand as a clear extension of the result of the (single) Fourier transform in (2.155).

The Double Discrete Fourier Transform

Similar to what is done for extending the Fourier integral transform in (2.50) to two dimensions in (2.161), the discrete Fourier transform of the sequence $G_n, 0 < n \leq N - 1$, in (2.100),

$$\tilde{\mathcal{F}}\{G_n\} = g_k = \frac{1}{N} \sum_{n=0}^{N-1} G_n \omega^{nk}, \quad 0 \leq k \leq N - 1 \quad (2.100)$$

can be extended to the *double* DFT of the sequence $G_{n,m}, \ 0 \leq n \leq N - 1; 0 \leq m \leq N - 1$ as,

$$g_{k,l} = \tilde{\mathcal{F}}_{(2)}\{G_{n,m}\} = \sum_{n=0}^{N-1} \sum_{m=0}^{N-1} G_{n,m} \omega^{-(nk+ml)}, \quad 0 \leq k, \ l \leq N - 1$$
$$(2.164)$$

with its inverse

$$G_{n,m} = \tilde{\mathcal{F}}_{(2)}^{-1}\{g_{k,l}\} = \frac{1}{N^2} \sum_{k=0}^{N-1} \sum_{l=0}^{N-1} g_{k,l}\omega^{-(nk+ml)}, \qquad 0 \le n, \ m \le N-1$$

$$(2.165)$$

Of course, it is easy to extend this definition to different limits on the two indices n and m of $G_{n,m}$, i.e., for $0 \le n \le N-1; 0 \le m \le M-1$, whereby we also have $g_{k,l}, 0 \le k \le N-1; 0 \le l \le M-1$.

One important operational property, which parallels that of the double Fourier transform in (2.163), should be very easy to see as

$$\tilde{\mathcal{F}}_{(2)}\{G_{n+2,m+2}\} = \omega^{-2(k+l)}g_{k,l} \qquad (2.166)$$

when we simply replace n and m in (2.165) by $n+2$ and $m+2$, respectively. This result (2.166) stands as an extension to two dimensions of (2.106) with $l = 2$. The other expected, simple to obtain property is that $G_{n,m}$ as well as $g_{k,l}$ are doubly periodic with period N in n, m, k and l, i.e., $G_{n+N,m} = G_{n,m}$, and $G_{n,m+N}$, and the same for $g_{k+N,l} = g_{k,l} = g_{k,l+N}$. As in the case of (2.106), this double periodicity is based on the typical periodicity with no jumps at the end points $n = 0$ and $m = N$ in the first direction of n, and the ends in the second direction, $m = 0$ and $m = N$. In order to bring about the involvement of the jump discontinuities, the summation by parts method should be used in both directions of n and m, as we did for the discrete Fourier transforms of one variable in this section. This subject will not be pursued further in this book.

Exercises 2.4

1. (a) Verify that $\tilde{G}(\frac{n}{NT})$ of (2.94) is the inverse discrete Fourier transform of $\tilde{g}(kT)$ in (2.93).

     Hint: Substitute $\tilde{G}(\frac{n}{NT})$, as a finite series of (2.94) inside the series of (2.93), interchange the summations, and use the discrete orthogonality in (2.97) for $k \ne l$. Notice that this series simply sums to N when $k = l$.

   (b) Do the same as part (a) of showing that $\tilde{g}(kT)$ in (2.93) is the inverse DFT of $\tilde{G}(\frac{n}{NT})$ in (2.94).

2. Let g_k and h_k be the discrete Fourier transforms of G_n and H_n, as given in (2.106) and (2.107), respectively. Show that:

   (a) $\mathcal{F}\{G_n + H_n\} = g_k + h_k$.

   Hint: Note that the finite summation operation of $\mathcal{F}$ on G_n is distributive over addition.

   (b) $\mathcal{F}\{cG_n\} = cg_k$, where c is a constant.

3. Prove the result in (2.106).

   Hint: As was done in (2.105), use the periodicity property of $g_m \, \omega^{-mn}$ with period N, where the resulting sum over m can be taken from $m = 0$ to $N - 1$ instead of $m = l$ to $N - 1 + l$.

4. Though longer than the method used in (2.109)–(2.110), use the method employed in deriving (2.107) with two summations by parts to establish (2.108).

5. Derive the general result-(2.109).

   Hint: Start with the special cases (2.107) and (2.108) and use mathematical induction, or better yet, follow the shorter route that we started with (2.110) to arrive at (2.108).

6. Establish the results (2.111) and (2.112) for the inverse discrete Fourier transform.

   Hint: See the method that led to (2.107) and (2.108).

7. (a) Show that the DST transform in (2.118) is linear, i.e., given $g_k = \mathcal{S}\{G_n\}$ and $h_k = \mathcal{S}\{H_n\}$, show that

   $$\mathcal{S}\{c_1 G_n + c_2 H_n\} = c_1 g_k + c_2 h_k \ .$$

   (b) Prove the same linearity property of part (a) for $\mathcal{C}$, the discrete Fourier cosine transform of (2.120).

Chapter 3

BASIC METHODS OF SOLVING LINEAR DIFFERENCE EQUATIONS

In this chapter we will present the basic methods of solving linear difference equations, and primarily with constant coefficients. The discrete Fourier Transforms method as well as the z-transform method will be covered in Chapters 4, 5, and 6, respectively. As this book covers mainly linear difference equations, some nonlinear equations are presented for merely exposing the reader to a very particular class of problems that are amenable to special methods which produce solutions in closed form. Such problems are presented as exercises with ample hints at the end of Section 3.6 exercises in Chapter 3.

3.1 Fundamentals of Linear Difference Equations

In Chapter 2 we introduced the basic elements and definitions of difference equations, and in particular the following nth order difference

139

equation in u_k,

$$b_0(k)u_{k+n} + b_1(k)u_{k+n-1} + \ldots + b_n(k)u_k = R_k, \qquad (3.1)$$

where the variable coefficients $b_0(k), b_1(k), \ldots, b_n(k)$; and the *non-homogeneous* term $R_k = R(k)$ are sequences defined over a set of integers $K_1 \leq k \leq K_2$. The above difference equation in (3.1) is clearly linear, and to make sure it is definitely of order n, we must insist that $b_0(k) \neq 0$ and $b_n(k) \neq 0$; or in other words $b_0(k)b_n(k) \neq 0$. The condition $b_0(k) \neq 0$ preserves the highest nth order term u_{k+n}, and the second condition $b_n(k) \neq 0$ prevent equation (3.1) from being equivalent to one of order $n-1$ in the resulting lowest order sequence $z_k = u_{k+1}$, where (3.1) becomes

$$
\begin{aligned}
b_0(k)u_{k+1+n-1} \quad &+ \quad b_1(k)u_{k+1+n-2} + \ldots + b_{n-1}u_{k+1} \\
&= \quad b_0(k)z_{k+n-1} + b_1(k)z_{k+n-2} + \ldots + b_{n-1}z_k \\
&= \quad R_k
\end{aligned}
\qquad (3.2)
$$

The development here will parallel that of linear ordinary differential equations. However, the nature of the methods and the solutions will vary. So, a familiarity with the basic elementary course is very helpful in following the mechanics of the procedures for the methods used here. However, we shall try to make the presentation here very much self contained.

For $b_0(k) \neq 0$, we can divide (3.1) by $b_0(k)$ to have a standardized form for (3.1) with 1 as the coefficient of the highest order term u_{k+n},

$$u_{k+n} + a_1(k)u_{k+n-1} + \ldots + a_n(k)u_k = r_k \qquad (3.3)$$

where

$$a_i(k) = \frac{b_i(k)}{b_0(k)}, \quad i = 1, 2, \ldots, n, \text{ and } r_k = \frac{R_k}{b_0(k)}.$$

Equation (3.3) is *nonhomogeneous* with nonhomogeneous term r_k, and it becomes *homogeneous* when $r_k = 0$ for all $k_1 \leq k \leq k_2$ over which (3.3) is defined,

$$u_{k+n} + a_1(k)u_{k+n-1} + \ldots + a_n(k)u_k = 0. \qquad (3.4)$$

As in the case of differential equations the typical direct methods of solving the nonhomogeneous equation (3.3) are very much dependent on solving the homogeneous equation (3.4), as we shall illustrate soon. Indeed the general solution of the homogeneous equation (3.4) is called a *complementary solution* of (3.3), and often written as $u_k^* \equiv u_k^{(c)}$. This means that with a *particular solution* $U_k \equiv u_k^{(p)}$ for the nonhomogeneous equation (3.4), its most general solution is

$$u_k = u_k^* + U_k \equiv u_k^{(c)} + u_k^{(p)}. \tag{3.5}$$

Because of the use here of the subscript for the sequence u_k, we shall need to use a superscript such as in the above $u_k^{(p)}$ and $u_k^{(c)}$ to differentiate between the two different sequences for the solution. This will become clear when we speak of the n expected solutions $u_k^{(1)}, u_k^{(2)}, \ldots, u_k^{(n)}$ of the homogeneous equation (3.4). Sometimes we may resort to using capital letters to distinguish between different solutions. For example we may use $U_k^{(p)}$ and $U_k^{(c)}$ instead of $u_k^{(p)}$ and $u_k^{(c)}$ to differentiate them from the above individual solutions $u_k^{(1)}, u_k^{(2)}, \ldots, u_k^{(n)}$ of the homogeneous equation (3.4). In an effort to avoid this notation difficulty, some authers resort to using the functional notations of a sequence $u(k)$ instead of the present indexed one u_k. Hence they may write $u_k^{(c)}, u_k^{(p)}$ and the individual solutions $u_k^{(1)}, u_k^{(2)}, \ldots$ as $u_c(k), u_p(k)$ and $u_1(k), u_2(k), \ldots$, respectively. For this book we shall stay with the traditional indexing for a sequence as indicated in (3.5), (3.7) and the rest of the book.

Next we present a few very basic results for the homogeneous linear equation (3.4), which are needed for constructing its general solution u_k of (3.5), then follow it by a result for (3.5). We shall state these results as theorems, whose proofs follow easily, and definitely parallel those done for the linear homogeneous differential equations.

Theorem 3.1   An Important Property of Linearity

For c as a constant, if u_k is a solution of (3.4), then the result cu_k of multiplying u_k by a constant amplitude c, is also a solution.

The proof is simple, since if we substitute cu_j instead of u_j on the left hand side of (3.4), we end up with the same value of zero for the right hand side,

$$cu_{k+n} + ca_1(k)u_{k+n-1} + \ldots + ca_n(k)u_k$$
$$= c[u_{k+n} + a_1(k)u_{k+n-1} + \ldots + a_n(k)u_k]$$
$$= c \cdot 0 = 0, \tag{3.6}$$

after using (3.4) for the zero value of the above parentheses.

Theorem 3.2   The Principle of Superposition

For c_1 and c_2 as arbitrary constants; let $u_k^{(1)}$ and $u_k^{(2)}$ be solutions of the homogeneous equation (3.4), then the linear combination,

$$u_k = c_1 u_k^{(1)} + c_2 u_k^{(2)} \tag{3.7}$$

of the two solutions $u_k^{(1)}$ and $u_k^{(2)}$, is also a solution.

The proof follows easily if we write the first line of (3.6) for $u_k^{(1)}$ with $c = c_1$ and for $u_k^{(2)}$ for $c = c_2$, then add the two resulting equations to have

$$\left[c_1 u_{k+n}^{(1)} + c_2 u_{k+n}^{(2)}\right] + \left[c_1 a_1(k)u_{k+n-1}^{(1)} + c_2 a_1(k)u_{k+n-1}^{(2)}\right]$$

$$+ \ldots + \left[c_1 a_n(k)u_k^{(1)} + c_2 a_n(k)u_k^{(2)}\right] = 0. \tag{3.8}$$

This is (3.4) with u_k replaced by $u_k = c_1 u_k^{(1)} + c_2 u_k^{(2)}$, which means that the linear combination $c_1 u_k^{(1)} + c_2 u_k^{(2)}$ of the solutions $u_k^{(1)}$ and $u_k^{(2)}$ of (3.4) is a also solution of the linear homogeneous equation (3.4).

The main result of this theorem is a very important property of the linearity of the equation (3.4), which is called the *principle of superposition*. Of course this principle of superposition can be applied to any number of the n solution $u_k^{(1)}, u_k^{(2)}, \ldots, u_k^{(n)}$ of equation (3.4).

The following theorem shows how a solution of the nonhomogeneous equation (3.3) can be a combination of any solution of (3.3), i.e., the *particular* solution $U_k \equiv u_k^{(p)}$, and a solution of the homogeneous equation $u_k^{(1)}$. We must stress here that we are not speaking of the most general solution of (3.3) as indicated in (3.5), but only about a step towards it.

Theorem 3.3   The Particular Solution

Let U_k be any solution of the nonhomogeneous equation (3.3), and $u_k^{(1)}$ be one of the solutions of the homogeneous equation (3.4), then

$$u_k = U_k + u_k^{(1)}$$

is also a solution of the nonhomogeneous equation (3.3).

The proof follows as we substitute this u_k in (3.3),

$$a_0(k)\left[U_{k+n} + u_{k+n}^{(1)}\right] + a_1(k)\left[U_{k+n+1} + u_{k+n+1}^{(1)}\right] + \ldots$$
$$+ a_n(k)\left[U_k + u_k^{(1)}\right]$$
$$= a_0(k)U_{k+n} + a_1(k)U_{k+n-1} + \ldots + a_n(k)U_k$$
$$+ \left[a_0(k)u_{k+n}^{(1)} + a_1(k)u_{k+n-1}^{(1)} + \ldots + a_n(k)u_k^{(1)}\right]$$
$$= r_k + 0 = r_k,$$

since the first part of the line before last gives r_k according to U_k being a particular solution of the nonhomogeneous equation (3.3), and the second line gives zero since $u_k^{(1)}$ is a solution of the homogeneous equation (3.4).

Example 3.1   The Particular Solution

Consider the second order linear nonhomogeneous difference equation with constant coefficients

$$u_{k+2} - 3u_{k+1} + 2u_k = 3^k, \qquad (E.1)$$

we can easily verify that $U_k = \frac{3^k}{2}$ is a particular solution, since

$$\frac{3^{k+2}}{2} - 3 \cdot \frac{3^{k+1}}{2} + 2 \cdot \frac{3^k}{2} = 3^k\left[\frac{9}{2} - \frac{9}{2} + 1\right] = 3^k.$$

Also $u_k^{(1)} = 2^k$ is a solution of the associated homogeneous equation of (E.1),

$$u_{k+2} - 3u_{k+1} + 2u_k = 0, \qquad (E.2)$$

since

$$2^{k+2} - 3 \cdot 2^{k+1} + 2 \cdot 2^k = 2^k \cdot (4 - 6 + 2) = 0.$$

We can also verify that $u_k^{(2)} = 1$ is a solution of (E.2),

$$u_{k+2} - 3u_{k+1} + 2u_k = 1 - 3 + 2 = 0.$$

Now, it is a simple matter to show that $u_k = U_k + u_k^{(1)}$, $u_k = U_k + u_k^{(2)}$, $u_k = U_k + u_k^{(1)} + u_k^{(2)}$, or even $u_k = U_k + c_1 u_k^{(1)} + c_2 u_k^{(2)} = \frac{3^k}{2} + c_1 2^k + c_2$, are all solutions of (E.1), which we shall leave as a very simple exercise.

We will return soon in Section 3.2 for the method of finding such $u_k^{(1)} = 2^k$, $u_k^{(2)} = 1$ solutions of the homogeneous second order equation (E.2).

Before attempting to solve an equation, we should first make sure that a solution *exists*, and that it is also *unique*. The following theorem states the (very reasonable) conditions for the existence of a unique solution of the linear difference equation (3.3), provided that we are given the n consecutive values of the solution $u_j, u_{j+1}, u_{j+2}, \ldots, u_{j+n-1}$ as $a_0, a_1, a_2, \ldots, a_{n-1}$ respectively. As we stated in Theorem 2.2, these n initial values of the sequence u_k along with the difference equation (3.3) constitute an *initial value problem*, and where the following theorem represents, the special case of Theorem 2.2 for the *linear* difference equation (3.3).

Theorem 3.4   Existence and Uniqueness of the Solution

For the linear difference equation (3.3), and the prescribed n initial conditions for n consecutive values of the solution $u_j = a_0$, $u_{j+1} = a_1$, $u_{j+2} = a_2$, $\ldots$, $u_{j+n-1} = a_{n-1}$, there exists a unique solution.

Since we have already presented the simple proof of Theorem 2.2, the proof here is done as a special case. For practice, of course, the simple iterative steps of proving Theorem 2.2 can be retraced easily for proving the present theorem.

Example 3.2   The Uniqueness of the Solution

In Example 3.1, we had a general solution

$$u_k = \frac{3^k}{2} + c_1 \cdot 2^k + c_2 \cdot 1 \tag{E.1}$$

to the second order linear difference equation

$$u_{k+2} - 3u_{k+1} + 2u_k = 3^k. \tag{E.2}$$

We note that the solution in (E.1) is not unique because of the two (if not yet determined) arbitrary constants c_1 and c_2. Also, in the same example, we assumed the existence of a solution. Furthermore, we were given a particular solution $U_k = \frac{3^k}{2}$ for the nonhomogeneous equation, and two solutions $u_k^{(1)} = 2^k$ and $u_k^{(2)} = 1$ for its corresponding homogeneous case. Here, according to Theorem 3.4, or its general case, we can be assured the existence of a unique solution, provided that for this second order linear difference equation, we are given two consecutive values of the solution u_k; for example, the two initial values

$$u_0 = 1, \tag{E.3}$$

and

$$u_1 = 2. \tag{E.4}$$

We can clearly see how the imposing of these two conditions on the solution

$$u_k = \frac{3^k}{2} + c_1 2^k + c_2 \tag{E.5}$$

determines the two arbitrary constants, c_1 and c_2, and hence we obtain a unique solution. If we let $k = 0$, then $k = 1$ in (E.5), and use the initial values $u_0 = 1$ and $u_1 = 2$, we obtain two linear simultaneous algebraic equations in c_1 and c_2,

$$u_0 = \frac{1}{2} + c_1 + c_2 = 1, \quad c_1 + c_2 = \frac{1}{2}, \tag{E.6}$$

$$u_1 = \frac{3}{2} + 2c_1 + c_2 = 2, \quad 2c_1 + c_2 = \frac{1}{2}. \tag{E.7}$$

From solving (E.6) and (E.7), we have $c_1 = 0$ and $c_2 = \frac{1}{2}$, which when substituted in (E.5), we obtain the unique solution

$$u_k = \frac{1}{2} + \frac{3^k}{2}.$$

With all the work in the example, there remains one important aspect to such solutions. That is, whether the nominated solution in (E.1) represents the *general* solution of (E.2), a question that will be addressed after the following example, and specially in Definition 3.1 and Theorem 3.5.

Example 3.3   An Initial Value Problem

Consider the following second order difference equation

$$u_{k+2} - 6u_{k+1} + 8u_k = 0 \qquad\qquad (E.1)$$

along with its two prescribed initial conditions

$$u_0 = 3, \qquad\qquad (E.2)$$

$$u_1 = 2. \qquad\qquad (E.3)$$

We can easily verify that $u_k^{(1)} = 2^k$, and $u_k^{(2)} = 4^k$, are the two solutions of (E.1), so their linear combination

$$u_k = c_1 2^k + c_2 4^k. \qquad\qquad (E.4)$$

is also a solution, but is not unique. Now, we use the initial conditions (E.2) and (E.3) in (E.4) to find the two arbitrary constants c_1, c_2,

$$u_0 = 3 = c_1 + c_2, \qquad\qquad (E.5)$$

$$u_1 = 2 = 2c_1 + 4c_2, \quad c_1 + 2c_2 = 1. \qquad\qquad (E.6)$$

Hence $c_2 = -2$, $c_1 = 5$, and the unique solution of the initial value problem (E.1)-(E.3) becomes

$$u_k = 5 \cdot 2^k - 2 \cdot 4^k = 2^k(5 - 2^{k+1}).$$

In the above example we verified that $u_k^{(1)} = 2^k$, and $u_k^{(2)} = 4^k$ are two solutions of the difference equation,

$$u_{k+2} - 6u_{k+1} + 8u_k = 0. \tag{3.9}$$

We also know from Theorem 3.1.2 that the linear combination $u_k = c_1 2^k + c_2 4^k$ of such two solutions is also a solution of (3.9). The question now is whether this last solution is the *general solution* of the homogeneous linear difference equation (3.9). The answer lies in whether the two solutions $u_k^{(1)} = 2^k$, and $u_k^{(2)} = 4^k$ are *linearly independent*, or in more simple terms that, for the special case of only two solutions, one of them cannot be obtained from the other by a simple change of its constant amplitude. With such an interpretation, the two solutions $u_k^{(1)} = 2^k$, and $u_k^{(2)} = 4^k$ are linearly independent, since $u_k^{(2)} = 4^k = 2^k \cdot 2^k = 2^k u_k^{(1)}$, which is not a constant multiple of $u_k^{(1)} = 2^k$. On the other hand, we can easily verify that $u_k^{(1)} = 2^k$ and $u_k^{(2)} = 2^{k+3}$ are solutions of the difference equation,

$$u_{k+2} - 4u_{k+1} + 4u_k = 0, \tag{3.10}$$

but they are not linearly independent. The reason is that $u_k^{(2)} = 2^{k+3} = 2^3 \cdot 2^k = 8 \cdot 2^k$ is a constant multiple of $u_k^{(1)} = 2^k$. In Section 3.2, we will discuss how $u_k^{(2)}$ is modified as $k \cdot 2^k$ to make it linearly independent of $u_k^{(1)}$.

Unfortunately, this last intuitive explanation of linear independence is very clear only for the case of the two solutions (of the special case of the second order equation). Next we give the definition of linear independence of n sequences, where we will follow it by a practical test for such linear independence. All these concepts run very parallel to the same concepts used for functions $f(x)$ that are covered in the basic course of elementary differential equations.

Definition 3.1   Linear Independence of Sequences

A set of sequences $f_1(k), f_2(k), \ldots, f_n(k)$ are said to be *linearly dependent* over the interval $K_1 \leq k \leq K_2$, if there exists a set of n constants $c_1, c_2, \ldots, c_n$ *not all zero* such that the following linear combination of the n sequences vanish, i.e.,

$$c_1 f_1(k) + c_2 f_2(k) + \ldots + c_n f_n(k) = 0.$$

If the set of sequences is *not* linearly dependent, then it is *linearly independent*.

Example 3.4   Linear Independence of Two Sequences

We can see that $f_1(k) = 2^k$ and $f_2(k) = 4^k$ are linearly independent, since $c_1 2^k + c_2 4^k$ cannot be made zero for all variable values of k, $K_1 \leq k \leq K_2$, without having to force both c_1 and c_2 to be zeros. On the other hand, the linear combination $c_1 2^k + c_2 2^{k+3}$ of $f_1(k) = 2^k$ and $f_2(k) = 2^{k+3}$ can easily be made to vanish by assuming $c_1 = -8c_2 \neq 0$, whence $c_1 \cdot 2^k + c_2 \cdot 2^{k+3} = c_1 \cdot 2^k + 8c_2 \cdot 2^k = -8c_2 \cdot 2^k + 8c_2 \cdot 2^k = 0$ for all (nonzero) constant c_2.

It is also easy to show that the two solutions $u_k^{(1)} = 2^k$, and $u_k^{(2)} = 1$ of Example 3.2 are linearly independent, since $c_1 2^k + c_2 \cdot 1 = 0$ requires that both c_1 and c_2 vanish.

While the case of two linearly independent sequences can be handled easily, it is much more difficult for the case of three or more sequences, as the following example illustrates. Thus we need a more practical mathematical test, which will be covered in Theorem 3.5, to follow.

Example 3.5   Linear Independence of Three Sequences

Consider the third order linear homogeneous (and constant coefficient) difference equation in u_k,

$$u_{k+3} - 6u_{k+2} + 11u_{k+1} - 6u_k = 0. \qquad (E.1)$$

In the next Section 3.2 we will discuss a simple method for finding the three solutions $u_k^{(1)} = 1$, $u_k^{(2)} = 2^k$, and $u_k^{(3)} = 3^k$ for (E.1), and which can be verified here very easily. According to the above definition of linear independence for these three sequences, the vanishing of their linear combination; i.e.,

$$c_1 \cdot 1 + c_2 \cdot 2^k + c_3 \cdot 3^k = 0 \qquad (E.2)$$

requires that all the three arbitrary constants c_1, c_2, and c_3 must vanish, so, $c_1 = c_2 = c_3 = 0$. This, we claim is not very easy to see as compared with the case of the two sequences in the above example. So, we will need the following test which involves the use of *determinants*.

Definition 3.2   The Casoratian Determinant of n Sequences

The *Casoratian determinant* (or just the *Casoratian*) of the n sequences $f_1(k), f_2(k), \ldots, f_n(k)$ is defined as the following determinant of an $n \times n$ matrix of these sequences evaluated at $k, k+1, \ldots, k+n$, and we denote it by $C(k)$,

$$
C(k) = \begin{vmatrix}
f_1(k) & f_2(k) & \cdots & f_n(k) \\
f_1(k+1) & f_2(k+1) & \cdots & f_n(k+1) \\
\vdots & \vdots & & \vdots \\
f_1(k+n) & f_2(k+n) & \cdots & f_n(k+n)
\end{vmatrix}. \tag{3.11}
$$

Now, we can state a theorem which serves as a test of the linear independence of n sequences, and we omit the proof.

Theorem 3.5   A Practical Test for Linear Independence of Sequences

Consider the set of n sequences $f_1(k), f_2(k), \ldots, f_n(k)$, they are *linearly dependent* if, and only if, their Casoration $C(k)$ as defined in (3.11) vanishes for all k, $K_1 \leq k \leq K_2$. As a very simple consequence of this theorem, we have a test for the linear independence of the above n sequences when their Casoratian does not vanish, i.e., $f_1(k), f_2(k), \ldots, f_n(k)$ are *linearly independent* if, and only if, $C(k) \neq 0$ for all $k, K_1 \leq k \leq K_2$.

When a set of solutions $u_k^{(1)}, u_k^{(2)}, \ldots, u_k^{(n)}$ of the nth order difference equation (3.4) are linearly independent (or their Casoration $C(k) \neq 0$, $K_1 \leq k \leq K_2$, then they are called a *fundamental set of solutions*. With such fundamental set of solutions, their linear combination

$$
u_k = c_1 u_k^{(1)} + c_2 u_k^{(2)} + \ldots + c_n u_k^{(n)} \tag{3.12}
$$

represents, finally, the *general solution* of the homogeneous nth order difference equation (3.4).

An important result concerning the Casoration $C(k)$ of (3.11) is that, except for a constant $C(k_0)$, it can be determined in terms of the coefficients $a_n(k)$ of the difference equation (3.4). This is in the sense that with the use of definition (3.11) we can show that $C(k)$ satisfies the difference equation

$$C(k+1) = (-1)^n a_n(k) C(k)$$

whose solution (as we shall cover in Section 3.6) is

$$C(k) = (-1)^{n(k-k_0)} C(k_0) \prod_{i=k_0}^{k-1} a_n(i), \quad k > k_0.$$

Example 3.6   Linear Independence and the General Solution

We can now use the Casoratian in the above Theorem 3.5, to show that the two solutions $u_k^{(1)} = 2^k$ and $u_k^{(2)} = 4^k$ of the difference equation

$$u_{k+2} - 6u_{k+1} + 8u_k = 0 \tag{E.1}$$

in Example 3.3, are indeed linearly independent, since their following Casoratian $C(k)$ does not vanish for all k, $K_1 \leq k \leq K_2$.

$$
\begin{aligned}
C(k) &= \begin{vmatrix} u_k^{(1)} & u_k^{(2)} \\ u_{k+1}^{(1)} & u_{k+1}^{(2)} \end{vmatrix} = \begin{vmatrix} 2^k & 4^k \\ 2^{k+1} & 4^{k+1} \end{vmatrix} \\
&= 2^k \cdot 4^{k+1} - 2^{k+1} \cdot 4^k = 2^k \cdot 2^{2k+2} - 2^{k+1} \cdot 2^{2k} \\
&= 2^{3k+2} - 2^{3k+1} = 2^{3k+1}(2-1) \\
&= 2^{3k+1} \neq 0 \text{ for all } K_1 \leq k \leq K_2.
\end{aligned}
$$

Hence $u_k^{(1)} = 2^k$ and $u_k^{(2)} = 4^k$ being solutions of equation (E.1), and now that they are also linearly independent makes them a set of two fundamental solutions for the homogeneous equation (E.1). Thus, the general solution of (E.1) is their linear combination

$$u_k = c_1 2^k + c_2 4^k. \tag{E.2}$$

We will also show here that the two solutions $u_k^{(1)} = 2^k$ and $u_k^{(2)} = 1$ of the homogeneous difference equation

$$u_{k+2} - 3u_{k+1} + 2u_k = 0 \tag{E.3}$$

in Example 3.2, are linearly independent, since

$$C(k) = \begin{vmatrix} u_k^{(1)} & u_k^{(2)} \\ u_{k+1}^{(1)} & u_{k+1}^{(2)} \end{vmatrix} = \begin{vmatrix} 2^k & 1 \\ 2^{k+1} & 1 \end{vmatrix}$$

$$= 2^k - 2^{k+1} = 2^k(1-2) = -2^k \neq 0 \text{ for all } K_1 \le k \le K_2. \tag{E.4}$$

Again, this makes the two solutions $u_k^{(1)} = 2^k$ and $u_k^{(2)} = 1$ of equation (E.3) as fundamental solutions, whence the general solution of (E.3) is

$$u_k = c_1 2^k + c_2. \tag{E.5}$$

Now we can show that the three solutions $u_k^{(1)} = 1$, $u_k^{(2)} = 2^k$, and $u_k^{(3)} = 3^k$ of the third order homogeneous equation (of Example 3.6)

$$u_{k+3} - 6u_{k+2} + 11u_{k+1} - 6u_k = 0 \tag{E.6}$$

are linearly independent, since their following Casoratian $C(k)$ does not vanish for all k, $K_1 \le k \le K_2$.

$$C(k) = \begin{vmatrix} u_k^{(1)} & u_k^{(2)} & u_k^{(3)} \\ u_{k+1}^{(1)} & u_{k+1}^{(2)} & u_{k+1}^{(3)} \\ u_{k+2}^{(1)} & u_{k+2}^{(2)} & u_{k+2}^{(3)} \end{vmatrix} = \begin{vmatrix} 1 & 2^k & 3^k \\ 1 & 2^{k+1} & 3^{k+1} \\ 1 & 2^{k+2} & 3^{k+2} \end{vmatrix}$$

$$= 1 \begin{vmatrix} 2^{k+1} & 3^{k+1} \\ 2^{k+2} & 3^{k+2} \end{vmatrix} - 2^k \begin{vmatrix} 1 & 3^{k+1} \\ 1 & 3^{k+2} \end{vmatrix} + 3^k \begin{vmatrix} 1 & 2^{k+1} \\ 1 & 2^{k+2} \end{vmatrix} \tag{E.7}$$

$$= 2^{k+1} \cdot 3^{k+2} - 2^{k+2} \cdot 3^{k+1} -$$

$$2^k(3^{k+2} - 3^{k+1}) + 3^k(2^{k+2} - 2^{k+1})$$

$$= 3 \cdot 6^{k+1} - 2 \cdot 6^{k+1} - 9 \cdot 6^k + 3 \cdot 6^k + 4 \cdot 6^k - 2 \cdot 6^k$$

$$= 6^{k+1} - 4 \cdot 6^k = 2 \cdot 6^k \neq 0, \text{ for all } K_1 \le k \le K_2.$$

Hence 1, 2^k, and 3^k constitute a *fundamental set* of solutions of the linear third order homogeneous equation (E.6), and its *general solution* is their linear combination,

$$u_k = c_1 + c_2 \cdot 2^k + c_3 \cdot 3^k. \tag{E.8}$$

Constructing the Homogeneous Difference Equation from its Linearly Independent Solutions

A similar determinant to that of the Casoratian in (3.11), of an $n + 1$ by $n + 1$ matrix of the sought solution u_k and the n linearly independent solutions $u_k^{(1)}, u_k^{(2)}, \ldots, u_k^{(n)}$ and their difference up to order n, can be used to find their nth order difference equation in u_k. The equation is obtained from

$$\begin{vmatrix} u_k & u_k^{(1)} & u_k^{(2)} & \cdots & u_k^{(n)} \\ u_{k+1} & u_{k+1}^{(1)} & u_{k+1}^{(2)} & \cdots & u_{k+1}^{(n)} \\ \vdots & \vdots & \vdots & & \vdots \\ u_{k+n} & u_{k+n}^{(1)} & u_{k+n}^{(2)} & \cdots & u_{k+n}^{(n)} \end{vmatrix} = 0. \qquad (3.13)$$

The proof lies in the fact that a determinant with two identical columns vanishes. This is the case, since the first column for the sought solution u_k is equal, by the definition of it being a solution, to all the other columns of the known linearly independent solutions $u_k^{(1)}, u_k^{(2)}, \ldots, u_k^{(n)}$ of the (same) sought homogeneous equation. We will illustrate this result by finding the (homogeneous) difference equation

$$u_{k+2} - 6u_{k+1} + 8u_k = 0$$

of Example 3.7 from the knowledge of its two linearly independent solutions $u_k^{(1)} = 2^k$, and $u_k^{(2)} = 4^k$; when used in (3.13).

Example 3.7   Constructing a Homogeneous Difference Equation from its Linearly Independent Solutions

For the two linearly independent solutions $u_k^{(1)} = 2^k$, and $u_k^{(2)} = 4^k$, we use (3.13) to have

$$\begin{vmatrix} u_k & 2^k & 4^k \\ u_{k+1} & 2^{k+1} & 4^{k+1} \\ u_{k+2} & 2^{k+2} & 4^{k+2} \end{vmatrix} = 0,$$

$$u_k(2^{k+1} \cdot 4^{k+2} - 2^{k+2} \cdot 4^{k+1}) - 2^k(u_{k+1} \cdot 4^{k+2} - u_{k+2} \cdot 4^{k+1})$$

$$+4^k(u_{k+1} \cdot 2^{k+2} - u_{k+2} \cdot 2^{k+1}) = 0,$$

$$u_k(2^{3k+5} - 2^{3k+4}) - 2^{3k+2}(4u_{k+1} - u_{k+2}) + 2^{3k+1}(2u_{k+1} - u_{k+2}) = 0$$

$$2^{3k+1} \cdot 8u_k - 2^{3k+1}(8u_{k+1} - 2u_{k+2}) + 2^{3k+1}(2u_{k+1} - u_{k+2}) = 0,$$

$$8u_k - 8u_{k+1} + 2u_{k+2} + 2u_{k+1} - u_{k+2} = 0,$$

$$u_{k+2} - 6u_{k+1} + 8u_k = 0,$$

which is the difference equation (E.1) of Example 3.7.

In the next sections we will discuss the usual methods of solving linear difference equations, primarily, with constant coefficients. The discrete Fourier transforms, and the z-transform methods of solving such equations will be left for Chapters 4 and 5 on the discrete Fourier Transforms, Chapter 6 on the z-transforms, and Chapter 7 on the applications.

Exercises 3.1

1. (a) In Example 3.1, show that $u_k = \frac{3^k}{2} + c_1 \cdot 2^k + c_2$ is a solution of the nonhomogeneous equation (E.1),

$$u_{k+2} - 3u_{k+1} + 2u_k = 3^k \qquad (E.1)$$

    in that example.

   (b) As we shall show in Section 3.2, the solution $u_k = \frac{3^k}{2} + c_1 \cdot 2^k + c_2$ happened to be the most *general solution* of (E.1), and to make it a *unique* solution, we need to have two initial conditions such as $u_1 = 1$ and $u_2 = 3$ for determining the two arbitrary constants c_1 and c_2 in this general solution. Use these initial conditions to find the unique solution.

2. (a) Verify that $u_k^{(1)} = 1$, $u_k^{(2)} = k$ and $u_k^{(3)} = k^2$ are solutions of the third order homogeneous equation

$$u_{k+3} - 3u_{k+2} + 3u_{k+1} - u_k = 0 \qquad (E.1)$$

(b) Show that the three solutions of (E.1) in part (a) are, indeed, linearly independent, thus they constitute a fundamental set of solutions to (E.1).

Hint: Consult Theorem 3.5 for a good test of linear independence, using the Casoratian determinant, and Example 3.7, in particular its last part in (E.6)-(E.8).

(c) Write the general solution for (E.1).

(d) Use the three solutions of (E.1) in part (a) to construct the difference equation (E.1) itself.

Hint: Consult the determinant in (3.13), and Example 3.8.

3. Do part (a)-(d) of problem 2 for the following difference equations and their corresponding solutions

(a)
$$u_{k+3} + u_{k+2} - u_{k+1} - u_k = 0,$$
$$u_k^{(1)} = (-1)^k, \quad u_k^{(2)} = (-1)^k k, \quad u_k^{(3)} = 1.$$

(b)
$$u_{k+3} - 3u_{k+2} + 3u_{k+1} - u_k = 0,$$
$$u_k^{(1)} = 1, \quad u_k^{(2)} = k, \quad u_k^{(3)} = k^2.$$

4. Do parts (a)-(c) of problem 2 for the following fourth order equation and its four solutions.

$$u_{k+4} - u_k = 0,$$
$$u_k^{(1)} = 1, \quad u_k^{(2)} = (-1)^k, \quad u_k^{(3)} = \cos\frac{k\pi}{2}, \quad u_k^{(4)} = \sin\frac{k\pi}{2}.$$

5. (a) Verify that $u_k = 2^k$ is a solution of the first order difference equation
$$u_{k+1} - 2u_k = 0$$

(b) Verify that $u_k = \sqrt{k+c}$ is a solution of the first order *nonlinear* difference equation
$$u_{k+1}^2 - u_k^2 = 1$$

(c) Verify that the two sequences $u_k^{(1)} = (-1)^k$ and $u_k^{(2)} = 1$ are two solutions of the second order difference equation

$$u_{k+1} - u_{k-1} = 0 \qquad (E.1)$$

(d) In part (c), verify that the linear combination of the two solutions

$$u_k = c_1 u_k^{(1)} + c_2 u_k^{(2)} = (-1)^k c_1 + c_2$$

is also a solution of (E.1). Can you give an explanation?

6. (a) Find a *first* order difference equation whose solution is the sequence $u_k = c2^k$ with one arbitrary constant c.
   Hint: Let $u_k = c2^k$, find u_{k+1} and relate to u_k.

(b) For the sequence $u_k = A2^k + B5^k$ with two arbitrary constants A and B, find a *second* order difference equation whose solution is this given u_k.
   Hint: As in part (a), find u_{k+1} and u_{k+2}, which give you two simultaneous equations in the two arbitrary constants A, B. Eliminate A and B to find the relation between u_{k+2}, u_{k+1} and u_k, which is your second order difference equation.

(c) Instead of a sequence with one or two arbitrary constants as in part (a) and (b), respectively, consider a sequence

$$u_k = f(k, c_1, c_2, \ldots, c_n)$$

with n arbitrary constants $c_1, c_2, \ldots, c_n$. Attempt to generalize the results of parts (a) and (b) as to the order of the resulting difference equation whose general solution is $u_k = f(k, c_1, c_2, \ldots, c_n)$.
   Hint: Note that

$$u_{k+1} = f(k+1, c_1, c_2, \ldots, c_n) \qquad (E.1)$$

$$\vdots$$

$$u_{k+n} = f(k+n, c_1, c_2, \ldots, c_n) \qquad (E.2)$$

as set of n equations. Now we try to eliminate the n arbitrary constants.

7. (a) Find the second order difference equation whose solution is $u_k = (A+Bk)2^k$ with its two arbitrary constants A and B.

   Hint: See the hint to problem 6(c).

   (b) In parallel to solutions of differential equations, what could the solution here remind us of?

8. Consider the third order homogeneous equation in Problem 3(b),

   $$u_{k+3} - 3u_{k+2} + 3u_{k+1} - u_k = 0. \qquad (E.1)$$

   Use the general solution as found in that problem and the following initial conditions to find the unique solution,

   $$u_1 = 1, \quad u_2 = 3, \quad u_4 = 5,$$

   then verify your answer.

9. Consider the nonhomogeneous difference equation of order 2,

   $$u_{k+1} - 3u_k + 2u_{k-1} = 1 + 5^k \qquad (E.1)$$

   (a) Verify that $u_k^{(1)} = 1$ and $u_k^{(2)} = 2^k$ are the two linearly independent solutions of its associated homogeneous equation

   $$u_{k+1} - 3u_k + 2u_{k-1} = 0 \qquad (E.2)$$

   (b) Find $u_k^{(c)}$, the general solution for (E.2).

   (c) Verify that $u_k^{(p)} = -1 - 12 \cdot 5^{k-1}$ is a particular solution of the nonhomogeneous equation (E.1), thus write the general solution of (E.1) with the help of the answer in part (b).

   (d) Use the following initial conditions

   $$u_0 = 3, \qquad (E.3)$$

   $$u_1 = 6, \qquad (E.4)$$

   to find the unique solution of this initial value problem, (E.1), (E.3)-(E.4).

3.2 Solutions of Homogeneous Difference Equations with Constant Coefficients

Consider the following homogeneous linear difference equation of order n in u_k,

$$a_0 u_{k+n} + a_1 u_{k+n-1} + \cdots + a_n u_k = 0, \qquad (3.14)$$

and its associated nonhomogeneous case

$$a_0 u_{k+n} + a_1 u_{k+n-1} + \cdots + a_n u_k = r_k \qquad (3.15)$$

as special case of (3.1), where $a_0, a_1, \ldots, a_n$ are *constants* as far as the variable k is concerned. As we had alluded to the method of solving this homogeneous equation (3.14) in Section 2.1, or as it can be seen by simple inspection, the sequence $u_k = \lambda^k$ would satisfy the homogeneous equation (3.14), a condition that would determine the value of λ from the following substitution of $u_k = \lambda^k$ in (3.14)

$$a_0 \lambda^{k+n} + a_1 \lambda^{k+n-1} + \cdots + a_n \lambda^k = 0,$$

$$= \lambda^k [a_0 \lambda^n + a_1 \lambda^{n-1} + \cdots + a_n] = 0$$

and since $\lambda^k \neq 0$, we must have

$$a_0 \lambda^n + a_1 \lambda^{n-1} + \cdots + a_n = 0 \qquad (3.16)$$

an nth degree algebraic equation in λ, from which we determine the n values of λ as the roots $\lambda = \lambda_1, \lambda_2, \ldots, \lambda_n$ of (3.16). Equation (3.16) is called the *auxiliary (or characteristic) equation* of the homogeneous difference equation (3.14), which parallels the characteristic equation we have in solving linear homogeneous differential equations that determines the values of λ in their (same type) nominated solution $u(x) = e^{\gamma x} = (e^{\gamma})^x = \lambda^x$, $(\lambda \equiv e^{\gamma})$.

We will limit ourselves here to the very basics of this method that are necessary to facilitate its very familiar use. We will have a chance in the latter chapters to compare it with the discrete transforms methods as a added feature of this book.

As we expect, for the solutions of the auxiliary (characteristic) algebraic equation, we have real or complex roots, which can be distinct or repeated.

The following summarizes such cases depending on the roots $\lambda_1, \lambda_2, \ldots, \lambda_n$ of equation (3.16). We will omit the proofs, and concentrate on the illustrations for the method.

Case 1. All the roots are real and distinct

For the real *distinct* roots $\lambda_1, \lambda_2, \ldots, \lambda_n$ of (3.16), we can show that the corresponding solutions $u_k^{(1)} = \lambda_1^k$, $u_k^{(2)} = \lambda_2^k, \ldots, u_k^{(n)} = \lambda_n^k$ are *linearly independent*, and as such they constitute the fundamental set of solutions to the homogeneous equation (3.14). Hence the general solution of (3.14) is their following linear combination

$$u_k = c_1 \lambda_1^k + c_2 \lambda_2^k + \cdots + c_n \lambda_n^k. \tag{3.17}$$

Example 3.8   Distinct Real Roots

Consider the second order homogeneous constant coefficient equation (E.1) of Example 3.3

$$u_{k+2} - 6u_{k+1} + 8u_k = 0. \tag{E.1}$$

We will use the present method of letting $u_k = \lambda^k$ to establish its two linearly independent solutions $u_k^{(1)} = 2^k$, $u_k^2 = 4^k$, which we accepted by inspection in Example 3.3. If we substitute $u_k = \lambda^k$ in (E.1), the auxiliary (second degree) equation (3.16) in a becomes

$$\lambda^2 - 6\lambda + 8 = 0,$$

$$(\lambda - 2)(\lambda - 4) = 0,$$

$$\lambda_1 = 2, \quad \lambda_2 = 4.$$

Hence we have two distinct real roots, whose corresponding solutions $u_k^{(1)} = 2^k$, $u_k^{(2)} = 4^k$ are linearly independent. Hence 2^k, 4^k constitute a fundamental set of solutions to (E.1), and the general solution to (E.1) is

$$u_k = c_1 2^k + c_2 4^k.$$

We may mention that the linear independence of the two solutions 2^k and 4^k was shown via the nonvanishing of their *Casoratian* in Example 3.6.

Next, we use the same method to solve the following third order equation that we used as (E.1) in Example 3.5.

$$u_{k+3} - 6u_{k+2} + 11u_{k+1} - 6u_k = 0. \qquad (E.2)$$

The auxiliary (characteristic) equation of (E.2) is

$$\lambda^3 - 6\lambda^2 + 11\lambda - 6 = 0$$

$$(\lambda - 1)(\lambda - 2)(\lambda - 3) = 0, \qquad (E.3)$$

with three real distinct roots $\lambda_1 = 1$, $\lambda_2 = 2$ and $\lambda_3 = 3$. The corresponding three linearly independent solutions $u_k^{(1)} = 1^k = 1$, $u_k^{(2)} = 2^k$ and $u_k^{(3)} = 3^k$ constitute a fundamental set of solutions, so the general solution to (E.2) is their following linear combination,

$$u_k = c_1 \cdot 1 + c_2 \cdot 2^k + c_3 \cdot 3^k$$
$$= c_1 + c_2 \cdot 2^k + c_3 \cdot 3^k \qquad (E.4)$$

Again, the direct verification of the linear independence of the above three solutions was done in the second part of Example 3.6, by showing that their Casoratian determinant does not vanish.

Case 2. Some of the real roots are repeated

If some of the n real roots of (3.16) are repeated, for example λ_1, $\lambda_2 = \lambda_3 = \lambda_4$, λ_5, $\lambda_6 = \lambda_7$, with the remaining roots being distinct, then there is a modification to $\lambda_2^k = \lambda_3^k = \lambda_4^k$ and $\lambda_6^k = \lambda_7^k$ to make them linearly independent. The modification process involves multiplying the solution λ_j^k, corresponding to a repeated root λ_k, by the lowest integer power of k to remove the linear dependence. For example, in the present case the modified set becomes λ_1^k, λ_2^k, $k\lambda_2^k$, $k^2\lambda_2^k$, λ_5^k, λ_6^k, $k\lambda_6^k$, λ_8^k, λ_9^k, ..., λ_n^k, which constitute a fundamental set of solutions, and the general solution to (3.16) becomes

$$\begin{aligned}
u_k = {} & c_1\lambda_1^k + c_2\lambda_2^k + c_3 k\lambda_2^k + c_4 k^2\lambda_2^k \\
& + c_5\lambda_5^k + c_6\lambda_6^k + c_7 k\lambda_6^k + c_8\lambda_8^k \\
& + c_9\lambda_9^k + \ldots + c_n\lambda_n^k. \qquad (3.18)
\end{aligned}$$

Example 3.9   Repeated Real Roots

The difference equation

$$u_{k+3} + u_{k+2} - u_{k+1} - u_k = 0 \qquad (E.1)$$

has the auxiliary equation,

$$\lambda^3 + \lambda^2 - \lambda - 1 = 0$$

$$= \lambda^2(\lambda + 1) - (\lambda - 1) = (\lambda + 1)(\lambda^2 - 1)$$

$$= (\lambda + 1)(\lambda + 1)(\lambda - 1) = (\lambda + 1)^2(\lambda - 1) = 0 \qquad (E.2)$$

which has two repeated roots $\lambda_1 = \lambda_2 = -1$ and $\lambda_3 = 1$. Hence, the linearly independent solutions are $u_k^{(1)} = (-1)^k$, $u_k^{(2)} = k(-1)^k$ and $u_k^{(3)} = 1^k = 1$, and the general solution to (E.1) is

$$u_k = c_1(-1)^k + c_2 k(-1)^k + c_3. \qquad (E.3)$$

As another illustration, the difference equation

$$u_{k+3} - 3u_{k+2} + 3u_{k+1} - u_k = 0. \qquad (E.4)$$

has the characteristic equation

$$\lambda^3 - 3\lambda^2 + 3\lambda - 1 = (\lambda - 1)^3 = 0 \qquad (E.5)$$

with three repeated roots $\lambda_1 = \lambda_2 = \lambda_3 = 1$, where the linearly independent solutions are $1^k = 1$, $k1^k = k$ and $k^2 1^k = k^2$, and the general solution to (E.4) becomes

$$u_k = c_1 + c_2 k + c_3 k^2. \qquad (E.6)$$

Case 3. Complex Roots

When we have a complex root $\lambda = \alpha + i\beta$, the corresponding solution is $\lambda^k = (\alpha + i\beta)^k$. Such expression can be simplified along with the anticipated algebraic operations, if we adopt the *polar representation* of the complex number $\alpha + i\beta$, which results in dealing with sine and cosine functions. The polar representation of the complex number $z = x + iy$ is

$$z = \lambda e^{i\theta} = \lambda(\cos\theta + i\sin\theta), \tag{3.19}$$

where

$$\lambda = \sqrt{x^2 + y^2}, \tag{3.20}$$

$$\tan\theta = \frac{y}{x}, \tag{3.21}$$

and that in (3.19) we used the *Euler identity*

$$e^{i\theta} = \cos\theta + i\sin\theta. \tag{3.22}$$

With such polar representation of complex numbers in (3.19), we can write the corresponding solution of the complex root $\lambda = x + iy = z = re^{i\theta}$ as

$$\lambda^k = z^k = \left(re^{i\theta}\right)^k = r^k e^{ik\theta} = r^k(\cos k\theta + \sin k\theta), \tag{3.23}$$

after using the Euler identity again for $e^{ik\theta}$ in the last step of (3.23). Here we see that the solution corresponding to the complex root $\lambda = z = x + iy$ has real and imaginary parts as $r^k \cos k\theta$ and $r^k \sin k\theta$, respectively.

For algebraic equations with *real* coefficients $b_0, b_1, b_2, \ldots, b_n$ as in (3.18), we can show that if (3.16) has a complex root $\lambda_1 = z_1 = x_1 + iy_1$, then it also has the *complex conjugate* of this root $\overline{\lambda_1} = \overline{z_1} = \overline{x_1 + iy_1} \equiv x_1 - iy_1$ as another root. For example, the characteristic equation of the difference equation

$$u_{k+2} - 2u_{k+1} + 5u_k = 0 \tag{3.24}$$

is

$$\lambda^2 - 2\lambda + 5 = 0, \tag{3.25}$$

which has two complex roots;

$$\lambda = \frac{2 \mp \sqrt{4 - 20}}{2} = \frac{2 \mp 4i}{2} = 1 \mp 2i,$$

with the first root as $\lambda_1 = 1 + 2i$, and the second one is its complex conjugate $\lambda_2 = \overline{1 + 2i} = 1 - 2i$. The polar representations of these two distinct complex roots are

$$\lambda_1 = 1 + 2i = \lambda(\cos\theta + i\sin\theta),$$

where $\lambda = \sqrt{1 + 4} = \sqrt{5}$, $\cos\theta = \frac{1}{\sqrt{5}}$, $\sin\theta = \frac{2}{\sqrt{5}}$,

$$\lambda_1 = \sqrt{5}\left(\frac{1}{\sqrt{5}} + i\frac{2}{\sqrt{5}}\right),$$

and

$$\lambda_2 = 1 - 2i = \sqrt{5}\left(\frac{1}{\sqrt{5}} - i\frac{2}{\sqrt{5}}\right).$$

So, for the complex roots $\lambda_1 = r(\cos\theta + i\sin\theta)$ and $\lambda_2 = r(\cos\theta - i\sin\theta)$, the corresponding solutions are

$$u_k^{(1)} = r^k(\cos k\theta + i\sin k\theta)$$

and

$$u_k^{(2)} = r^k(\cos k\theta - i\sin k\theta),$$

which can be shown to be linearly independent. Hence they constitute a fundamental solution, and the general solution of their second order difference equation becomes

$$u_k = c_1 r^k(\cos k\theta + i\sin k\theta) + c_2 r^k(\cos k\theta - i\sin k\theta), \qquad (3.26)$$

$$
\begin{aligned}
u_k &= (c_1 + c_2)r^k \cos k\theta + i(c_1 - c_2)r^k \sin k\theta \\
&= Ar^k \cos k\theta + Br^k \sin k\theta \qquad\qquad (3.27)
\end{aligned}
$$

after renaming the arbitrary constants as $(c_1 + c_2) = A$ and $i(c_1 - c_2) = B$. So, for the difference equation (3.24), we have

$$\lambda_1 = 1 + 2i = \sqrt{5}(\cos\theta + i\sin\theta)$$

with $\tan\theta = 2$, and

$$\lambda_2 = \overline{\lambda_1} = \sqrt{5}(\cos\theta - i\sin\theta),$$

and the solution of (3.24) in the form (3.27) becomes

$$u_k = A(5)^{\frac{k}{2}} \cos k\theta + B(5)^{\frac{k}{2}} \sin k\theta, \tag{3.28}$$

where θ is known from $\tan \theta = 2$.

Example 3.10   Complex Roots

Consider the fourth order difference equation

$$u_{k+4} - u_k = 0. \tag{E.1}$$

The characteristic equation is

$$\begin{aligned} \lambda^4 - 1 = 0 &= (\lambda^2 + 1)(\lambda^2 - 1) \\ &= (\lambda + i)(\lambda - i)(\lambda + 1)(\lambda - 1) \end{aligned} \tag{E.2}$$

Hence we have one complex root $\lambda_1 = i$, its complex conjugate $\lambda_2 = \overline{\lambda_1} = \overline{i} = -i$, and two distinct real roots $\lambda_3 = 1$ and $\lambda_4 = -1$. The corresponding solutions are i^k, $(-i)^k = (-1)^k i^k$, $1^k = 1$ and $(-1)^k$, which are linearly independent as we have two distinct real roots 1, -1, and, obviously two distinct complex roots i, $-i$. The (formal) general solution to the equation (E.1) is

$$u_k = c_1 i^k + c_2 (-i)^k + c_3 \cdot 1 + c_4 \cdot (-1)^k, \tag{E.3}$$

which is not very desirable due to the complex expressions i^k and $(-i)^k$. The polar form of these complex numbers offers a more desirable and familiar expression for them in terms of sine and cosine functions, since

$$i = e^{i\frac{\pi}{2}} = \cos \frac{\pi}{2} + i \sin \frac{\pi}{2} = 0 + i,$$

and

$$-i = e^{-i\frac{\pi}{2}} = \cos \left(-\frac{\pi}{2}\right) - i \sin \left(-\frac{\pi}{2}\right) = 0 - i = -i.$$

In this form, their corresponding solutions become $\left(e^{i\frac{\pi}{2}}\right)^k = e^{\frac{i\pi k}{2}}$ and $\left(e^{-i\frac{\pi}{2}}\right)^k = e^{-\frac{i\pi k}{2}}$. Now the general solution to (E.1) becomes

$$u_k = c_1 1^k + c_2 (-1)^k + c_3 e^{\frac{i\pi k}{2}} + c_4 e^{-\frac{i\pi k}{2}} \tag{E.4}$$

and if we use the polar representation of complex numbers as in (3.26)-(3.27) expression for combining the last two complex terms, we have

$$u_k = c_1 + c_2(-1)^k + c_3 1^k \left(\cos\frac{k\pi}{2} + i\sin\frac{k\pi}{2}\right)$$

$$+ c_4 1^k \left(\cos\frac{k\pi}{2} - i\sin\frac{k\pi}{2}\right) \qquad (E.5)$$

$$= c_1 + c_2(-1)^k + A\cos\frac{k\pi}{2} + B\sin\frac{k\pi}{2},$$

where $\lambda = \left| e^{\frac{ik\pi}{2}} \right| = \sqrt{e^{\frac{ik\pi}{2}} \cdot e^{-\frac{ik\pi}{2}}} = \sqrt{1} = 1$, and we let $c_3 + c_4 = A$ and $i(c_3 - c_4) = B$.

Two more interesting illustrations that use the method of solution in this section are Examples 7.3 and 7.5 in Chapter 7. They model, respectively, the population growth and the evaluation of determinants.

Exercises 3.2

1. Consider the homogeneous equation

$$u_{k+3} - 3u_{k+1} + 2u_k = 0$$

associated with the nonhomogeneous equation (E.1) of problem 1 in the Exercises of Section 3.1.

(a) Find the characteristic equation.

(b) Find the two linearly independent solutions, and write the general solution to verify what was given in problem 1.

2. Solve the difference equation

$$u_{k+3} - 3u_{k+2} + 3u_{k+1} - u_k = 0,$$

of problem 2 in Exercises 3.1, by first solving the characteristic equation, then write its general solution.

3. Find the characteristic equation of the homogeneous difference equation

$$u_{k+3} + u_{k+2} - u_{k+1} - u_k = 0,$$

of problem 3 in Exercises 3.1, then write the general solution.

4. Find the general solution of the difference equation

$$u_{k+2} - 5u_{k+1} + 6u_k = 0.$$

5. Find the general solution of the difference equation,

$$u_{k+3} - 8u_{k+2} - 21u_{k+1} - 18u_k = 0.$$

6. Solve the difference equation

$$u_{k+4} - 16u_k = 0$$

by first solving its characteristic equation, then write the general solution.

Hint: See Example 3.11.

7. Find the general solution of the difference equation

$$u_{k+2} - u_{k+1} + u_k = 0$$

by first solving its characteristic equation.

8. Find the general solution of the difference equation

$$u_{k+4} + 12u_{k+2} - 64u_k = 0$$

by first solving its characteristic equation.

9. Solve the difference equation by first solving its characteristic equation

$$u_{k+4} - 2u_{k+3} + 3u_{k+2} - 2u_{k+1} + u_k = 0$$

Hint: Note that the characteristic equation of this problem has complex repeated roots. Also, see problem 7.

10. Solve the difference equation

$$u_{k+3} - 8u_k = 0$$

$$u_0 = 3 \ , u_1 = 2.$$

11. Solve the initial value problem

$$u_{k+2} - 6u_{k+1} + 8u_k = 0$$

$$u_0 = 3, \quad u_1 = 2.$$

Hint: Use the initial conditions $u_0 = 3$ and $u_1 = 2$ to determine the two arbitrary constants in the general solution

$$u_k^{(c)} = c_1 u_k^{(1)} + c_2 u_k^{(2)}.$$

12. The *method of generating functions* for solving a linear homogeneous difference equation in u_k, assumes the following infinite series with the sought solution u_k as its coefficient,

$$G(t) = \sum_{k=0}^{\infty} u_k t^k. \qquad (E.1)$$

The method evolves around multiplying the difference equation by t^k, summing over each term of the equation from $k = 0$ to $k = \infty$, then attempting to recognize terms related to the *generating function* in (E.1), and solving for this $G(t)$. Finally we look at the coefficient in the series of $G(t)$ as the desired solution u_k. We will illustrate this method in the form of an exercise with leading hints.

For example, consider the following homogeneous equation in u_k,

$$u_{k+2} - 3u_{k+1} + 2u_k = 0 \qquad (E.2)$$

with the initial conditions

$$u_0 = 2 \text{ and } u_1 = 3. \qquad (E.3)$$

(a) Multiply the equation (E.2) by t^k then sum over from $k = 0$ to $k = \infty$ to show that (E.2) is transformed to the following simple *algebraic equation* in the generating function $G(t)$,

$$\frac{G(t) - u_0 - u_1 t}{t^2} - 3\frac{G(t) - u_0}{t} + 2G(t) = 0. \qquad (E.4)$$

Hint: You should add and subtract $u_0 + u_1$ and u_0 for the terms associated with u_{k+2} and u_{k+1}, respectively, in order to generate $G(t)$.

(b) Use the initial conditions of (E.3) in (E.4), then show that the generating function is

$$G(t) = \frac{2 - 3t}{(1 - t)(1 - 2t)}. \qquad (E.5)$$

(c) Use partial fraction expansion for $G(t)$ in (E.5), then expand each of the resulting terms in an infinite series to arrive at the following series of $G(t)$,

$$G(t) = \sum_{k=0}^{\infty} (1 + 2^k) t^k = \sum_{k=0}^{\infty} u_k t^k, \qquad (E.6)$$

and that the solution to the initial value problem (E.2)-(E.3) is the coefficient in (E.6),

$$u_k = 1 + 2^k. \qquad (E.7)$$

Hint: $\frac{2-3t}{(1-t)(1-2t)} = \frac{1}{1-t} + \frac{1}{1-2t}$, then expand $\frac{1}{1-t}$ and $\frac{1}{1-2t}$ as geometric series.

(d) Verify that u_k in (E.7) is indeed the solution to the initial value problem (E.2)-(E.3).

13. Consider the difference equation

$$u_{k+2} - u_{k+1} + u_k = 0 \qquad (E.1)$$

of problem 7.

(a) Use the generating function method of problem 12 above to solve (E.1).

(b) Compare the two answers.

14. Use the generating function method of problem 12 to solve the difference equation

$$u_{k+4} - 16u_k = 0$$

of problem 6, then compare the two answers.

15. For all the examples and exercises in this chapter, we have covered difference equations with auxiliary conditions as *initial conditions* such as $u_0 = A$ and $u_1 = B$, where these conditions are given for the initial values u_0 and u_1, and along with the difference equation we have an *initial value problem*. If we consider a sequence in u_k, which has only a *finite number* of elements N, i.e., $k = 1, 2, \ldots, N$, then we expect conditions at or around the two end points, for example, $u_1 = A$ and $u_N = B$, which are called *boundary conditions*, and along with the difference equation we have a *boundary value problem* in u_k. Chapters 4 and 5 and Section 2.4 deal with such boundary value problems. The following boundary value problem may illustrate the nature of such type of problems.

Consider the difference equation with the *parameter* μ,

$$u_{k+1} - (2 - \mu)u_k + u_{k-1} = 0, \quad 1 \le k \le N - 1 \qquad (E.1)$$

and the two (homogeneous) boundary conditions

$$u_0 = 0 \qquad\qquad\qquad (E.2)$$

$$u_N = 0 \qquad\qquad\qquad (E.3)$$

(a) Use the discriminant $\mu(\mu-4)$ of the characteristic equation to show that when $0 \le \mu \le 4$, we have two complex roots $\lambda_1 = \cos\theta + i\sin\theta$ and $\lambda_2 = \cos\theta - i\sin\theta$ after using the trigonometric substitution $2 - \mu = 2\cos\theta$ to obtain these roots in their simplified form.

Hint: For the resulting complex roots, for example $\lambda_1 = \cos\theta + i\sin\theta \equiv e^{i\theta}$, note that its corresponding solution is $u_k = \lambda_1^k = (e^{i\theta})^k = e^{ik\theta} = \cos k\theta + i\sin k\theta$, after using the Euler identity for $e^{ik\theta}$, and expect the two arbitrary constants c_1 and c_2 of the general solution to (E.1) to involve the pure imaginary number i.

(b) For $0 < \mu < 4$, use the results in part (a) to show that the solution to the difference equation (E.1) is

$$u_k = c_1 \sin k\theta + c_2 \cos k\theta, \quad \theta = \arccos\left(\frac{2-\mu}{2}\right) \qquad (E.4)$$

and that the solution of (E.1), which satisfies the first boundary condition $u_0 = 0$ in (E.2) is

$$u_k = c_1 \sin k\theta, \quad \theta = \arccos\left(\frac{2-\mu}{2}\right), \qquad (E.5)$$

noting that this solution still depends on the parameter μ, $0 < \mu < 4$ via the θ dependence on it in (E.5).

(c) Show that for the solution in (E.5) to satisfy the second boundary condition $u_N = 0$ in (E.3), the parameter μ must take only *discrete* values, namely,

$$\mu_n = 4\sin^2\frac{n\pi}{2N}, \quad n = 1, 2, \ldots, N-1 \qquad (E.6)$$

Hint: For a nontrivial solution of $\sin N\theta = 0$, we must have $N\theta = n\pi$, $n = 1, 2, \ldots, N-1$, $\theta = \frac{n\pi}{N}$, Hence $\mu_n = 2(1 - \cos\frac{n\pi}{N}) = 4\sin^2\frac{n\pi}{2N}$.

(d) Show that a solution to the boundary value problem (E.1)-(E.3) is the following set of N sine functions

$$u_k = c_2 \sin\frac{n\pi k}{N}, \quad n = 1, 2, \ldots, N-1 \qquad (E.7)$$

which also satisfies the restriction of $0 < \mu_n < 4$.

(e) Verify that the sequence in (E.7) is a solution to the boundary value problem (E.1)-(E.3).

Hint: Note how the boundary conditions (E.2) and (E.3) are verified at a glance with the proper form of the solution (E.7), where $u_0 = c_2 \sin 0 = 0$ and $u_N = c_2 \sin \frac{n\pi N}{N} = c_2 \sin n\pi = 0$. Such a useful property, of satisfying the given boundary conditions at the outset, is a main goal of the direct discrete Fourier transform method of Section 2.4 (in Chapter 2) and Chapters 4 and 5, for solving difference equations with boundary conditions, i.e., discrete boundary value problems.

(f) As opposed to the case $0 < \mu < 4$ of part (a), that resulted in a set of sine functions solution in (E.7) for the boundary value problem (E.1)-(E.3), consider the cases of $\mu = 0$ and $\mu = 4$ and show that the only solution to the same boundary value problem is the trivial solution $u_k = 0$ for $k = 1, 2, \ldots, N - 1$.

Hint: Here we have two equal real roots $\lambda_1 = \lambda_2$ for the characteristic equation, with the solution to (E.1) becoming $u_k = (c_1 + c_2 k)\lambda_1^k$, where the two boundary conditions (E.2) and (E.3) result in $c_1 = c_2 = 0$.

(g) Instead of $\mu = 0$, and 4 of part (f), consider the other values $\mu < 0$ and $\mu > 4$ outside the range $0 < \mu < 4$ of part (a), to show also that the trivial solution is the only solution to the boundary value problem (E.1)-(E.3).

Hint: Here we have two distinct real roots $\lambda_1 \neq \lambda_2$ with the solution $u_k = c_1 \lambda_1^k + c_2 \lambda_2^k$, where the first boundary condition (E.1) gives $u_0 = c_1 + c_2 = 0$, $c_2 = -c_1$. The second boundary condition gives $u_N = c_1(\lambda_1^N - \lambda_2^N) = 0$, and for $\lambda_1 \neq \lambda_2$, $c_1 = 0$, hence $u_k = 0$ for $k = 1, 2, \ldots, N - 1$, a trivial solution.

(h) What do you notice about the difference between a boundary value problem and an initial value problem that are associated with the difference equation?

3.3   Nonhomogeneous Difference Equations with Constant Coefficients

In Section 3.1 we presented the general linear nonhomogeneous difference equation (3.3) of order n,

$$u_{k+n} + a_1(k)u_{k+n-1} + \cdots + a_n(k)u_k = r_k, \qquad (3.3)$$

and its associated homogeneous case (3.4),

$$u_{k+n} + a_1(k)u_{k+n-1} + \cdots + a_n(k)u_k = 0, \qquad (3.4)$$

and showed that the general solution of the nonhomogeneous equation (3.3) consists of two parts,

$$u_k = U_k + u_k^* \equiv u_k^{(p)} + u_k^{(c)} \qquad (3.29)$$

The second part is the *complementary* solution $u_k^{(c)}$, which is the general solution of the homogeneous equation (3.4), and which we studied in the last section,

$$u_k^{(c)} = c_1 u_k^{(1)} + c_2 u_k^{(2)} + \cdots + c_n u_k^{(n)} \qquad (3.30)$$

where $u_k^{(1)}, u_k^{(2)}, \cdots, u_k^{(n)}$ are the n linearly independent solutions of the homogeneous equation (3.4). The first part in (3.5) is $u_k^{(p)}$, the *particular* solution of the nonhomogeneous equation (3.3), which is *any* solution of (3.3), and is what we are about to discuss.

The Method of Undetermined Coefficients

The first, and most known method that we shall cover is the method of the *undetermined coefficients*, which is limited to equations with *constant* coefficients $a_0, a_1, a_2, \cdots, a_n$,

$$a_0 u_{k+n} + a_1 u_{k+n-1} + \cdots + a_n u_k = r_k, \qquad (3.15)$$

and *for only a very special class of sequences for the nonhomogeneous term r_k in (3.5)*. The method, basically, depends on "guessing" at a solution of (3.30), allowing arbitrary constant coefficients for the different terms of the proposed (guessed) expression of the particular solution $u_k^{(p)}$. The required particular form of the nonhomogeneous single term r_k, for example, is that all its differences must result in no more than a finite set of sequences, each one of which, aside from a constant amplitude, is a special case, or very much related to that term r_k in the way that we will discuss in detail next. In case the nonhomogeneous part r_k in (3.30) consists of more than one term, then this essential condition must be applied to each one of its terms. An example is $r_k = a^k$ whose difference of nth order

$$E_n a^k = a^{k+n} = a^n \cdot a^k = Ca^k, \quad C = a^n$$

is a constant multiple of itself $r_k = a^k$. The other example of such Un-determined Coefficients (UC) method – type nonhomogeneous terms is

$$r_k = \cos bk \quad \text{and} \quad r_k = \sin bk.$$

For each of these terms, the differences result in terms as constant multiples of $\cos bk$ and $\sin bk$. For example,

$$\begin{aligned}
E(\sin bk) &= \sin b(k+1) = \sin bk \cos b + \cos bk \sin b \\
&= C \sin bk + D \cos bk,
\end{aligned}$$

with constant amplitudes $C = \sin b$ and $D = \cos b$. The same happens for $E(\cos bk)$, and for higher differences of $\sin bk$ or $\cos bk$. The third form is the polynomial $P_m(k)$ of order m

$$r_k = P_m(k) = b_0 k^m + b_1 k^{m-1} + \cdots + b_m \tag{3.31}$$

which is here with $m+1$ terms of the form $r_k = a_j k^m$. For example, the difference of $r_k = 3k^2 + 5k + 1$, as a polynomial $P_2(k)$ of degree 2 is

$$\begin{aligned}
E(r_k) &= P_2(k+1) = 3(k+1)^2 + 5(k+1) + 1 \\
&= 3(k^2 + 2k + 1) + 5k + 5 + 1 \\
&= 3k^2 + 6k + 3 + 5k + 5 + 1 \\
&= 3k^2 + 11k + 9
\end{aligned} \tag{3.32}$$

where this result is only a special case of $P_2(k) = b_0 k^2 + b_1 k + b_2$, which is of the same form as that of the above nonhomogeneous term $r_k = 3k^2 + 5k + 1$.

So, these special forms

$$r_k = a^k \tag{3.33}$$

$$r_k = \sin bk \tag{3.34}$$

$$r_k = \cos bk \tag{3.35}$$

$$r_k = P_m(k) \tag{3.36}$$

are called the Undetermined Coefficients (UC) method – type forms (UC) sequences.

It can be shown that the following products of the above UC sequences:

$$r_k = a^k P_m(k) \tag{3.37}$$

$$r_k = a^k \sin bk \tag{3.38}$$

$$r_k = a^k \cos bk \tag{3.39}$$

are also UC sequences.

The other main condition we made for this UC method to work is that the difference equation is with *constant coefficients*, which will become clear as we see how this method works.

The purpose here is to find *any* solution for the nonhomogeneous equation (3.3). So, we start by nominating a sequence whose differencing on the left hand side of (3.3) results, aside from arbitrary constant amplitude, in the same form as that of the results of all differencing of the nonhomogeneous term r_k on the right hand side of (3.3). For two such equal sides of (3.3), the combination of the resulting amplitudes for each term on the left hand side of (3.3) are matched with those on the right hand side. This results in a system of algebraic equations in the unknown (undetermined) constants whose solution determines the arbitrary constants used in the nominated sequence for the particular solution.

Of course, we will not have any part of the nominated sequence to be one of the solutions of the associated homogeneous equation,

$$a_0 u_{k+n} + a_1 u_{k+n-1} + \cdots + a_n u_k = 0 \qquad (3.40)$$

The reason is that such part will by definition, result in a zero value on the left hand side. Thus, it contributes nothing to our strategy of attempting to have the left hand side end up with terms related to the term (or terms) r_k, the nonhomogeneous part on the right hand side of (3.30).

We shall return to this matter and the full procedure of the method shortly. But, for now, we will illustrate the essentials of the method in finding the particular solution of the difference equation in Example 3.1.

Example 3.11   A Particular Solution via the UC Method

Consider the nonhomogeneous difference equation

$$u_{k+2} - 3u_{k+1} + 2u_k = 3^k \qquad (E.1)$$

of Example 3.1, where we were given the particular solution as $u_k^{(p)} = \frac{3^k}{2}$, and we only verified that, indeed it was. Here we illustrate the UC method by finding this solution. In (E.1) the nonhomogeneous term is $r_k = 3^k$, which is a UC sequence as special case of a^k in (3.33) with $a = 3$. Since all the differencing of 3^k will end up with $C3^k$, a constant multiple of 3^k, then we may nominate a constant multiple of this result for the particular solution as $u_k^{(p)} = A3^k$. Furthermore, the two linearly independent solutions of the associated homogeneous equation

$$u_{k+2} - 3u_{k+1} + 2u_k = 0 \qquad (E.2)$$

are

$$u_k^{(1)} = 1^k = 1 \ \text{ and } \ u_k^{(2)} = 2^k,$$

which is clear from solving its characteristic equation,

$$\lambda^2 - 3\lambda + 2 = (\lambda - 1)(\lambda - 2) = 0, \qquad (E.3)$$

with $\lambda_1 = 1$ and $\lambda_2 = 2$ for $u_k = \lambda^k$,

$$u_k^{(c)} = c_1 + c_2 2^k. \qquad (E.4)$$

So, the third condition for the success of the nominated particular solution $u_k^{(p)} = A3^k$, namely, that *no term in it intersects with any term of the complementary solution*, is satisfied. So $u_k^{(p)} = A3^k$ is the final *correct* form in this case, and what remains is to determine the coefficient A. This is done by substituting $u_k^{(p)} = A3^k$ for u_k on the left hand side of (E.1) to have

$$A3^{k+2} - A3^{k+1} + 2A3^k = 3^k$$

$$9A3^k - 9A3^k + 2A3^k = 3^k$$

$$(9A - 9A + 2A)3^k = 3^k$$

$$2A3^k = 3^k \qquad (E.5)$$

Now, the above matching of the coefficients (of the same UC term 3^k) on both sides of (E.2) determines the coefficient $A = \frac{1}{2}$ from (E.5), and we obtain the particular solution

$$u_k^{(p)} = A3^k = \frac{3^k}{2} \qquad (E.6)$$

With the *complementary* solution $u_k^{(c)}$ in (E.4), and this *particular* solution of (E.6), the *general solution* to the nonhomogeneous difference equation (E.1) becomes

$$u_k = u_k^{(c)} + u_k^{(p)} = c_1 + c_2 2^k + \frac{3^k}{2} \qquad (E.7)$$

We must note here that because of the two arbitrary constants c_1 and c_2 of $u_k^{(c)}$, u_k in (E.7) represents an infinite number of solutions. However, in order to have a unique solution, we must supply two auxiliary conditions on u_k in this second order difference equation (E.1) such as the following two *initial conditions*

$$u_0 = 1 \qquad (E.8)$$

$$u_1 = 5 \qquad (E.9)$$

If we impose these initial conditions (E.8), (E.9) on the general solution of (E.7), we obtain two simultaneous algebraic equations in the two coefficients c_1 and c_2,

$$u_0 = c_1 + c_2 2^0 + \frac{3^0}{2} = c_1 + c_2 + \frac{1}{2} = 1 \qquad (E.10)$$

and

$$u_1 = c_1 + 2c_2 + \frac{3}{2} = c_1 + 2c_2 + \frac{3}{2} = 5, \qquad (E.11)$$

If we subtract (E.10) from (E.11), we obtain

$$c_2 + 1 = 4, \quad c_2 = 3.$$

With this $c_2 = 3$, substituted in (E.1)), we obtain c_1,

$$c_1 = 1 - \frac{1}{2} - c_2 = 1 - \frac{1}{2} - 3 = -\frac{5}{2}.$$

Hence the final *unique* solution to the *initial value problem* (E.1), (E.8)-(E.9) is

$$u_k = -\frac{5}{2} + 3 \cdot 2^k + \frac{3^k}{2} \qquad (E.12)$$

The UC Method – Various Illustrations

The next few examples illustrate finding the forms of the nominated sequences (with undetermined coefficients) for particular solutions associated with a number of nonhomogeneous UC terms r_k.

Example 3.12   Determining the Form of a Particular Solution

Consider the nonhomogeneous term

$$r_k = 7k^2 + 2 - 8 \cdot 3^k \qquad (E.1)$$

Here, we see that r_k consists of two basic UC terms, the first $7k^2 + 2$ is in the polynomial of degree 2 form $P_2(k)$, and the second is in the form a^k with $a = 3$. The differencing of $7k^2$ will clearly result, aside from constant amplitude, in terms involving k^2, k, and 1, while the term 2 results in itself. So, for the part $7k^2 + 2$, we nominate the linear combination of k^2, k and 1 as $Ak^2 + Bk + C$. For the second part 3^k, the differencing end up with 3^k, and we nominate $D3^k$ for it. So, the final particular solution is

$$u_k^{(p)} = Ak^2 + Bk + C + D3^k. \qquad (E.2)$$

Of course, this is to be substituted in the left hand side of the given nonhomogeneous difference equation. The similar terms are grouped, then their coefficients are matched with those of their corresponding terms in $r_k = 7k^2 + 2 - 8 \cdot 3^k$ on the right hand side. This, in general, will result in four linear algebraic equations in the four (undetermined) coefficients A, B, C, and D.

Of course, as we mentioned earlier, what is missing in the above example is the difference equation itself. This is very important, since we have to make sure that the above nominated form of the particular solution $u_k^{(p)}$ in (E.2) does not contain a term (or terms) which is a solution of the associated homogeneous equation. In this case such term, by definition, will result in a zero on the left hand side, and there is nothing to be matched with its corresponding term on the right hand side of the nonhomogeneous equation. If that happens, such term is to be modified by multiplying it by k^m with high enough m to remove this difficulty, which is what we shall illustrate in the next example.

Example 3.13   The UC Method Difficulty with the Complementary Solution

Consider the difference equation

$$u_{k+2} - 5u_{k+1} + 6u_k = 7k^2 + 2 - 8 \cdot 3^k \qquad (E.1)$$

with the same nonhomogeneous term as we discussed in the above Example 3.12. Here we have $u_k^{(1)} = 3^k$ and $u_k^{(2)} = 2^k$ as the two linearly independent solutions of the associated homogeneous equation of (E.1),

$$u_{k+2} - 5u_{k+1} + 6u_k = 0. \qquad (E.2)$$

In this case, the above nominated form of a particular solution in (E.2) in Example 3.12,

$$u_k^{(p)} = Ak^2 + Bk + C + D3^k \qquad (E.3)$$

contains the last term $D3^k$, which is the first solution $u_k^{(1)} = 3^k$ of the homogeneous equation (E.2), and which shall, by definition, contribute nothing to the left hand side of (E.1). Thus we have to modify this term by multiplying it by k as $k3^k$. With such modification, the correct suggested form of a particular solution for (E.1) becomes

$$u_k^{(p)} = Ak^2 + Bk + C + Dk3^k \qquad (E.4)$$

Sometimes multiplying by k, to remove the difficulty of having one solution of homogeneous equation as part of the particular solution, is not enough. This happened when we had, for example, $\lambda = 3$ as a *repeated root* of the homogeneous equation. In this case the term in $u_k^{(p)}$ corresponding to the result of differencing a term in the nonhomogeneous part r_k of (3.30) must be multiplied by enough power k^j (with the lowest possible j) to remove this solution, as we shall illustrate in the following example.

Example 3.14   The UC Method and the Repeated Roots

In the case of the nonhomogeneous equation

$$u_{k+3} - 8u_{k+2} + 21u_{k+1} - 18u_k = r_k = 7k^2 + 2 - 8 \cdot 3^k \qquad (E.1)$$

the roots of the characteristic equation

$$\lambda^3 - 8\lambda^2 + 21\lambda - 18 = 0$$

$$(\lambda - 3)^2(\lambda - 2) = 0,$$

of the associated homogeneous equation

$$u_{k+3} - 8u_{k+2} + 21u_{k+1} - 18u_k = 0 \qquad (E.2)$$

are $\lambda_1 = \lambda_2 = 3$, and $\lambda_3 = 2$ with the first two roots as the repeated roots $a_1 = a_2 = 3$. Hence the three linearly independent solutions of the homogeneous equation (E.2) are

$$u_k^{(1)} = 3^k, \quad u_k^{(2)} = k3^k \quad \text{and} \quad u_k^{(3)} = 2^k$$

In this case, if we multiply the term 3^k, of the suggested particular solution $u_k^{(p)}$ of (E.2) in Example 3.13,

$$u_k^{(p)} = Ak^2 + Bk + C + D3^k \qquad (E.3)$$

by k to have $k3^k$, this is still a solution of the homogeneous equation (E.2), and will not contribute to our goal of finding the particular solution. This difficulty is removed, for the case of doubly repeated roots, if we multiply 3^k by k^2, and the correct form of the particular solution for (E.1) becomes

$$u_k^{(p)} = Ak^2 + Bk + C + Dk^2 3^k, \qquad (E.4)$$

the k^2 factor used to remove the difficulty of the two repeated roots, represents the lowest power of k in k^j that is necessary to do the job.

Example 3.15   More on the Particular Solution

1. For the nonhomogeneous term

$$r_k = 5 \cdot 2^k + 7 \cdot 4^k, \qquad (E.1)$$

and if we assume that neither 2^k nor 4^k are solutions of the given associated homogeneous equation, it becomes clear that the form of the particular solution is

$$u_k^{(p)} = A2^k + B4^k. \qquad (E.2)$$

2. For the case
$$r_k = 9k + 2, \qquad (E.3)$$

and if neither k nor 1 are solutions of the associated homogeneous equation, then the form of the particular solution should be the linear combination of all the results of differencing k and 1, namely k and 1,

$$u_k^{(p)} = Ak + B \qquad (E.4)$$

In case that $u_k^{(1)} = 1$, $u_k^{(2)} = k$ and $u_k^{(3)} = k^2$ happened to be three solutions of the associated homogeneous equation, then the two above terms in (E.4) corresponding to the differencing of k and 1, must be multiplied by enough power k^j to remove this situation. In this case all the terms in (E.4) must be multiplied by k^3 to have the correct suggested form of the particular solution as

$$u_k^{(p)} = Ak^4 + Bk^3. \qquad (E.5)$$

3. Now, consider the case of

$$r_k = k + \cos 2k.$$

Here, the result of differencing k and $\cos 2k$ are k, 1, $\cos 2k$ and $\sin 2k$. Hence, in the absence of any solution of the associated homogeneous equation being one of these sequences 1, k, $\cos 2k$ or $\sin 2k$, the form of the particular solutions becomes the linear combination of these results of differencing k and $\cos 2k$, namely, k, 1, $\sin 2k$ and $\cos 2k$,

$$u_k^{(p)} = Ak + B + C \sin 2k + D \cos 2k.$$

Example 3.16   A High Order Nonhomogeneous Equation

Consider the nonhomogeneous difference equation

$$u_{k+4} - u_k = 2 + \cos \frac{k\pi}{2} \qquad (E.1)$$

whose associated homogeneous case is

$$u_{k+4} - u_k = 0 \qquad (E.2)$$

which is (E.1) in Example 3.11. The initial suggested particular solution of (E.1), as we discussed in Example 3.16 above, should consist of terms (with arbitrary coefficients) that are the results of differencing $r_k^{(1)} = 1$ (in the term 2), which gives 1 itself, and the differencing of $r_k^{(2)} = \cos\frac{k\pi}{2}$ in (E.1), which results in $\cos\frac{k\pi}{2}$ and $\sin\frac{k\pi}{2}$, aside from constant amplitudes. Thus, the initial form of a particular solution of (E.1) becomes the linear combination of these three resulting linearly independent terms 1, $\cos\frac{k\pi}{2}$ and $\sin\frac{k\pi}{2}$,

$$u_k^{(p)} = A + B\cos\frac{k\pi}{2} + C\sin\frac{k\pi}{2} \qquad (E.3)$$

However, if we consult Example 3.11 we have the complementary solution (of the associated homogeneous equation) as

$$u_k^{(c)} = c_1 \cdot 1 + c_2(-1)^k + c_3\cos\frac{k\pi}{2} + c_4\sin\frac{k\pi}{2} \qquad (E.4)$$

where we see that three of the solutions of the homogeneous equation, namely, 1 (in the first term c_1), $\cos\frac{k\pi}{2}$ and $\sin\frac{k\pi}{2}$ are included in our initial guess of the particular solution (E.3). To remedy this, we must multiply each term of (E.3) by a large enough power of k to remove this difficulty. For the first term 1 in (E.3), multiplying it by k will do, and the same is for the two terms of $\cos\frac{k\pi}{2}$ and $\sin\frac{k\pi}{2}$. Hence the *correct* form for the particular solution of (E.1) becomes

$$u_k^{(p)} = Ak + Bk\cos\frac{k\pi}{2} + Ck\sin\frac{k\pi}{2}. \qquad (E.5)$$

Example 3.17   The UC Method Difficulty with the Complementary Solution

Consider the equation

$$u_{k+3} + u_{k+2} - u_{k+1} - u_k = k^2 + \sin\frac{k\pi}{2} \qquad (E.1)$$

whose associated homogeneous case was considered in Example 3.10,

$$u_{k+3} + u_{k+2} - u_{k+1} - u_k = 0 \qquad (E.2)$$

where the complementary solution was found as

$$u_k^{(c)} = c_1(-1)^k + c_2 k(-1)^k + c_3 \qquad (E.3)$$

To find a form for the particular solution of (E.1), we find the terms resulting from differencing $r_k^{(1)} = k^2$ in (E.1), which are k^2, k, and 1, and the terms resulting from differencing $r_k^{(2)} = \sin\frac{k\pi}{2}$, which are $\cos\frac{k\pi}{2}$, and $\sin\frac{k\pi}{2}$. Here, it is more efficient to do the modification on these resulting terms, so as to avoid the presence of any term of the complementary solution (E.3) from being among them. From the last term c_3 in (E.3), we note that the term 1 is a part of the complementary solution, which is also among the terms k^2, k, and 1 resulting from differencing k^2 in r_k of (E.1). So all these three terms must be multiplied by enough power of k to remove this intersection. In this case, multiplying each one by k resolves the problem, and we suggest $Ak^3 + Bk^2 + Ck$ as the part of the particular solution related to the term k^2 in r_k. The terms related to the part $\sin\frac{k\pi}{2}$ of r_k consist of the results of differencing $\sin\frac{k\pi}{2}$, which are $\sin\frac{k\pi}{2}$ and $\cos\frac{k\pi}{2}$. Since these two sequences do not intersect any of the terms of the complementary solution $u_k^{(c)}$ in (E.3), their suggested part of the particular solution is $D\sin\frac{k\pi}{2} + E\cos\frac{k\pi}{2}$. Hence the correct form of the particular solution for (E.1) is

$$u_k^{(p)} = Ak^3 + Bk^2 + Ck + D\sin\frac{k\pi}{2} + E\cos\frac{k\pi}{2}. \qquad (E.4)$$

As we did in the case of Example 3.12, what remains after having the correct form of the particular solution with its arbitrary constant coefficients, is to determine these "undetermined" coefficients. This is done by plugging, for example, $u_k^{(p)}$ of (E.1) in the above Example 3.17, gather the similar (linearly independent) terms, and equate their coefficients with their corresponding ones on the right hand side of (E.1). In case one of the terms is missing on the right hand side of (E.1), then we equate the coefficients of its corresponding term on the left hand side of (E.1) to zero. In general, this process results, for the Example 3.18, in five linear simultaneous algebraic equations in the coefficients (to be determined) A, B, C, D, and E to have the final unique form of the particular solution $u_k^{(p)}$ of (E.1). In the case of Example 3.12, we had the very simple case of one term $A3^k$ for the suggested particular solution, where the above described procedure resulted in a single algebraic equation (E.2) of Example 3.12,

$$2A \cdot 3^k = 3^k$$

which determined $A = \frac{1}{2}$.

Example 3.18   The Solution for an Initial Value Problem

Let us consider the same nonhomogeneous difference equation of Example 3.11, except that the nonhomogeneous part consists of two terms

$$u_{k+1} - 3u_k + 2u_{k-1} = 1 + 5^k \qquad (E.1)$$

We can easily find that the solution of the associated homogeneous equation is

$$u_k^{(c)} = c_1 + c_2 2^k \qquad (E.2)$$

For the particular solution, the first term $r_k^{(1)} = 1$ in (E.1), which we note is the first solution of the homogeneous equation as seen in (E.2). Hence we must modify it to remove this situation, which requires multiplying it by k to have $1 \cdot k = k$. The result of differencing $r_k^{(2)} = 5^k$ in (E.1) is 5^k (aside from a constant amplitude), and this needs no modification, as it has no intersection with the parts of the complementary solution 1 and 2^k in (E.2). So the suggested form for the particular solution becomes

$$u_k^{(p)} = Ak + B5^k. \qquad (E.3)$$

To determine the unknown coefficients A and B, as described above, we substitute $u_k^{(p)}$ of (E.3) in the left hand side of (E.1), collect the similar terms, then equate their coefficients to their corresponding ones in $r_k = 1 + 5^k$ on the right hand side of (E.1),

$$A(k+1) + B5^{k+1} - 3(Ak + B5^k) + 2[A(k-1) + B5^{k-1}] = 1 + 5^k,$$

and after collecting the similar terms on the left hand side, we have

$$A(1-2) + (A - 3A + 2A)k + (5B - 3B + \frac{2}{5}B)5^k = 1 + 5^k,$$

$$-A + \frac{12}{5}B5^k = 1 + 5^k$$

So, if we equate the coefficients of 1 as $-A = 1$, we have $A = -1$. From equating the coefficients of 5^k we obtain $\frac{12}{5}B = 1$, $B = \frac{5}{12}$. Hence the particular solution $u_k^{(p)}$ in (E.3), after determining its coefficients $A = -1$ and $B = \frac{5}{12}$, becomes

$$u_k^{(p)} = -k + \frac{5}{12} \cdot 5^k = -k + \frac{1}{12} \cdot 5^{k+1}.$$

With the complementary solution $u_k^{(c)}$ in (E.2), and this particular solution $u_k^{(p)}$, the general solution of the nonhomogeneous equation (E.1) becomes

$$u_k = c_1 + c_2 2^k - k + 12 \cdot 5^{k+1} \qquad (E.4)$$

Again, because of the two *arbitrary* constants c_1 and c_2 (of $u_k^{(c)}$), this solution is not unique, and we need two auxiliary conditions for this second order difference equation (E.1) to determine c_1 and c_2 in (E.4). We may supply these two conditions as the following two initial conditions

$$u_0 = 3 \qquad (E.5)$$

and

$$u_1 = 6 \qquad (E.6)$$

to make u_k in (E.4) as the unique solution of the *initial value problem* (E.1), (E.5)-(E.6). If we impose these two initial conditions (E.5) and (E.6) on u_k in (E.4), we obtain two simultaneous linear algebraic equations in c_1 and c_2,

$$u_0 = c_1 + c_2 + \frac{5}{12} = 3 \qquad (E.7)$$

$$u_1 = c_1 + 2c_2 - 1 + \frac{25}{12} = 6 \qquad (E.8)$$

Subtracting (E.7) from (E.8), we obtain

$$c_2 + \frac{2}{3} = 3, \quad c_2 = \frac{7}{9},$$

then from (E.7) we have

$$c_1 = 3 - \frac{5}{12} - c_2 = 3 - \frac{5}{12} + \frac{7}{9} = \frac{65}{36}.$$

With $c_1 = \frac{65}{36}$ and $c_2 = -\frac{7}{9}$ in (E.4), the unique solution to the initial value problem (E.1), (E.5)-(E.6), (associated with the nonhomogeneous equation (E.1)) becomes,

$$u_k = \frac{65}{36} + \frac{7}{9}2^k - k + \frac{1}{12}5^{k+1} \qquad (E.9)$$

It should be a simple exercise to show that this u_k satisfies the initial value problem (E.1), (E.5)-(E.6).

The Operator Approach of the UC Method

The above undetermined coefficients method of finding a particular solution of the nonhomogeneous constant coefficient equation (3.15) with its very special class of sequences (UC) for r_k,

$$a_0 u_{k+n} + a_1 u_{k+n-1} + \cdots + a_n u_k = r_k \qquad (3.15)$$

may be facilitated with the help of simple results concerning the difference operator E.

We have already seen, for example, how the sequence a^k or λ^k served for solving the homogeneous equation (3.14), and how it also made one of the main special sequences in the nonhomogeneous term r_k of (3.15), for which the UC method works. The real reason behind this is the special effect the E operator has on such a sequence as b^k, namely,

$$Eb^k = b^{k+1} = b \cdot b^k, \qquad (3.41)$$

i.e., it leaves b^k intact, aside from a multiplicative constant. Of course, this process can be repeated to have

$$E^n b^k = b^n b^k = C b^k, \qquad (3.42)$$

where $C = b^n$ is a constant.

Now, equation (3.30) may be written in the following *operator form*:

$$(a_0 E^n + a_1 E^{n-1} + \cdots + a_n)u_k = r_k$$

$$\equiv \Phi(E)u_k = r_k. \tag{3.43}$$

So, if we can define an inverse $\Phi^{-1}(E) \equiv \frac{1}{\Phi(E)}$ of this operator $\Phi(E)$, which means $\Phi^{-1}(E)\Phi(E)u_k = u_k$, then we can formally apply the operator Φ^{-1} on both sides of equation (3.43) to find a particular solution (u_k) for it (or (3.30)),

$$\Phi^{-1}(E)\Phi(E)u_k = \Phi^{-1}(E)r_k,$$

$$u_k = \Phi^{-1}(E)r_k = \frac{1}{\Phi(E)}r_k. \tag{3.44}$$

For example, the difference equation (E.1) of Example 3.1

$$u_{k+2} - 3u_{k+1} + 2u_k = 3^k \tag{3.45}$$

may be written in the operator form as

$$(E^2 - 3E + 2)u_k \equiv \Phi(E)u_k = 3^k,$$

$$u_k = \frac{1}{\Phi(E)}3^k = \frac{1}{E^2 - 3E + 2}3^k. \tag{3.46}$$

As with any operational method which transforms the problem from differencing, differential or even integral operations to simple algebraic manipulations, we need to establish results (or pairs) that will help us in transforming to the desired algebraic equation. Also such pairs will help us in transforming the found solution back to the original space. In this case, the solution of the difference equation. This is no more than a parallel to the transforms (or Heaviside) methods used first in algebraizing differential equations, where the Laplace transform is the most known among many other transforms (or operational) methods. The next well known transforms are the Fourier transforms. Indeed, a related discrete version to the Laplace transform, namely, the z-transform will be covered in Chapter 6 for solving difference equations associated with *initial value problems*. The discrete Fourier transforms will be covered in Chapters 4, 5 and

7, and they will be used in Chapter 7 for algebraizing difference equations associated, mainly, with boundary conditions, i.e., for discrete *boundary value problems*. The first pair, among a number of other pairs to follow, used in facilitating the solving of $\frac{1}{\Phi(E)}$ in (3.44) is related to the basic result (3.42).

1.

$$E^n b^k = b^n b^k \tag{3.42}$$

If we use this result for each term in (3.43), we obtain

$$
\begin{aligned}
\Phi(E)b^k &= a_0 b^{n+k} + a_1 b^{n-1+k} + \cdots + a_n b^k \\
&= (a_0 b^n + a_1 b^{n-1} + \cdots + a_n)b^k \\
&= \Phi(b)b^k,
\end{aligned}
$$

and we have our first result or pair,

$$\Phi(E)b^k = \Phi(b)b^k,$$

$$\frac{1}{\Phi(E)}b^k = \frac{1}{\Phi(b)}b^k, \quad \Phi(E)\Phi(b) \neq 0, \tag{3.47}$$

after dividing both sides by $\Phi(E)\Phi(b) \neq 0$. We note that this condition warns us against, for example, the use of b_1^k, where $\Phi(b_1) = 0$ in the possible method of finding a particular solution of (3.30) as in (3.44) where in case b_1^k is part of the nonhomogeneous term r_k. This is so, since $\Phi(b) = 0$ is no more than the *characteristic* equation

$$a_0 b^n + a_1 b^{n-1} + \cdots + a_n = 0 \tag{3.48}$$

of the associated homogeneous equation. So, in case the nonhomogeneous term r_k of (3.30) has a term b_1^k, where b_1 is one of the solutions of $\Phi(b_1) = 0$, then (3.47) can not be used, and some modified approach has to be followed as we shall present the pair (3.56) and illustrate this case in Example 3.20 at the end of this section. In equation (E.1) of Example 3.1 as shown in the above equation (3.46),

$$\Phi(b) = b^2 - 3b + 2 = (b - 2)(b + 1)$$

with both of its two zeros $b_1 = 2$ and $b_2 = 1$ not equal to 3 in the only (UC) term 3^k of the nonhomogeneous term in (3.45) or (3.46), and we have, in this case, the condition $\Phi(b) \neq 0$, $b = 1, 2$, of (3.47) being satisfied. So, from (3.44) and (3.47), we have

$$
\begin{aligned}
u_k &= \frac{1}{\Phi(E)} r_k = \frac{1}{E^2 - 3E + 2} 3^k \\
&= \frac{1}{\Phi(3)} 3^k = \frac{1}{9 - 3 \cdot 3 + 2} 3^k = \frac{1}{2} 3^k
\end{aligned} \qquad (3.49)
$$

as the particular solution of (3.45), which is the same answer we obtained in Example 3.1 by using the direct UC method.

The result of the final pair (3.47) can be easily generalized to

$$
\Phi(E) b^k F(k) = b^k \Phi(bE) F(k) \qquad (3.50)
$$

where $F(k)$ is any sequence, and we leave its proof for an exercise (see Exercise 13).

The pair (3.47) and this simple example illustrates how important is the UC sequence form of r_k for the success of this method.

The following essential pairs will complement the one in (3.47), and also illustrate the dependence of this method on r_k being a UC sequence.

2.

$$
\frac{1}{\Phi(E)} \sin \alpha k = \frac{1}{\Phi(\alpha)} \sin \alpha k \qquad (3.51)
$$

$$
\frac{1}{\Phi(E)} \cos \alpha k = \frac{1}{\Phi(\alpha)} \cos \alpha k \qquad (3.52)
$$

These can be proved with the help of the following *Euler identities*

$$
\cos \alpha k = \frac{e^{i\alpha k} + e^{-i\alpha k}}{2} \qquad (3.53)
$$

and

$$
\sin \alpha k = \frac{e^{i\alpha k} - e^{-i\alpha k}}{2i} \qquad (3.54)
$$

after applying the pairs (3.47) on each of $e^{i\alpha k} = (e^{i\alpha})^k = a^k$ and $e^{-i\alpha k} = (e^{-i\alpha})^k = c^k$ in (3.53) and (3.54).

3.

$$
\begin{aligned}
\frac{1}{\Phi(E)} P_m(k) &= \frac{1}{\Phi(1+\Delta)} P_m(k) \\
&= (b_0 + b_1 \Delta + \cdots + b_m \Delta^m + \cdots) P_m(k) \\
&= (b_0 + b_1 \Delta + \cdots + b_m \Delta^m) P_m(k)
\end{aligned} \qquad (3.55)
$$

where $\Delta = E - 1$, and $P_m(k)$ is a polynomial of degree m as given in (3.32). The terms after Δ^m in the line before last of (3.55) vanish since $\Delta^j P_m(k) = 0$ for $j > m$.

4.

$$\frac{1}{\Phi(E)} b^k P_m(k) = b^k \frac{1}{\Phi(bE)} P_m(k) \tag{3.56}$$

We note that this result may be resorted to in case b^k, as a term in the nonhomogeneous part r_k of (3.30), happened to be in the complementary solution as we shall illustrate in Example 3.20. Also, we can show that the above result (3.56) is valid for any sequence $F(k)$ instead of just the polynomial $P_m(k)$, whose proof we shall leave for the exercises (see problem 14). We may add that this fact, of having (3.56) valid for any sequence $F(k)$, may enlarge the scope of this method to that of not necessarily UC functions–an advantage that other (indirect) operational methods do enjoy. Examples are the methods of using the Laplace or Fourier transform for solving differential equations with, essentially, constant coefficients.

Since this chapter is meant to present a clear introduction of the familiar methods of solving difference equations, we shall be satisfied with this brief exposure of the operational form of the UC method. However, we will return to its same (operational) spirit in our main emphasis of using (discrete) transform methods for solving difference equations, which we shall cover in Chapters 4 and 5 and for the discrete Fourier transforms, and in Chapter 6 for the z-transform.

For completeness, we give the following example, which illustrates the use of the pair (3.55) in case the UC sequence r_k has a term that is a part of the complementary solution. We will do that with the direct UC method of the last section as well as the present operational method.

Example 3.19   The Direct and the Operational UC Method Application

Consider the nonhomogeneous difference equation

$$u_{k+2} - 4u_{k+1} + 4u_k = 2^k \tag{E.1}$$

with its nonhomogeneous term $r_k = 2^k$ as a UC sequence. The characteristic equation is

$$\lambda^2 - 4\lambda + 4 = (\lambda - 2)^2 = 0, \qquad (E.2)$$

with its *double* roots $\lambda_1 = \lambda_2 = 2$. Consequently, the complementary solution becomes

$$u_k^{(c)} = c_1 2^k + c_2 k 2^k \qquad (E.3)$$

after modifying 2^k to make the second linearly independent solution $k2^k$. Here we face the difficulty of the nonhomogeneous term $r_k = 2^k$ being a part of the complementary solution. So, in order to prepare for the particular solution, we must multiply 2^k by k^2 to have $u_k^{(p)} = Ak^2 2^k$ that has no intersection with either of the two parts 2^k and $k2^k$ of $u_k^{(c)}$.

(1). We shall first illustrate finding this $u_k^{(p)}$ by the UC method of the last section, where we determine the coefficient A, and also by the present operational method with the help of the pair in (3.56) in the second part of this example. If we substitute $u_k^{(p)} = Ak^2 2^k$ in (E.1), we obtain

$$A(k+2)^2 2^{k+2} - 4A(k+1)^2 2^{k+1} + 4Ak^2 2^k = 2^k,$$

$$A2^k[4(k+2)^2 - 8(k+1)^2 + 4k^2] = 2^k,$$

$$A[4k^2 + 16k + 16 - 8k^2 - 16k - 8 + 4k^2] = 1,$$

$$8A = 1, \quad A = \frac{1}{8}.$$

Hence, we have the particular solution

$$u_k^{(p)} = Ak^2 2^k = \frac{1}{8} k^2 2^k \qquad (E.4)$$

as the result of using the direct UC method. Thus, from $u_k^{(c)}$ in (E.3) and the above $u_k^{(p)}$ in (E.4), obtained via the direct UC method, the general solution to the nonhomogeneous equation (E.1) becomes

$$u_k = u_k^{(c)} + u_k^{(p)} = c_1 2^k + c_2 k 2^k + \frac{1}{8} k^2 2^k. \qquad (E.5)$$

This is to be compared with the answer in (E.12), which is obtained with the help of the operational method.

(2). Now we use the operational method, where (E.1) can be written as

$$(E^2 - 4E + 4)u_k = 2^k, \qquad (E.1)$$

where, unfortunately, the simple pair in (3.47),

$$\frac{1}{\Phi(E)}2^k = \frac{1}{\Phi(2)}2^k \qquad (E.6)$$

cannot be used here, since $\Phi(2) = 4 - 8 - 4 = 0$. Hence we may resort to the pair (3.56),

$$\frac{1}{\Phi(E)}b^k P_m(k) = b^k \frac{1}{\Phi(bE)}P_m(k) \qquad (3.56)$$

where we take $b^k P_m(k)$ in (3.56) as $2^k P_0(1) = 2^k$ to accommodate the left hand side of (E.6). Thus, the left hand side of (E.6) may now be written different from its (forbidden) right hand side as

$$u_k = \frac{1}{\Phi(E)}2^k \cdot 1 = 2^k \frac{1}{\Phi(2E)} \cdot 1 = 2^k \frac{1}{4E^2 - 8E + 4}(1). \qquad (E.7)$$

With this form, we shall resort to expressing E in terms of Δ, $E = \Delta + 1$ as written in (3.55), hence (E.7) becomes

$$
\begin{aligned}
u_k &= 2^k \frac{1}{4E^2 - 8E + 4}(1) = 2^k \frac{1}{4(1+\Delta)^2 - 8(1+\Delta) + 4}(1) \\
&= 2^k \frac{1}{4\Delta^2}(1) = \frac{2^k}{4}(\Delta^2)^{-1}(1) \qquad (E.8) \\
&= \frac{2^k}{4}(\Delta^{-1})^2(1) = \frac{2^k}{4}(\Delta^{-1})\left(\Delta^{-1}(1)\right).
\end{aligned}
$$

Now, we need to use the pair (1.37) which we developed for this inverse difference operator in Chapter 1,

$$\Delta^{-1}(1) = k^{(1)} = k \qquad (E.9)$$

to have

$$u_k = \frac{2^k}{4}\Delta^{-1}(k^{(1)}) \qquad (E.10)$$

and if we use the pair (1.52) for $\Delta^{-1}(k^{(1)}) = \frac{1}{2}k^{(2)} = \frac{1}{2}k(k-1)$, we obtain

$$u_k^{(p)} = \frac{2^k}{4}\frac{k^{(2)}}{2} = \frac{2^k}{8}k(k-1)$$

$$= \frac{2^k}{8}k^2 - \frac{2^k}{8}k$$

$$(E.11)$$

as the particular solution of (E.1)

We must note that this sought (particular solution) $u_k^{(p)}$ is different from the particular solution $u_k^{(p)} = \frac{2^k}{8}k^2$ of (E.4) that we obtained by the direct UC method. In (E.11) we see that there is an extra term $-\frac{2^k}{8}k$ compared to what we have for $u_k^{(p)}$ in (E.4). This term is a part of the complementary solution $u_k^{(c)}$ in (E.3). However, this should not make a difference when we know that, ultimately this $u_k^{(p)}$ in (E.11) has to be added to the $u_k^{(c)}$ solution to obtain the general solution for (E.1). Thus, the extra term $-k\frac{2^k}{8}$ will be incorporated with the part $c_2 k 2^k$ of $u_k^{(c)}$ in (E.3), which results in only a different arbitrary constant $c_3 = \frac{1}{8}(c_2 - \frac{1}{8})$, and the general solution to (E.1) becomes

$$u_k = u_k^{(c)} + u_k^{(p)}$$

$$= c_1 2^k + c_2 k 2^k + \frac{k^2}{8}2^k - k\frac{2^k}{8}$$

$$= c_1 2^k + (c_2 - \frac{1}{8})k\frac{2^k}{8} + \frac{k^2}{8}2^k$$

$$(E.12)$$

$$= c_1 2^k + c_3 k 2^k + \frac{k^2}{8}2^k$$

where $c_3 = \frac{1}{8}(c_2 - \frac{1}{8})$. This general solution in (E.12) is equivalent to the one we had in (E.5),

$$u_k = c_1 2^k + c_2 k 2^k + \frac{1}{8}k^2 2^k \qquad (E.5)$$

where the particular solution was obtained via the direct UC method in part (1) of this example.

———————

Two interesting applications that use the method of solution in this section are those of Example 7.1 and 7.2 in Chapter 7. They model, respectively, the chemical concentration problem with the input pipe injecting drops at regular time intervals, and the problem of the compounding of interest on investments.

Exercises 3.3

1. (a) In Example 3.11, verify that

$$u_k = -\frac{5}{2} + 3 \cdot 2^k + \frac{3^k}{2}$$

of (E.12), is the solution to the initial value problem (E.1), (E.8)-(E.9).

   (b) In Example 3.17, verify that $u_k^{(p)}$ in (E.5) is indeed a particular solution of the nonhomogeneous equation (E.1).

   (c) In Example 3.19, verify that the general solution u_k in (E.9) is the solution of the initial value problem (E.1), (E.5)-(E.6).

2. (a) Use the UC method to find the particular solution $u_k^{(p)}$ of the nonhomogeneous equation

$$u_{k+2} - 3u_{k+1} + 2u_k = 4^k + 3k^2$$

   Hint: Consult Example 3.1, for the complementary solution (of its associated homogeneous equation).

   (b) Find the general solution.

3. Use the UC method to find a particular solution, then write the general solution of the equation

$$u_{k+2} - u_{k+1} - 2u_k = 6.$$

4. Use the UC method to find a particular solution, then find the general solution of the equation

$$u_{k+2} - 5u_{k+1} + 6u_k = 3^k.$$

5. Use the UC method to find a particular solution, then find the general solution of the equation

$$u_{k+3} - 3u_{k+2} + 3u_{k+1} - u_k = 48 + 24k$$

Hint: Watch for three repeated roots of the characteristic equation.

6. Solve the initial value problem (E.1)-(E.3),

$$u_{k+2} - 8u_{k+1} + 16u_k = 3 \cdot 4^k \qquad (E.1)$$

$$u_0 = 0 \qquad (E.2)$$

$$u_1 = 0. \qquad (E.3)$$

7. (a) Use the UC method to solve the nonhomogeneous first order equation
$$u_{k+1} - u_k = k. \qquad (E.1)$$

(b) Use the following initial condition (E.2) to determine the unique solution of this initial value problem (E.1)-(E.2),

$$u_0 = 1. \qquad (E.1)$$

8. Find the general solution of the equation

$$u_{k+3} - u_{k+2} - 4u_{k+1} + 4u_k = 1 + k + 2^k.$$

9. Solve the following difference equation

$$u_{k+2} - 4u_{k+1} + 4u_k = 3 \cdot 2^k + 5 \cdot 4^k.$$

Hint: Watch for the repeated roots $\lambda_1 = \lambda_2 = 2$. Prepare all the terms resulting from the action of $\Phi(E)$ of the characteristic equation.

10. Use the UC method to find the particular solution then find the general solution of the following difference equations

(i) $u_{k+2} + u_k = 4 \cos 2^k$

(ii) $u_{k+3} + u_k = 2^k \cos 3^k.$

11. Use the UC method to find a particular solution, then find the general solution of the equation

$$u_{k+2} - 6u_{k+1} + 8u_k = 2 + 3k^2 - 5 \cdot 3^k.$$

12. Prove the two operational pairs in (3.51) and (3.52).

    Hint: See the outline of the proofs that followed these equations with the help of the Euler identities (3.53)-(3.54) and the pair (3.42).

13. Prove the following generalization (of $\Phi(E)b^k = \Phi(b)b^k$ that lead to (3.47)),

$$\Phi(E)b^k F(k) = b^k \Phi(bE) F(k),$$

    where $F(k)$ is any sequence.

    Hint: Prepare all the terms resulting from the action of $\Phi(E)$ on $b^k F(k)$, i.e.,

$$
\begin{aligned}
Eb^k F(k) = b^{k+1} F(k+1) &= b^k \left[bEF(k) \right] \\
E^2 b^k F(k) = b^{k+2} F(k+2) &= b^k \left[b^2 F(k+2) \right] \\
&= b^k \left[b^2 E^2 F(k) \right]
\end{aligned}
$$

$$\vdots$$

$$E^n b^k F(k) = b^k \left[b^n E^n F(k) \right]$$

    then write $\Phi(E) \left[b^k F(k) \right]$ as was done in deriving (3.47).

14. (a) Use the result in problem 13 to generalize the result

$$\frac{1}{\Phi(E)} b^k P_m(k) = b^k \frac{1}{\Phi(bE)} P_m(k)$$

    of (3.56) for any sequence $F(k)$ instead of just the polynomial $P_m(k)$, i.e., prove that

$$\frac{1}{\Phi(E)} b^k F(k) = b^k \frac{1}{\Phi(bE)} F(k) \qquad (E.1)$$

Hint: Let $H(k) = \frac{1}{\Phi(bE)}F(k)$, i.e., $\Phi(bE)H(k) = F(k)$, then on the right hand side of (E.1), apply $\Phi(E), \Phi(E)b^k H(k) = b^k F(k)$), whence use the result of problem 13.

(b) Use the operational method to find the particular solution of the following difference equation, then find the general solution

$$u_{k+2} - 6u_{k+1} + 8u_k = 3k^2 + 2 \qquad (E.1)$$

Hint: See Example 3.18, and note to write

$$\frac{1}{\Phi(E)} = \frac{1}{E^2 - 6E + 8} = \frac{1}{(1+\delta)^2 - 6(1+\delta) + 8}$$

$$= \frac{1}{3 - 4\delta + \delta^2} = \frac{1}{3}\left[1 + \frac{4}{3}\delta + \frac{13}{9}\delta^2 + \cdots\right]$$

(c) Use the direct UC method to solve the difference equation (E.1) in part (a).

15. (a) Use the operational method to solve the difference equation in problem 9, and compare your answer with that of problem 9. *Hint:* See Example 3.18 for comparing the results obtained for a particular solution.

(b) Use the direct UC method to solve the equation in part (a).

16. Do the same as in problem 15 to solve the two nonhomogeneous equations in part (i) and (ii) of problem 10.

Other Linear Equations

Before we start to cover difference equations with *variable coefficients* in Sections 3.6 and 3.7, we will cover other types of difference equations with constant coefficients, since their methods of solution, generally, reduce to the methods that we have already discussed in the past two sections 3.2 and 3.3 for linear difference equations with constant coefficients. These topics include *system of linear first order difference equations*, that we shall cover in the next Section 3.4, and the *partial difference equations* in Section 3.5. The latter involves difference equations in more than one variable, hence the partial differencing that parallels partial differentiation in infinitesimal calculus.

3.4 Linear System of Difference Equations with Constant Coefficients

The methods of the last two Sections 3.2 and 3.3 will be used in solving system of linear difference equations. For simplicity, we will limit ourselves to a system of two linear difference equations of first order. Also, because of the limitation on the methods of Section 3.2 and 3.3, of constant coefficients and a UC sequence for the nonhomogeneous term, we shall adhere in our examples to the same conditions. However, what we are about to illustrate can be generalized to variable coefficients, which may be looked at in light of the method that covers such general case, namely, the method of *variation of parameters*, that we shall discuss in Sections 3.6 and 3.7.

Example 3.20 Two Simultaneous Linear Homogeneous Equations

Consider the system of two linear first order homogeneous difference equations with constant coefficients in the two sequences u_k and v_k

$$4u_{k+1} - 17u_k + v_{k+1} - 4v_k = 0 \qquad (E.1)$$

and

$$2u_{k+1} - u_k + v_{k+1} - 2v_k = 0 \qquad (E.2)$$

The usual method in solving such simultaneous equations is to eliminate one unknown first to result in one equation for the other unknown. However, here we have the difference operator E acting on both u_k and v_k in both equations. So, it is best to write the two equations in the operator form, then try to apply the appropriate combination of the operator E to eliminate, for example, v_k and its related terms that are usually operated on by E. The two equations (E.1), (E.2) with the operator E form are

$$(4E - 17)u_k + (E - 4)v_k = 0 \qquad (E.3)$$

$$(2E - 1)u_k + (E - 2)v_k = 0 \qquad (E.4)$$

In order to prepare for eliminating the terms related to the second unknown v_k, we apply the operators $(E - 2)$ and $(E - 4)$ on the equations (E.3) and (E.4), respectively to have

$$(E - 2)(4E - 17)u_k + (E - 2)(E - 4)v_k = 0 \qquad (E.5)$$

$$(E - 4)(2E - 1)u_k + (E - 4)(E - 2)v_k = 0 \qquad (E.6)$$

Since, in the present situation of constant coefficients, the difference operators here are commutative, then the second terms in both equations (E.5) and (E.6) are equal, i.e., $(E - 2)(E - 4)v_k = (E - 4)(E - 2)v_k$. Thus in subtracting (E.6) from (E.5) we eliminate them, and end up with one second order (homogeneous) equation in u_k,

$$(E - 2)(4E - 17)u_k - (E - 4)(2E - 1)u_k = 0$$

$$(4E^2 - 25E + 34)u_k - (2E^2 - 9E + 4)u_k = 0$$

$$(2E^2 - 16E + 30)u_k = 0$$

$$(E^2 - 8E + 15)u_k = 0. \qquad (E.7)$$

For this homogeneous equation, the characteristic equation is

$$\lambda^2 - 8\lambda + 15 = (\lambda - 3)(\lambda - 5) = 0, \qquad (E.8)$$

$$\lambda_1 = 3, \quad \lambda_2 = 5,$$

and the two linearly independent solutions are $u_k^{(1)} = 3^k$ and $u_k^{(2)} = 5^k$. Hence the general solution for the first unknown u_k in (E.1) is

$$u_k = c_1 3^k + c_2 5^k \qquad (E.9)$$

where c_1 and c_2 are arbitrary constants. The typical next step for solving algebraic system of equations is to substitute this solution u_k in one of the original equations (E.3) or (E.4) (or (E.1), (E.2)) to determine the second solution v_k. However in both equations (E.3) and (E.4) we have v_k (in this example) operated on by $(E - 4)$ and $(E - 2)$, respectively, which means that we will end up solving first order difference equation in v_k only. Moreover, this resulting equation will be *nonhomogeneous*. To illustrate this rather costly step compared to the other alternative to be discussed very shortly, let us substitute u_k of (E.9) in (E.1), where we obtain the following first order nonhomogeneous equation in v_k,

$$4(c_1 3^{k+1} + c_2 5^{k+1}) - 17(c_1 3^k + c_2 5^k) + v_{k+1} - 4v_k = 0$$

$$= -5c_1 3^k + 3c_2 5^k + v_{k+1} - 4v_k = 0 \qquad (E.10)$$

along with the arbitrary constants c_1 and c_2. Of course, this situation becomes much more simple (and preferable) if v_k in one of the equations is not operated on by E or its combination. In this case, for example, let us take another such equation instead of (E.1),

$$(4E - 17)u_k + 7v_k = 0$$

where, upon the same above substitution of a known first solution like u_k of (E.9), we have a direct answer for v_k instead of having it involved in a first order equation,

$$
\begin{aligned}
v_k &= -\frac{1}{7}(4E - 17)u_k = -\frac{1}{7}(4E - 17)(c_1 3^k + c_2 5^k) \\
&= -\frac{1}{7}(-5c_1 3^k + 3c_2 5^k) = \frac{5}{7}c_1 3^k - \frac{3}{7}c_2 5^k.
\end{aligned}
$$

The other preferable way of finding the second unknown v_k in (E.1)-(E.2) after having the first one u_k as in (E.9), is to try the same above method anew by eliminating the second unknown v_k in (E.3) and (E.4). As we shall see, the process will end up with another second order *homogeneous* difference equation in v_k with the *same* form as that of (E.7) for u_k. Thus it will have the same characteristic equation, whence v_k will have the same form as that of u_k in (E.9)

$$v_k = b_1 3^k + b_2 5^k, \qquad (E.11)$$

We will also show soon that even these arbitrary constants b_1 and b_2 of (E.11) are related in a simple way to c_1 and c_2 of u_k in (E.9). So, in order to eliminate u_k from (E.3) and (E.4) this time we apply the operators $(2E - 1)$ and $(4E - 17)$ on the equation (E.3) and (E.4), respectively,

$$(2E - 1)(4E - 17)u_k + (2E - 1)(E - 4)v_k = 0 \qquad (E.12)$$

$$(4E - 17)(2E - 1)u_k + (4E - 17)(E - 2)v_k = 0, \qquad (E.13)$$

then subtract (E.12) from (E.13) to have

$$(4E - 17)(E - 2)v_k - (2E - 1)(E - 4)v_k = 0$$
$$= (4E^2 - 25E + 34)v_k - (2E^2 - 9E + 4)v_k = 0$$
$$= (2E^2 - 16E + 30)v_k = 0,$$

$$(E^2 - 8E + 15)v_k = 0 \qquad (E.14)$$

which is the same difference equation as that of (E.7) in u_k. Hence, in the same way done for u_k of (E.7), we have the general solution of (E.14) as

$$v_k = b_1 3^k + b_2 5^k \qquad (E.15)$$

Here we are facing the solution, of the system of the two linear difference equations (E.3) and (E.4), with four arbitrary constants c_1, c_2, b_1 and b_2 as shown in the solutions of the system u_k and v_k in (E.9) and (E.15). But, in practice, we think of the system of the two linear first order difference equations in (E.3)-(E.4), as equivalent to second order system, which needs only two arbitrary constants. This brings the idea that the arbitrary constants b_1 and b_2 of v_k in (E.15) must be related to c_1 and c_1 of u_k in (E.9). This happens to be the case as we will see next, when we substituted the solution u_k of (E.9) and v_k of (E.15) in either of the original equations, say (E.3), or its equivalent (E.1),

$$4u_{k+1} - 17u_k + v_{k+1} - 4v_k = 0$$
$$= 4[c_1 3^{k+1} + c_2 5^{k+1}] - 17[c_1 3^k + c_2 5^k]$$
$$+ [b_1 3^{k+1} + b_1 5^{k+1}] - 4[b_1 3^k + b_2 5^k] = 0$$
$$= -5c_1 3^k + 3c_2 5^k - b_1 3^k + b_2 5^k = 0$$
$$= (-5c_1 - b_1)3^k + (3c_2 + b_2)5^k = 0$$

Now, we must equate the coefficients of the two linearly independent terms 3^k and 5^k, to zero,

$$-5c_1 - b_1 = 0, \quad b_1 = 5c_1 \qquad (E.16)$$

$$3c_2 + b_2 = 0, \quad b_2 = -3c_2 \qquad (E.17)$$

Thus, as seen in (E.16)-(E.17), the arbitrary constants b_1 and b_2 of v_k in (E.15) are indeed related to c_1 and c_2 of u_k in (E.9), and the final general solution to the system of two linear difference equations (E.1)-(E.2) is the pair

$$u_k = c_1 3^k + c_2 5^k, \qquad (E.18)$$

$$v_k = -5c_1 3^k - 3c_2 5^k \qquad (E.19)$$

Nonhomogeneous Simultaneous Equations

Of course, in the case of the two linear simultaneous first order equations being nonhomogeneous, the resulting equation in u_k , after eliminating the part involving the second unknown v_k will, in general, be nonhomogeneous second order equation. So, in addition to the complementary solution

$$u_k^{(c)} = c_1 u_k^{(1)} + c_2 u_k^{(2)} \qquad (3.57)$$

we must find its corresponding particular solution $u_k^{(p)}$ to have the general solution u_k,

$$u_k = c_1 u_k^{(1)} + c_2 u_k^{(2)} + u_k^{(p)} \qquad (3.58)$$

The same is done for the second unknown v_k,

$$v_k = b_1 v_k^{(1)} + b_2 v_k^{(2)} + v_k^{(p)} \qquad (3.59)$$

where again, the arbitrary constants b_1 and b_2 of v_k are related to c_1 and c_2 of u_k. This relation can be obtained, as we did in the above example, from insisting that u_k and v_k in (3.58) and (3.59) do indeed satisfy either one of the two original nonhomogeneous first order equations in u_k and v_k. We shall leave the illustration of this and other cases to the exercises (see Exercise 1, in particular), which are supported by detailed hints.

We may conclude this section with a general note that just as in the case of the nth order linear difference equation, the *nth order linear difference equation with constant coefficients*,

$$u_{k+n} + a_1 u_{k+n-1} + \cdots + a_{n-1} u_{k+1} + a_n u_k = r_k \qquad (3.60)$$

can be reduced to the following system of n linear simultaneous equations of *first order* in $u_j(k) \equiv u_k^{(j)}$, $j = 1, 2, \cdots, n$, where $u_{k+1} = Eu_k = u_k^{(2)}$:

$$Eu_1(k) = u_2(k)$$

$$Eu_2(k) = u_3(k)$$

$$\vdots$$

$$Eu_n(k) + a_1 u_n(k) + \cdots + a_{n-1} u_2(k) + a_n u_1(k) = r_k \qquad (3.61)$$

Exercises 3.4

1. Consider the two simultaneous linear nonhomogeneous equation in u_k and v_k

$$u_{k+1} - 3u_k + v_k = k \qquad (E.1)$$

$$3u_k + v_{k+1} - 5v_k = 4^k \qquad (E.2)$$

(a) In the attempt of solving this system of two (nonhomogeneous) equations, as we did in Example 3.20 for the system of two homogeneous equations, and as we discussed afterward for the nonhomogeneous case, start by eliminating the parts related to v_k in (E.1)-(E.2) to show that we obtain the following second order nonhomogeneous equation in u_k,

$$(E^2 - 8E + 12)u_k = 1 - 4k - 4k^2. \qquad (E.3)$$

(b) Since this equation (E.3) is with constant coefficients, and with a nonhomogeneous term as a UC sequence, $r_k = 1 - 4k - 4k^2$, attempt the UC method to find its particular solution as

$$u_k^{(p)} = \frac{1}{4}4^k - \frac{4}{5}k - \frac{19}{25}, \qquad (E.4)$$

and thus the general solution of (E.3) as

$$u_k = c_1 2^k + c_2 6^k + \frac{1}{4}4^k - \frac{4}{5}k - \frac{19}{25} \qquad (E.5)$$

Hint: Start with $u_k^{(p)} = A + Bk + C4^k$, noting that this does not have an intersection with 2^k and 6^k, the parts of the complementary solution.

(c) Find the general solution v_k,

$$v_k = b_1 v_k^{(1)} + b_2 v_k^{(2)} + v_k^{(p)}, \qquad (E.6)$$

Hint: Note that (E.1) does not involve differencing of v_k, so it is better to substitute u_k of part (b) in it to find v_k.

(d) Relate the arbitrary constants b_1, b_2 of v_k in (E.6) to c_1 and c_2 of u_k in (E.5).

Hint: Substitute v_k of (E.6) and u_k of (E.5) in (E.1), and equate the coefficients of 2^k and 6^k to zero.

(e) Use the initial conditions $u_1 = 2$ and $v_1 = 0$, to find the unique solution (as a pair u_k and v_k) for the two simultaneous equations (E.1), (E,2).

2. Solve the following system of first order linear nonhomogeneous equations in u_k and v_k

$$2u_{k+1} - 3u_k + 5v_k = 2 \qquad (E.1)$$

$$2u_k + v_{k+1} - 2v_k = 7 \qquad (E.2)$$

Hint: Note that these equations are nonhomogeneous, so follow the detailed steps of Exercise 1, and also of Example 3.20.

3. Solve

$$(E - 1)u_k + 2v_k = 0 \qquad (E.1)$$

$$-2u_k + (E - 1)v_k = a^k \qquad (E.2)$$

Hint: Attempt here to eliminate the terms involving u_k, so that when you finally find v_k, it becomes a simple matter to find u_k from (E.2), since it does not involve any differencing of u_k.

3.5   Linear Partial Difference Equations with Constant Coefficients

What we did until now is consider sequences of one discrete variable $u_k = u(k)$ and (ordinary) difference equations. This is in the sense that there is one differencing with respect to the only *independent variable k*, such as $Eu_k = u_{k+1}$, and $E^2 u_k = u_{k+2}$. However, as in the case of functions, we may consider sequences $w(k,l)$ as functions of the *two independent (discrete) variables k* and *l*. This parallels what we have in multivariable calculus, where we speak of $z = g(x,y)$ as a function of the two independent variables x and y. In the latter case we have partial differentiation $\frac{\partial g}{\partial x}$ with respect to the first variable x as well as a partial differentiation $\frac{\partial g}{\partial y}$ with respect to the second variable y. In this case we speak of the two first order *partial differential operators*, $\frac{\partial}{\partial x}$ and $\frac{\partial}{\partial y}$ on $g(x,y)$ of $z = g(x,y)$ instead of the (total) differential operator $\frac{d}{dx}$ on $f(x)$ in $\frac{df}{dx}$, the derivative of the function of one variable $y = f(x)$, the latter parallels what we have been doing for differencing the sequence u_k of one variable k via the difference operator E (or Δ).

For the sequence $w(k,l)$ of the two (discrete) variables k and l, we have to define two first order *partial differencing operators E_1* and E_2, the first is with respect to the first variable k,

$$E_1 w(k,l) = w(k+1,l) \equiv w_{k+1,l} \qquad (3.62)$$

and the second is with respect to the second variable l,

$$E_2 w(k,l) = w(k,l+1) \equiv w_{k,l+1} \qquad (3.63)$$

Just as in the case of higher order partial derivatives of $g(x,y)$, such as the second partial derivatives $\frac{\partial^2 f}{\partial x^2}$, $\frac{\partial^2 f}{\partial y^2}$, $\frac{\partial^2 f}{\partial x \partial y}$ and $\frac{\partial^2 f}{\partial y \partial x}$, we can also apply E_1 and E_2 more than one time on $w(k,l)$ to result in higher order partial differencing. For example

$$E_1^2 w(k,l) = E_1 \left(E_1 w(k,l) \right) = E_1 w(k+1,l) \qquad (3.64)$$

$$\equiv E_1 w_{k+1,l} = w(k+2,l) \equiv w_{k+2,l},$$

$$E_2^2 w(k, l) = E_2 \left(E_2 w(k, l) \right) = E_2 w(k, l + 1) \tag{3.65}$$

$$\equiv E_2 w_{k,l+1} = w(k, l + 2) \equiv w_{k,l+2}$$

and

$$E_2 E_1 w(k, l) = E_2 \left(E_1 w(k, l) \right) = E_2 w(k + 1, l)$$

$$\equiv E_2 w_{k+1,l} = w_{k+1,l+1}. \tag{3.66}$$

It is easy to see that the two difference operators E_1 and E_2 are commutative,

$$E_1 E_2 w(k, l) = w(k + 1, l + 1) = w_{k+1,l+1}$$

$$= E_2 E_1 w(k, l) = w(k + 1, l + 1) = w_{k+1,l+1}. \tag{3.67}$$

An equation in $w(k, l)$ that involves the differencing operator of E_1 and/or E_2 is called a *partial difference equation* in $w(k, l)$. An example

$$w_{k+2,l} - 3w_{k+1,l+1} + 2w_{k,l+2} = 0$$

$$= E_1^2 w_{k,l} - 3E_1 E_2 w_{k,l} + 2E_2^2 w_{k,l} = 0$$

$$= (E_1^2 - 3E_1 E_2 + 2E_2^2) w_{k,l} = 0, \tag{3.68}$$

which is a linear second order partial difference equation is $w_{k,l}$, that is homogeneous with constant coefficients.

This concept of partial difference equations can be extended to sequences of n discrete variables with the help of the corresponding difference operators E_1, E_2, E_3, $\cdots$, E_n of these variables.

From the previous sections we note that one of the most important properties of the difference operator E, that facilitate solving the *constant coefficient* homogeneous equations as well as the non-homogeneous ones with a UC sequence as a nonhomogeneous term, was the following property:

$$E a^k = a^{k+1} = a \cdot a^k,$$

which can be easily extended to

$$E^n a^k = a^{k+n} = a^n \cdot a^k$$

This operation leaves the sequence a^k intact except for a multiplicative constant a^n. This same property is also valid for the partial sequence of two variables $\lambda^k \mu^l$, where λ and μ are constants,

$$E_1 E_2(\lambda^k \mu^l) = \lambda^{k+1} \mu^{l+1} = (\lambda \mu) \lambda^k \mu^l,$$

$$E_1^n E_2^m(\lambda^k \mu^l) = \lambda^{k+n} \mu^{l+m} = (\lambda^n \mu^m) \lambda^k \mu^l. \tag{3.69}$$

This result (3.69) is no more than a constant multiple of $\lambda^k \mu^l$, where the constant is $\lambda^n \mu^m$.

The Lagrange Method of Solving Homogeneous Partial Difference Equations

One of the main methods of solving homogeneous partial difference equations with constant coefficients, namely, the *Lagrange method*, assumes the above form as a solution, i.e.,

$$w_{k,l} = w(k,l) = \lambda^k \mu^l.$$

Upon substituting this form for $w_{k,l}$ in the homogeneous difference equation, a relation between the constants λ and μ is obtained. For example, letting $w_{k,l} = \lambda^k \mu^l$ in the above partial difference equation (3.68), we have

$$
\begin{aligned}
(E_1^2 - 3E_1 E_2 + 2E_2^2)\lambda^k \mu^l &= \lambda^{k+2} \mu^l - 3\lambda^{k+1} \mu^{k+1} \\
&\quad + 2\lambda^k \mu^{l+2} \\
&= \lambda^k \lambda^l (\lambda^2 - 3\lambda\mu + 2\mu^2) = 0,
\end{aligned}
\tag{3.70}
$$

and since, in general, $\lambda^k \mu^l \neq 0$, we have a *characteristic* equation in λ and μ,

$$(\lambda^2 - 3\lambda\mu + 2\mu^2) = (\lambda - \mu)(\lambda - 2\mu) = 0. \tag{3.71}$$

Hence we either have $\mu = \lambda$ or $\mu = \frac{1}{2}\lambda$, and the two corresponding solutions become

$$w_{k,l}^{(1)} = \lambda^k \mu^l = \lambda^k \lambda^l = \lambda^{k+l} \tag{3.72}$$

and

$$w_{k,l}^{(2)} = \lambda^k \mu^l = \lambda^k \left(\frac{\lambda}{2}\right)^l = \left(\frac{1}{2}\right)^l \lambda^{k+l}, \tag{3.73}$$

We should note here the dependence of the two solutions $w_{k,l}^{(1)}$ and $w_{k,l}^{(2)}$ on one of the parameters λ.

Example 3.21 A Homogeneous Partial Difference Equation

The first order difference equation in $w_{k,l}$

$$w_{k+1,l} - 2w_{k,l+1} - 3w_{k,l} = 0 \qquad (E.1)$$

is homogeneous with constant coefficients

$$(E_1 - 2E_2 - 3)w_{k,l} = 0 \qquad (E.2)$$

So, if we let $w_{k,l} = \lambda^k \mu^l$, we have

$$\lambda^{k+1}\mu^l - 2\lambda^k\mu^{l+1} - 3\lambda^k\mu^l = 0$$

$$\lambda^k\mu^l(\lambda - 2\mu - 3) = 0, \quad \lambda^k\mu^l \neq 0, \qquad (E.3)$$

$$\lambda - 2\mu - 3 = 0, \quad \lambda = 2\mu + 3, \quad \mu = \frac{\lambda - 3}{2} \qquad (E.4)$$

with the one relation $\mu = \frac{\lambda-3}{2}$ in (E.4). Hence, the corresponding solution becomes

$$w_{k,l} = \lambda^k \mu^l = \lambda^k \left(\frac{\lambda - 3}{2}\right)^l = (2\mu + 3)^k \mu^l. \qquad (E.5)$$

As can be seen here, and was noted earlier, the above solution depends on one of the parameters, here, μ. So, for the general solution $w_{k,l}$ that is independent of μ, we may sum over this, generally continuous parameter μ, in an integral,

$$w_{k,l} = \int_{-\infty}^{\infty} C(\mu)(2\mu + 3)^k \mu^l \, d\mu \qquad (E.6)$$

Of course, we need some condition that enables us to find the unknown amplitude $C(\mu)$ inside the above integral. The subject may be revisited in the exercises.

We may note again that the Lagrange method for homogeneous equations works for constant coefficients. For example it does not work for the following homogeneous equation

$$w_{k+1,l+1} - (k+1)w_{k+1,l} - w_{k,l} = 0$$

$$= (E_1E_2 - (k+1)E_1 - 1)w_{k,l} = 0, \tag{3.74}$$

which has its second term with variable coefficient.

The Separation of Variables Method

There is another method that may be applied, which again assumes a very special form of the solution $w_{k,l} = C(k)D(l) \equiv C_kD_l$, i.e., it assumes the solution is a *product* of a sequence $C_k = C(k)$ as a function of the first variable k only, and another sequence $D_l = D(l)$ as a function of the second variable l only. This is called the method of the *products* or the method of *separation of variables*, which has its origin in facilitating the solution of many partial differential equations in physics and engineering. The main condition for this method is that after substituting $w_{k,l} = C_kD_l$ in the homogeneous partial difference equation, the action of the operator E_1 on C_k along with C_k itself and may be some known factor as a function of k, must separate on one side of the equation, while the operation of E_2 on D_l and D_l, with possibly a factor involving l, be on the other side. For example, in the partial difference equation

$$w_{k+1,l+1} - (k+1)w_{k+1,l} - w_{k,l} = 0$$

$$= (E_1E_2 - (k+1)E_1 - 1)w_{k,l} = 0, \tag{3.74}$$

with its *variable coefficient*, the Lagrange method would not work, but the present method of separation of variables does work. Upon substituting the product from the solution $w_{k,l} = C_kD_l$ in (3.74), we have

$$C_{k+1}D_{l+1} - (k+1)C_{k+1}D_l - C_kD_l = 0 \tag{3.75}$$

Now if we divide by $C_kD_l \neq 0$, we obtain

$$\frac{C_{k+1}}{C_k}\frac{D_{l+1}}{D_l} - (k+1)\frac{C_{k+1}}{C_k} - 1 = 0,$$

$$\frac{D_{l+1}}{D_l} = \frac{(k+1)C_{k+1} + C_k}{C_{k+1}}, \tag{3.76}$$

where now the equation *separates* to a left hand side that depends on D_l and its E_2 differencing only, and a right hand side that depends on C_k and its E_1 differencing only. Now comes a main point that would enable this process to reduce the partial difference equation (3.74) to two ordinary difference equations, one in C_k and the other in D_l. The idea is that k and l are the two *independent variables* of $w(k, l)$, thus they cannot be related. But the left hand side of (3.76) depends on l only, while the right hand side depends on k only, however k and l cannot be related! So, the only way out is for the two sides of equation (3.76) to be independent of both k and l, i.e., we equate both sides to a *constant* β,

$$\frac{D_{l+1}}{D_l} = \frac{(k+1)C_{k+1} + C_k}{C_{k+1}} = \beta \qquad (3.77)$$

This results in the following *two ordinary (homogeneous)* difference equations in C_k and D_l,

$$\frac{D_{l+1}}{D_l} = \beta,$$

$$D_{l+1} - \beta D_l = 0 \qquad (3.78)$$

and

$$\frac{(k+1)C_{k+1} + C_k}{C_{k+1}} = \beta,$$

$$(k+1)C_{k+1} + C_k - \beta C_{k+1} = 0 \qquad (3.79)$$

Since (3.78) is with constant coefficients, it is easy to see its characteristic equation, after letting $D_k = \lambda^k$,

$$(\lambda - \beta) = 0, \quad \lambda = \beta$$

where the solution $D_l = \lambda^l$ becomes

$$D_l = A\beta^l \qquad (3.80)$$

after allowing a constant amplitude A in this solution (3.80) of the homogeneous equation (3.78). The second equation (3.79) is also a first order, but with *variable coefficients*, which requires the methods of the next Section 3.6, specifically, the method of *variation of parameters* and the method of *summation factor*. It can be shown then that the solution is

$$C_k = B \cdot \frac{(-1)^k}{\Gamma(k - \beta + 1)} \qquad (3.81)$$

where Γ is the gamma function defined by the following integral

$$\Gamma(\mu) = \int_0^\infty e^{-x} x^{\mu-1} dx, \quad \mu \neq 0, -1, -2, \ldots, \tag{3.82}$$

$$\Gamma(m+1) \equiv m!, \quad m = 0, 1, 2, \ldots. \tag{3.83}$$

With the two separate results in (3.80) and (3.81), their product $w_{k,l} = C_k D_l$ gives

$$w_{k,l}(\beta) = AB\beta^l \frac{(-1)^k}{\Gamma(k-\beta+1)} \tag{3.84}$$

as a function of the parameter β. This $w_{k,l}(\beta)$ is then summed over β with coefficient $A(\beta)$ instead of the constant AB to have the solution

$$w_{k,l} = (-1)^k \int \frac{A(\beta)\beta^l}{\Gamma(k-\beta+1)} d\beta \tag{3.85}$$

As was the case for the solution of the (homogeneous) partial difference equation, via the Lagrange method in (E.6) of Example 3.21, extra condition (or conditions) are needed to determine the unknown amplitude $A(\beta)$ inside the above integral.

The condition of separatiality for the method of the products is not always easy to meet. For example, the partial difference equation

$$w_{k+2,l} + w_{k,l+2} - 2w_{k+1,l+1} = 0 \tag{3.86}$$

does not separate to one side that depends on k only, and another side that depends on l only, since if we let $w_{k,l} = C_k D_l$ in (3.86), then divide by $C_k D_l \neq 0$, we have

$$\frac{C_{k+2}D_l}{C_k D_l} + \frac{C_k D_{l+2}}{C_k D_l} - \frac{2C_{k+1}D_{l+1}}{C_k D_l} = 0,$$

$$\frac{C_{k+2}}{C_k} = -\frac{D_{l+2}}{D_l} + 2\frac{C_{k+1}}{C_k} \cdot \frac{D_{l+1}}{D_l} \tag{3.87}$$

Here, the second term on the right hand side is dependent on both k and l, which prevents the separation of variables process. On the other hand, the Lagrange method of letting $w_{k,l} = \lambda^k \mu^l$ works here, since upon the substitution in (3.86) (with its constant coefficients), we have

$$\lambda^{k+2}\mu^l + \lambda^k\mu^{l+2} - 2\lambda^{k+1}\mu^{l+1} = 0,$$

$$\lambda^k\mu^l[\lambda^2 + \mu^2 - 2\lambda\mu] = 0,$$

$$(\lambda - \mu)^2 = 0, \quad \lambda = \mu$$

So, with $\lambda = \mu$, a solution $w_{k,l} = \lambda^k\mu^l$ of (3.71) becomes

$$w_{k,l} = \lambda^{k+l}. \tag{3.88}$$

The Operational Sum Calculus Method

As mentioned above, the *separation of variables* method has its origin in solving partial differential equations in functions of two (or more) variables $G(x,y)$. Even though it applies to many problems, nevertheless, it does assume the special product form $G(x,y) = X(x)Y(y)$ on the outset. The operational integral calculus method, as discussed in Section 2.3 does not assume such a product form of the solution. In addition, with the use of the integration by parts, the auxiliary conditions associated with the differential equation are also automatically satisfied when the proper, or compatible, transform is chosen. As we mentioned very briefly at the end of Section 2.4, the Fourier transforms are extended to two (or higher) dimensions, as shown in (2.161)-(2.162), to solve partial differential equations associated with boundary conditions. There, we mentioned the possibility of extending the discrete Fourier transform to *two dimensions* as shown in (2.164)-(2.165) for the purpose of solving partial difference equations associated with boundary conditions. Of course, the summation by parts method would be used in the two dimensions of the two variables n and m of the sequence $G_{n,m}$, for example, to bring about the jumps at the ends in the two directions of n and m. For this we need many operational tools to be developed. However, we do not have the space in this book in which to pursue it further.

It may be appropriate to mention, as we shall discuss in Chapter 6, that the z-transform, as the discrete version of the Laplace transform, is used to solve ordinary difference equations associated with

initial conditions, i.e., initial value problems. So, for a partial difference equation in $w_{n,m,j}$ with two spacial variables n, m and time variable j, we can employ a double discrete Fourier transform followed
by a z-transform to completely algebraize the partial difference equation. This, of course, is well developed operational integral calculus
for solving partial differential equations associated with boundary
conditions on the several spatial variables, and initial conditions for
the time variable, i.e., for boundary and initial value problems.

Symbolic Methods

We may note in our above examples of partial difference equations
with constant coefficients such as (3.68), (E.1) of Example 3.21, and a
third equation $(E_1 - 4E_2)w_{k,l} = 0$ that we shall use in the illustration
of the last part of this section,

$$(E^2 - 3E_1E_2 + 2E_2)w_{k,l} = 0 \qquad (3.68)$$

$$= (E_1 - E_2)(E_1 - 2E_2)w_{k,l} = 0$$

$$(E_1 - 2E_2 - 3)w_{k,l} = 0 \qquad (3.89)$$

$$(E_1 - 4E_2)w_{k,l} = 0 \qquad (3.90)$$

have particular constant coefficient operators that are *factorizable*
with factors of the form $(E_1 - (aE_2 + b))$. For such case a symbolic
method may be employed, which depends on the formality that, with
E_1 and E_2 being commutative operators and operate on two different
(independent) variables k and l, then an operation E_1 (or $c_1E_1 + c_2$)
has no effect on the variable l, and the same is true for $(cE_2 + d)$
with regards to the first variable k. For example in the case of the
simple equation (3.90), we may write

$$E_1w_{k,l} = 4E_2w_{k,l} \qquad (3.91)$$

and concentrate only on the operation of E_1, considering for the
moment, symbolically, as if the operator E_2 on the right hand side
is a constant, i.e., as if we have

$$E_1w_{k,l} = 4E_2w_{k,l} = \alpha w_{k,l},$$

where $\alpha = 4E_2$. In this case we can solve, formally, for $w_{k,l}$ by letting $w_{k,l} = \lambda^k A(l)$, where, in general we must have the arbitrary amplitude be a function $A(l)$ of the other <u>ignored</u> variable l,

$$E_1 w_{k,l} - \alpha w_{k,l} = \lambda^{k+1} A(l) - \alpha \lambda^k A(l)$$

$$= \lambda^k A(l)[\lambda - \alpha] = 0, \quad \lambda = \alpha = 4E_2 \tag{3.92}$$

Hence, symbolically,

$$\begin{aligned} w_{k,l} &= \alpha^k A(l) = (4E_2)^k A(l) \\ &= 4^k E_2^k A(l) = 4^k A(l+k), \end{aligned} \tag{3.93}$$

where only at the last step we saw the action of the operator E_2^k on its function $A(l)$. This is the same type of solution we can obtain with using the Lagrange method as we shall show very shortly in (3.92).

This symbolic method can be extended to consider the general cases of $(aE_1 + b)$ and $(cE_2 + d)$, and also their repeated cases $(aE_1 + b)^n$ and $(cE_2 + d)^m$, which we shall leave their illustration for the exercises that are supported by detailed hints. In the same vein we shall also include an exposure to other methods via the exercises.

Next we will discuss methods of finding the *particular* solution of an important special class of *nonhomogeneous* partial difference equations.

Some Nonhomogeneous Partial Difference Equations-The Particular Solution

We will first consider an important special class of nonhomogeneous partial difference equations, that allows finding the particular solution in a relatively simple way. Such class of equations requires that the partial difference operator is in a *polynomial form*, and that the nonhomogeneous term is also in a polynomial form. An example is the following equation in $w_{k,l}$

$$3w_{k+1,l} + 2w_{k,l+2} = k^2 + l + 1$$

$$= [3E_1 + 2E_2^2]w_{k,l} = k^2 + l + 1, \tag{3.94}$$

where both of the partial difference operator $3E_1 + 2E_2^2$ and the nonhomogeneous term $R_{k,l} = k^2 + l + 1$ are in a polynomial form.

A general nth order nonhomogeneous partial difference equation in $w_{k,l}$, of the special form described above, may be written as

$$\Phi(E_1, E_2)w_{k,l} = R_{k,l}, \tag{3.95}$$

where Φ is a *polynomial* in E_1 and E_2 of degree n, and $R_{k,l}$ is the nonhomogeneous term, which is also a polynomial in k and l. In parallel to the theory of linear (ordinary) difference equations, the general solution $w_{k,l}$ here also consists of the *complementary* part $w_{k,l}^{(c)}$ as the solution of the associated homogeneous equation,

$$\Phi(E_1, E_2) = 0, \tag{3.96}$$

and the *particular* part $w_{k,l}^{(p)}$, which is any solution of the nonhomogeneous equation (3.95),

$$w_{k,l} = w_{k,l}^{(c)} + w_{k,l}^{(p)}. \tag{3.97}$$

In general, the good success of the method of finding the particular solution here, depends on the partial difference operator $\Phi(E_1, E_2)$ as well as the nonhomogeneous term $R_{k,l}$ in (3.95) being in polynomial forms. For example, the partial difference equation

$$w_{k+1,l} - 4w_{k,l+1} = 6k^2l + 4$$

$$= (E_1 - 4E_2)w_{k,l} = 6k^2l + 4 \tag{3.98}$$

has $\Phi(E_1, E_2) = E_1 - 4E_2$, a polynomial of degree 1 in both E_1 and E_2, and the nonhomogeneous term $R_{k,l} = 6k^2l + 4$ is a polynomial of degree 2 in k and of degree 1 in l as a special case of (3.95)

With such polynomial forms, the method of finding the particular solution is much simplified via the following rather simple operation. If we assume the existence of the inverse $\frac{1}{\Phi(E_1,E_2)}$ for the difference operator $\Phi(E_1, E_2)$ in (3.95), we can write the solution of (3.95), formally, as

$$w_{k,l} = \frac{1}{\Phi(E_1, E_2)} R_{k,l} \tag{3.99}$$

Since, we assume that $\Phi(E_1, E_2)$ is a polynomial in E_1 and E_2, then, after writing it as $\Phi(1+\Delta_1, 1+\Delta_2)$, we can expand $\frac{1}{\Phi(1+\Delta_1, 1+\Delta_2)}$ on the right hand side in powers of E_1 and E_2 to have

$$
\begin{aligned}
w_{k,l} &= \frac{1}{\Phi(a+\Delta_1, 1+\Delta_2)} R_{k,l} \\
&= [b_0 + b_1\Delta_1 + b_2\Delta_2 + b_{11}\Delta_1^2 + b_{12}\Delta_1\Delta_2 \\
&\qquad + b_{22}\Delta_2^2 + b_{112}\Delta_1^2\Delta_2 + \cdots] R_{k,l}
\end{aligned}
\tag{3.100}
$$

where b_0, b_1, b_2, b_{11}, b_{112}, ... are constants. Now, if $R_{k,l}$ is also polynomial in k and l, say with the highest degree (in either k or l) is m, then the result of the operator in the parenthesis of (3.100) will only have a *finite number* of operations up to Δ_1^m or Δ_2^m. For example, in the case of the nonhomogeneous equation (3.98), $R_{k,l} = 6k^2l + 4$ is a polynomial of degree 2 in k, and of degree 1 in l. So, only the Δ_1 operations up to Δ_1^2, and Δ_2 operations up to Δ_2 and their mixed term $\Delta_1^2\Delta_2$ will contribute, while the remaining infinite number of terms of the operators Δ_1 and Δ_2 will produce zeros upon acting on the polynomial $6k^2l + 4$. Hence according to (3.100), the particular solution of (3.79) will via (3.94), formally, becomes

$$
w_{k,l} = [b_0 + b_1\Delta_1 + b_2\Delta_2 + b_{11}\Delta_1^2 + b_{112}\Delta_1^2\Delta_2] R_{k,l}
\tag{3.101}
$$

where the constants b_0, b_1, b_2, b_{11} and b_{112} here are evaluated from the actual expansion,

$$
\begin{aligned}
\frac{1}{E_1 - 4E_2} &= \frac{1}{\Delta_1 + 1 - 4(\Delta_2 + 1)} = \frac{1}{\Delta_1 - 4\Delta_2 - 3} \\
&= -\frac{1}{3[1 - \frac{1}{3}(\Delta_1 - 4\Delta_2)]} \\
&= -\frac{1}{3}\left[1 + \frac{\Delta_1 - 4\Delta_2}{3} + \left(\frac{\Delta_1 - 4\Delta_2}{3}\right)^2 \right. \\
&\qquad \left. + \left(\frac{\Delta_1 - 4\Delta_2}{3}\right)^3 + \cdots\right] \\
&= -\frac{1}{3}[1 + \frac{1}{3}\Delta_1 - \frac{4}{3}\Delta_2 + \frac{1}{9}\Delta_1^2 \\
&\qquad - \frac{8}{9}\Delta_1\Delta_2 - \frac{4}{9}\Delta_1^2\Delta_2 + \cdots]
\end{aligned}
\tag{3.102}
$$

after using, formally, the geometric series expansion,

$$\frac{1}{1-\alpha} = 1 + \alpha + \alpha^2 + \alpha^3 + \cdots \tag{3.103}$$

and keeping only the operators that contribute when acting on $(6k^2l + 4)$.

Because of the polynomial form of the nonhomogeneous term $R_{k,l} = 6k^2l + 4$, only the first five terms plus the seventh term of this expansion will contribute to the solution in (3.100) as indicated in (3.101) and shown below with $b_0 = -\frac{1}{3}$, $b_1 = -\frac{1}{9}$, $b_2 = \frac{4}{9}$ and $b_{11} = -\frac{1}{27}, b_{12} = \frac{8}{27}$, and $b_{112} = \frac{4}{27}$. Hence the particular solution of (3.94), according to (3.99) and (3.102) becomes

$$w_{k,l} = -\frac{1}{3}[1 + \frac{1}{3}\Delta_1 - \frac{4}{3}\Delta_2 + \frac{1}{9}\Delta_1^2 - \frac{8}{9}\Delta_1\Delta_2 - \frac{4}{9}\Delta_1^2\Delta_2](6k^2l + 4)$$

$$= -\frac{1}{3}\left[6k^2l + 4 + 4kl + 2l - 8k^2 + \frac{4}{3}l\right.$$

$$\left. - \frac{32k}{3} - \frac{16}{3} - \frac{16}{3}\right],$$

$$w_{k,l} = w_{k,l}^{(p)} = -2k^2l - \frac{4}{3}kl + \frac{8}{3}k^2 + \frac{32}{9}k - \frac{10l}{9} + \frac{20}{9} \tag{3.104}$$

In (3.104) we found the particular solution of the nonhomogeneous equation (3.98), and to have the general solution we need the complementary solution $w_{k,l}^{(c)}$ of its associated homogeneous equation,

$$(E_1 - 4E_2)w_{k,l} = 0. \tag{3.105}$$

As we have already illustrated in Example 3.21, the method of establishing $w_{k,l}^{(c)} = 4^k H(k + l)$ for this homogeneous constant coefficient equation (3.105), may start with the Lagrange method of assuming the form of solution $w_{k,l} = \lambda^k \mu^l$ in (3.105)

$$\lambda^{k+1}\mu^l - 4\lambda^l\mu^{k+l} = 0,$$

$$\lambda^k\mu^l(\lambda - 4\mu) = 0, \quad \lambda = 4\mu,$$

$$w_{k,l} = \lambda^k\mu^l = (4\mu)^k\mu^l = 4^k\mu^k\mu^l = 4^k\mu^{k+l} \tag{3.106}$$

Since this solution is dependent on the parameter μ, we can sum over this parameter μ, to have

$$w_{k,l}^{(c)} = \int w_{k,l}(\mu)\, d\mu = 4^k \int \mu^{k+l} d\mu = 4^k H(k+l). \qquad (3.107)$$

where the integral is clearly a function of $k + l$. We may note that this result was also established via the symbolic method for solving (3.75) as seen in (3.78).

With this result of $w_{k,l}^{(c)}$ in (3.107) and $w_{k,l}^{(p)}$ in (3.104), we have now the *general solution* to the *nonhomogeneous* equation (3.98),

$$
\begin{aligned}
w_{k,l} &= w_{k,l}^{(c} + w_{k,l}^{(p)} \\
&= 4^k H(k+l) - 2k^2 l - \tfrac{4}{3} kl + \tfrac{8}{3} k^2 \qquad (3.108) \\
&+ \tfrac{32}{9} k - \tfrac{10l}{9} + \tfrac{20}{9}.
\end{aligned}
$$

We note here that, in this case, no term of $w_{k,l}^{(p)}$ belongs to $w_{k,l}^{(c)}$. In case there is one, then we can discard it, since it will, in essence, not affect $w_{k,l}^{(c)}$.

To verify that $w_{k,l}^{(c)}$ is, indeed, a solution of the homogeneous equation (3.105), we merely substitute to have

$$
\begin{aligned}
w_{k+1,l} \quad - \quad 4w_{k,l+1} &= 4^{k+1} H(k+1+l) - 4^k H(k+l) \\
&= 4^{k+1}[H(k+l+1) - H(k+l+1)] = 0
\end{aligned}
$$

The verification of $w_{k,l}^{(p)}$ in (3.104) as a particular solution of (3.98) is left as a simple exercise that involves the substitution of $w_{k,l}^{(p)}$ (with its six terms) of (3.104) in (3.98).

Since this example of (3.98) is a linear nonhomogeneous equation with constant coefficients, and with a nonhomogeneous term $R_{k,l} = 6k^2 l + l$ in a UC form in both of the variables k and l, we shall in the following example try the extension of the UC method, discussed in the previous Section 3.3, for this partial difference equation.

Example 3.22   The UC Method for Nonhomogeneous
Partial Difference Equations

To find the particular solution for the nonhomogeneous equation (3.83), using the UC method, we first note that the nonhomogeneous term,

$$R_{k,l} = 6k^2l + 4 \qquad (E.1)$$

is a UC function in both k and l. So, if we extend the method of UC functions, to include differencing this $R_{k,l}$ with all possible powers of the operators E_1 and E_2, the result of such differencing will, clearly, end up with the terms k^2l, kl, k^2, k, l and one that we shall illustrate at the end of this example. So, assuming that none of these terms is a part of the complementary solution, which is the case for $w_{k,l}^{(c)} = 4^k H(k+l)$ in (3.107), the nominee for a particular solution becomes a linear combination of the above six terms,

$$w_{k,l}^{(p)} = Ak^2l + Bkl + Ck^2 + Dk + El + F. \qquad (E.2)$$

This form is then substituted in (3.83), and the coefficients of the similar terms are equated, which results in $w_{k,l}^{(p)}$ of (3.89), and we leave the details of the computations for an exercise (see Exercise 4). An illustration of how some of the above six terms in (E.2) are generated, by the differencing via E_1 and E_2 and their combination, is

$$E_1 k^2 l = (k+1)^2 l = k^2 l + 2kl + l, \qquad (E.3)$$

which produces the first, second and the fifth terms in (E.2). On the other hand,

$$E_1 E_2 k^2 l = (k+1)^2(k+1) = (k^2 + 2k + 1)(l+1)$$

$$= k^2 l + 2kl + l + k^2 + 2k + 1 \qquad (E.4)$$

produces the six of them.

We will return in Chapter 4, 5, 6 and 7 to solve partial difference equations by the transforms method, which include the discrete Fourier transform and/or the z-transform for solving boundary and initial value problems associated with partial difference equations.

Laplace Method

The Laplace method for solving partial difference equation is limited to a particular class of equations, where the sum or the difference of the indices of the solution $w(x_k, y_k) \equiv w_{k,l}$ is the same. For example, equation (3.86) is with the sum of the indices being the same $k + 2 + l$ in $w_{k+2,l}$, $k + l + 2$ in $w_{k,l+2}$, and $k + 1 + l + 1 = k + l + 2$ in $w_{k+1,l+1}$,

$$w_{k+2,l} + w_{k,l+2} - 2w_{k+1,l+1} = 0 \qquad (3.86)$$

The method is based on letting the same (fixed) sum (or difference) of the indices be, say $k + l + 2 = m$ with $w_{k,l+2}$ becoming $v_{k,m-k}$, and equation (3.86) becoming, for the moment, an equation in $v_{k,m-k}$ of one variable k, leaving m as constant.

The basic details and illustration of this method is left as a few exercises with very detailed hints or directions, especially in problem 12 (see also problems 13-15.)

Exercises 3.5

1. (a) Verify that $w_{k,l} = \lambda^{k+l}$ is a solution of the first order homogeneous partial difference equation in $w_{k,l}$

$$w_{k+1,l} - w_{k,l+1} = 0 \qquad (E.1)$$

(b) Since the solution in part (a) depends on the parameter λ, assume an amplitude as a function of λ for this solution, i.e., $w_{k,l} = A(\lambda) \cdot \lambda^{k+l}$, then sum over λ as in the following integral, to show that the result is a function of k and l, namely $H(k+l)$, which also satisfies the difference equation (E.1) in part (a)

$$w_{k,l} = \int_{-\infty}^{\infty} A(\lambda) \cdot \lambda^{k+l} d\lambda \equiv H(k+l).$$

Hint: See the verification of the solution in (3.92) for equation (3.105), which was done after (3.108).

2. Consider the homogeneous equation of the second order in $w_{k,l}$,

$$w_{k+2,l} - 4w_{k,l+1} = 0. \qquad (E.1)$$

Verify that $w_{k,l} = 2^{-2l} H(k + 2l)$, where $H(j)$ is an arbitrary function of the one variable j, is a solution of (E.1).

3. Consider the first order homogeneous equation in $w_{k,l}$

$$w_{k+1,l} - 4w_{k,l+1} = 0. \qquad (E.1)$$

with its *auxiliary* condition for $k = 0$,

$$w_{0,l} = l^2 \qquad (E.2)$$

Verify that $w_{k,l} = 2^{2k}(k+l)^2$ is the solution to (E.1)-(E.2).

4. (a) Verify that

$$w_{k,l} = -2k^2 l - \frac{4}{3}kl + \frac{8}{3}k^2 + \frac{32}{9}k - \frac{10}{9}l + \frac{20}{9} \qquad (E.1)$$

is a *particular* solution of the *nonhomogeneous* equation

$$w_{k+1,l} - 4w_{k,l+1} = 6k^2 l + 4. \qquad (E.2)$$

(b) Since the nonhomogeneous term in (E.2) is of the UC *(double)* sequence form, use the UC method, as we had started in Example 3.22 to find the expression (E.1) of the particular solution of (E.2).

5. (a) Use the Lagrange method to solve equation (E.1) in problem 3.

(b) Sum over the parameter μ in the solution of part (a), which is $w_{k,l} = 2^{2k}\mu^{k+l}$, as an integral like the one suggested in (3.107) and (3.85), to find the solution of (E.1) as $w_{k,l} = 2^{2k}H(k+l)$, where $H(j)$ is an arbitrary function of the single variable j.

(c) Use the auxiliary condition $w_{0,l} = l^2$ in (E.2) of problem 3 on the resulting solution of (E.1) as the integral in part (b), to find the final solution of (E.1)-(E.2) of problem 3, namely,

$$w_{k,l} = 2^{2k}(k+l)^2.$$

Hint: From part (b), you would have

$$w_{k,l} = 2^{2k} \int C(\mu) \cdot \mu^{k+l} d\mu \equiv 2^{2k} H(k+l).$$

If we impose the condition (E.2) on this general solution, we have

$$w_{0,l} = l^2 = \int C(\mu)\mu^l d\mu \equiv H(l),$$

where we determine the function $H(j) = j^2$. Hence the final solution of (E.1)-(E.2) is $w_{k,l} = 2^{2k}H(k+l) = 2^{2k}(k+l)^2$.

6. Attempt the method of separation of variables to solve problem 3

   Hint: Replace step (a) of problem 5 by using the method of separation of variables, then follow steps (b)-(c) in problem 5.

7. (a) Attempt the symbolic method to show that the solution of the equation (E.1) of problem 3 is $w_{k,l} = 2^{2k}H(k+l)$.
   Hint: See (3.90)-(3.93).

   (b) Use the result in part (a) to find the final solution of (E.1)-(E.2) in problem 3.
   Hint: See steps (b)-(c) in problem 5.

8. Use the Lagrange method to solve the equation

$$w_{k+1,l} - 2w_{k,l+1} - 3w_{k,l} = 0.$$

9. Use the Lagrange method with $w_{k,l} = \lambda^k \mu^l$ to solve the equation

$$w_{k+2,l} - 4w_{k,l+1} = 0$$

   to find two special solutions.
   Hint: Note that you get $\lambda^2 = 4\mu$, $\lambda_1 = 2\mu^{\frac{1}{2}}$, $\lambda_2 = -2\mu^{\frac{1}{2}}$.

10. Solve the second order equation

$$w_{k+2,l+2} - 3w_{k+1,l+1} + 2w_{k,l} = 0.$$

11. (a) Consider the second order homogeneous equation

$$w_{k+1,l+2} - w_{k,l} = 0. \qquad (E.1)$$

   Use the Lagrange and the separation of variables method to solve this equation.
   Hint: Consult problem (5) for summing over the parameter in order to get a general solution.

(b) Consider now the nonhomogeneous equation, associated with (E.1)

$$w_{k+1,l+1} - w_{k,l} = 2^{k-l}. \tag{E.2}$$

Verify that the nonhomogeneous term $R_{k,l} = 2^{k-l}$ is a UC sequence in both k and l, and find its UC set after all (partial) differencing with respect to both k and l.

(c) Use the UC method to find a particular solution $u_{k,l}^{(p)}$ for (E.2), then use the result of part (a) to write the general solution of (E.2).

Hint: $w_{k,l}^{(p)} = A2^{k-l}$ may very well be the obvious nominee!

12. The *Laplace method* is a special method for solving a very particular class of partial difference equations. The method depends entirely on the fact that the *sum* or the *difference* of the indices $k + r$ and $l + s$ of the two arguments x_{k+r} and y_{l+s} of the function of the two variables $w(x_{k+r}, y_{l+s}) \equiv w_{k+r,l+s}$, are constants. For example, the difference equation

$$w_{k,l} - kw_{k-1,l-l} = kl \tag{E.1}$$

has a constant difference of the indices, i.e., $k - l$ in $w_{k,l}$ and also the same $k - 1 - (l - 1) = k - l$ in $w_{k-1,l-1}$.

In such a case, we use the substitution of $k - l = m$, where m is considered, for now, as constant, and we have $w_{k,l} = w_{k,k+m} \equiv v_k$ as if it is a function of one variable v_k in k, while m is considered constant. This substitution reduces (E.1) momentarily to an ordinary difference equation in v_k where it can be solved for v_k. As we expect, the solution is a summation with *arbitrary constant* of the summation that, in general, should be a function $C(m)$ of the *neglected* other index (or variable) m during the main summation operation with respect to k.

(a) Use the substitution $k - l = m$ in (E.1) to show that it reduces to the following equation in $v_k \equiv w_{k,k-m}$,

$$v_k - kv_{k-1} = k(k - m). \tag{E.2}$$

(b) Consider the associated homogeneous equation of (E.1),

$$w_{k,l} - kw_{k-1,l-1} = 0. \qquad (E.3)$$

Use the substitution $k - l = m$ of part (a) to reduce this equation, for the moment, to an ordinary equation in v_k, then solve for v_k, remembering the arbitrary constant $C(m)$ of the neglected index m, to find the solution $w_{k,l}$ of (E.3) as

$$w_{k,l} = k! \cdot g(k - l)$$

where $g(k-l) \equiv g(m)$ is an arbitrary function of $k-l = m$. *Hint:* The solution of the first order homogeneous equation

$$v_k - kv_{k-1} = 0 \qquad (E.4)$$

is $v_k = k!C_1 = k!C_1(m) = k!C_1(k - l)$.

13. Consider the partial difference equation in $w_{k,l}$,

$$w_{k,l} - bw_{k+1,l-1} + (1 - b)w_{k-1,l+1} = 0 \qquad (E.1)$$

(a) Show that the Laplace method of problem 1 is applicable to (E.1) by showing that the *sum*, of the indices of the two arguments of the solution, is constant.
Hint: Note $k + l$ in $w_{k,l}$ is the same as $k + 1 + l - 1 = k + 1$ of $w_{k+1,l-1}$ and $k - 1 + l + 1 = k + l$ of $w_{k-1,l+1}$.

(b) Follow the steps in part (a) and (b) of the preceding problem to reduce (E.1) to the following equation in $v_k \equiv w_{k,m-k}$ by letting $k + l = m$

$$v_k - bv_{k+1} + (1 - b)v_{k-1} = 0. \qquad (E.2)$$

(c) Solve this second order equation in v_k, remembering that the two arbitrary constants are functions of the (neglected) index m, i.e., $C_1(m)$ and $C_2(m)$, to show the solution as

$$v_k = C_1(m) + C_2(m) \cdot \left(\frac{1 - b}{b}\right)^k, \qquad (E.3)$$

$$w_{k,l} = g(k+l) + h(k+l) \cdot \left(\frac{1-b}{b}\right)^k, \qquad (E.4)$$

where g and h are arbitrary functions of $k+l$, the sum of the indices k and l.

(d) Verify that $w_{k,l}$ in (E.4) is indeed a solution of (E.1).

14. (a) Verify that the Laplace method of problems 13 and 14 is applicable to the following equation

$$w_{k+3,l} - 3w_{k+2,l+1} + 3w_{k+1,l+2} - w_{k,l+2} = 0, \qquad (E.1)$$

then reduce it to the equation in v_k,

$$v_{k+3} - 3v_{k+2} + 3v_{k+1} - v_k = 0. \qquad (E.2)$$

Hint: Note that the sum of the arguments is constant $k+l+3$, so let $k+l = m$, $v_k = w_{k,m-k}$.

(b) Solve (E.2) to show that the solution of (E.1) is

$$w_{k,l} = f(k+l) + kg(k+l) + k^2 h(k+l)$$

where f, g, and h are arbitrary functions of the sum $k+l$.

Hint: Use the method in Section 3.2, and watch for three repeated roots $\lambda_1 = \lambda_2 = \lambda_3 = 1$. Also watch for the three arbitrary constants as functions of m, $C_1(m) = C_1(k+l)$, $C_2(m) = C_2(k+l)$ and $C_3(m) = C_3(k+l)$ of the general solution of (E.2).

15. Consider the nonhomogeneous equation

$$w_{k,l} + 2w_{k-1,l-1} = l \qquad (E.1)$$

(a) Verify that the Laplace method of problem 12 is applicable.

Hint: See that the difference in the two arguments is a constant $k-l$.

(b) Use the Laplace method to find the solution of the resulting equation of part (a) in v_k, after letting $k - l = m$,

$$v_k + 2v_{k-1} = k - m, \qquad (E.2)$$

as

$$v_k = (-2)^k f(k - l) + (\frac{2}{9} - \frac{1}{3}m) + \frac{1}{3}k, \qquad (E.3)$$

$$w_{k,l} = (-1)^k f(k - l) + \frac{2}{9} + \frac{1}{3}l \qquad (E.4)$$

Hint: For solving the first order homogeneous equation in v_k, remember the arbitrary constant $C(m) = C(k - l) \equiv f(k - l)$. For the particular solution of (E.2), you may use the UC method.

16. Solve problem 15 by the usual method of this section as illustrated in the examples.

17. Solve problem 14 by the method of this section.

3.6 First Order Difference Equations with Variable Coefficients

In the previous sections, we considered only difference equations with constant coefficients, and with, primarily, a particular class of nonhomogeneous terms, namely, the UC sequences. In this section and the following Section 3.7 we will illustrate methods for solving *variable coefficient* equations, which, in principle, can also cover nonhomogeneous terms that are not necessarily UC sequences. We will limit ourselves to applying such methods to first then second order equations, since these are the most used ones, noting that these methods can be extended, without much trouble, to higher order equations.

A somewhat central or very useful idea for such methods, is that of *reducing the order* of the equation. Hence in dealing with a second order equation, we should have the good ground for solving the *first order* nonhomogeneous equation with variable coefficients, which is what we shall start with for this section.

Consider the most general linear first order nonhomogeneous equation in u_k,

$$u_{k+1} - p_k u_k = E u_k - p_k u_k = r_k \qquad (3.109)$$

with the variable coefficient $p_k = p(k)$, and the nonhomogeneous term r_k. We shall discuss two methods of solving this equation, the first parallels that of the *integrating factor method* used for linear first order differential equations. The second will represent a prelude to the *variation of parameters* method as one of the most general methods for solving second order (and higher) equations with variable coefficients, and for nonhomogeneous terms that are not necessarily of the UC form. This goes for both linear differential as well as the linear difference equations that we are considering here.

The Summation Factor Method for Linear First Order Equations

For the first order equation (3.109)

$$u_{k+1} - p_k u_k = (E - p_k) u_k = r_k, \qquad (3.109)$$

we note that we can introduce the difference operator Δ on the left hand side if we multiply both sides of (3.109) by the *summation factor* $\frac{1}{p_1 p_2 \cdots p_{k-1} p_k}$, which parallels the integrating factor used for first order linear differential equations,

$$\frac{u_{k+1}}{p_1 p_2 \cdots p_{k-1} p_k} - \frac{p_k u_k}{p_1 p_2 \cdots p_{k-1} p_k} = \frac{r_k}{p_1 p_2 \cdots p_{k-1} p_k} \qquad (3.110)$$

$$= \frac{u_{k+1}}{p_1 p_2 \cdots p_{k-1} p_k} - \frac{u_k}{p_1 p_2 \cdots p_{k-1}} = \frac{r_k}{p_1 p_2 \cdots p_{k-1} p_k} \qquad (3.111)$$

So, if we let $v_k = \frac{u_k}{p_1 p_2 \cdots p_{k-1}}$, we can see clearly that the left hand side is $\Delta v_k = v_{k+1} - v_k$, and (3.110) becomes

$$\Delta v_k = \Delta \left(\frac{u_k}{p_1 p_2 \cdots p_{k-1}} \right) = \frac{r_k}{p_1 p_2 \cdots p_{k-1} p_k} \qquad (3.112)$$

This operation, of multiplying by the summation factor that resulted in (3.111), parallels that of putting the first order differential equation in the 6 th exact differential form, whence be ready for one step of integration to obtain the solution. In our present case of (3.111), we only need one operation of the *inverse difference operator* (or the *summation operator*) Δ^{-1} on the middle part of (3.112), to free the solution u_k,

$$\Delta^{-1}\Delta\left(\frac{u_k}{p_1 p_2 \cdots p_{k-1}}\right) = \Delta^{-1}\left(\frac{r_k}{p_1 p_2 \cdots p_k}\right), \qquad (3.113)$$

$$\frac{u_k}{p_1 p_2 \cdots p_{k-1}} = \Delta^{-1}\left(\frac{r_k}{p_1 p_2 \cdots p_k}\right) \qquad (3.114)$$

Here, we see that all we need is the effect of the summation operator Δ^{-1} on the sequence $\frac{r_k}{p_1 p_2 \cdots p_k}$, a subject that we covered in Section 1.3, to have the final solution u_k

$$u_k = p_1 p_2 \cdots p_{k-1}\Delta^{-1}\left(\frac{r_k}{p_1 p_2 \cdots p_k}\right) \qquad (3.115)$$

after, simply, multiplying the result of the summation operation in (3.94) by $p_1 p_2 \cdots p_k$. But from (1.55) in Section 1.3, we know that the Δ^{-1} summation operation is defined by

$$\begin{aligned}\Delta^{-1}w_k &= \sum_{j=1}^{k-1} w_j + c_1 \\ &= w_1 + w_2 + \cdots + w_{k-1} + c_1,\end{aligned} \qquad (3.116)$$

where c_1 is an arbitrary constant (of this summation operator), which can be determined from the given initial condition on the desired sequence such as $u_1 = 3$ to obtain the unique solution of (3.94).

Hence, if we use (3.116) for Δ^{-1} in (3.112), we have the solution

$$\begin{aligned}u_k &= p_1 p_2 \cdots p_{k-1}\left[\sum_{j=1}^{k-1}\frac{r_j}{p_1 p_2 \cdots p_j} + c_1\right] \\ &= p_1 p_2 \cdots p_{k-1}\sum_{j=1}^{k-1}\frac{r_j}{p_1 p_2 \cdots p_j} + c_1 p_1 p_2 \cdots p_{k-1}\end{aligned} \qquad (3.117)$$

We note here that the final solution of (3.116) consists of two parts, the first that does not involve an arbitrary constant, which makes the *particular* solution $u_k^{(p)}$,

$$u_k^{(p)} = p_1 p_2 \cdots p_{k-1}\sum_{j=1}^{k-1}\frac{r_j}{p_1 p_2 \cdots p_j}$$

and the second, which involves the arbitrary constant c_1, this stands for the *complementary* solution $u_k^{(c)}$,

$$u_k^{(c)} = c_1 p_1 p_2 \cdots p_{k-1} \qquad (3.118)$$

which is the solution of the associated *homogeneous* equation

$$(E - p_k) u_k = u_{k+1} - p_k u_k = 0 \qquad (3.119)$$

of (3.109). Hence the solution u_k in (3.117) represents the general solution $u_k = u_k^{(c)} + u_k^{(p)}$ to (3.109), where $u_k^{(c)}$ and $u_k^{(p)}$ are as in (3.119) and (3.118), respectively. We should note again that this represents our first encounter with a *variable coefficient* homogeneous equation in (3.120) as well as the nonhomogeneous one in (3.109). We should also note, that this summation factor method does present us the with complementary solution $u_k^{(c)}$ as a sort of a bonus, via the arbitrary constant of the summation operation of the inverse difference operator Δ^{-1} as seen in (3.116) and (3.117).

Example 3.23   Variable Coefficient First Order Equation–The Summation Factor Method

Consider the first order linear (nonhomogeneous) equation with variable coefficient

$$u_{k+1} - k u_k = \frac{1}{2} k^2 - \frac{1}{2} k, \quad 2 \leq k \leq K_2 \qquad (E.1)$$

with its initial condition

$$u_1 = 1. \qquad (E.2)$$

This is a special case of (3.109) with $p_k = k$ and $r_k = \frac{1}{2} k^2 - \frac{1}{2} k = \frac{1}{2} k(k-1)$. So, if we consult (3.114), we need $p_1 p_2 \cdots p_k = 1 \cdot 2 \cdots k = k!$ and, where $p_1 p_2 \cdots p_{k-1} = (k-1)!$ and, where the solution to (E.1), according to (3.114) is

$$u_k = p_1 p_2 \cdots p_{k-1} \Delta^{-1} \left(\frac{r_k}{p_1 p_2 \cdots p_k} \right) - (k-1)! \Delta^{-1} \left(\frac{\frac{1}{2} k(k-1)}{k!} \right)$$

$$\qquad (E.3)$$

$$= (k-1)! \Delta^{-1} \left(\frac{1}{2} \frac{1}{(k-2)!} \right), \quad 2 \leq k \leq K_2 \qquad (E.4)$$

But, as we have seen in Section 1.3 and (3.115), the inverse operator Δ^{-1}, as it is called the summation operator, is defined to be the following simple finite sum, which is the parallel of integration in differential calculus,

$$\Delta^{-1}w_k = \sum_{j=1}^{k-1} w_j = w_1 + w_2 + \cdots + w_{k-1} + c_1 \qquad (E.5)$$

If we examine (E.4) in light of this definition, we see that this sum involves a last term $w_{k-1} = \frac{1}{2} \cdot \frac{1}{(k-3)!}$ in (E.5), which is not defined for $k = 3$. Hence, u_2 in (E.4) becomes undefined, and we may resort to the initial condition $u_1 = 1$ in (E.2) and the difference equation (E.1) to generate u_2. So we may look for the solution in (E.4) only for $k \geq 3$. With $k = 1$ in the difference equation (E.1) and $u_1 = 1$ from (E.2) we obtain

$$u_2 - u_1 = \frac{1}{2} - \frac{1}{2} = 0, \quad u_2 = u_1 = 1. \qquad (E.6)$$

Hence

$$\Delta^{-1}\left(\frac{1}{2}\frac{k(k-1)}{k!}\right) = \frac{1}{2}\sum_{j=1}^{k-1}\frac{j(j-1)}{j!} + c_1$$

$$= \frac{1}{2}\sum_{j=2}^{k-1}\frac{1}{(j-2)!} + c_1, \quad 3 \leq k \leq K_2 \qquad (E.7)$$

since the term corresponding to $j = 1$ vanishes because of the $(j-1)$ factor in the numerator,

$$\Delta^{-1}\left(\frac{1}{2}\frac{k(k-1)}{k!}\right) = \frac{1}{2}\sum_{j=2}^{k-1}\frac{1}{(j-2)!} + c_1,$$

$$= \frac{1}{2}\left[\frac{1}{0!} + \frac{1}{1!} + \frac{1}{2!} + \cdots + \frac{1}{(k-3)!}\right] + c_1, \quad 3 \leq k \leq K_2 \qquad (E.8)$$

where $0! = 1$. Hence from (3.114) and (E.8), the solution to (E.1) is

$$u_k = (k-1)!\frac{1}{2}\left[\frac{1}{0!} + \frac{1}{1!} + \frac{1}{2!} + \cdots + \frac{1}{(k-3)!}\right] + c_1(k-1)!$$

$$= w_k^{(p)} + w_k^{(c)}, \quad 3 \leq k \leq K_2 \qquad (E.9)$$

where the first and the last two terms of the solution u_k represent the particular and the complementary solutions, respectively.

For the solution in (E.9) to be *unique*, we must determine the arbitrary constant c_1. But, we note that we cannot use the initial value $u_1 = 1$, or what we have already computed $u_2 = 1$ (with the help of $u_1 = 1$ and the difference equation (E.1)), since u_k in (E.9) is now not defined for $k = 1$ and $k = 2$. So, we must appeal again to the difference equation (E.1) and the initial values $u_1 = 1$ and $u_2 = 1$, and find u_3, to use it in (E.9) for determining the arbitrary constant c_1. Letting $k = 2$ in the difference equation (E.1), we have

$$u_3 - 2u_2 = 2 - 1, \quad u_3 = 2u_2 + 1 = 3$$

after using $u_2 = 1$. So, if we substitute $k = 3$ in (E.9) with $u_3 = 3$, and note that the series in the part of the particular solution in (E.9) stops at $\frac{1}{0!} = 1$ when $k = 3$ makes its last term of $\frac{1}{(k-3)!}$ as $\frac{1}{0!} = 1$, we have

$$u_3 = \frac{1}{2} \cdot 2! \left[\frac{1}{0!}\right] + c_1 \cdot 2! = 1 + 2c_1 = 3,$$

Hence $c_1 = 1$, and the unique solution to the difference equation (E.1) along with its initial condition $u_1 = 1$ of (E.2), becomes

$$u_k = \frac{1}{2}(k-1)! \left[\frac{1}{0!} + \frac{1}{1!} + \frac{1}{2!} + \cdots + \frac{1}{(k-3)!}\right]$$
$$+ (k-1)!, \quad \text{for } 3 \le k \le K_2,$$

and

$$u_k = 1, \quad \text{for } 1 \le k \le 2. \tag{E.10}$$

The linear first order difference equation with *constant* coefficient p,

$$u_{k+1} - pu_k = r_k \tag{3.120}$$

is clearly a special case of (3.119), where its solution can be obtained from (3.117) with $p_1 p_2 \cdots p_k = \overbrace{pp \cdots p}^{k} = p^k$ to have the solution

$$u_k = p^{k-1} \sum_{j=1}^{k-1} \frac{r_j}{p^j} + c_1 p^{k-1}$$
$$= u_k^{(p)} + u_k^{(c)}, \tag{3.121}$$

which consists of the *particular* and the *complementary* solutions $u_k^{(p)}$ and $u_k^{(c)}$.

We may note that if we had not made much mention of the solution of first order equations with constant coefficients in Section 3.3, the reason may be that this is the proper place for covering it.

Example 3.24   A Constant Coefficient First Order Equation–
The Summation Factor Method

Let us consider the first order equation with constant coefficient

$$u_{k+1} - 3u_k = (E - 3)u_k = k^2, \quad 1 \le k \le K_2 \qquad (E.1)$$

According to (3.121) we have $p = 3$ and $r_k = k^2$, and we obtain the general solution of (E.1) from (3.117) as

$$u_k = 3^{k-1} \sum_{j=1}^{k-1} \frac{j^2}{3^j} + c_1 3^{k-1}, \quad 1 \le k \le K_2 \qquad (E.2)$$

$$= 3^{k-1} \left(-\frac{1}{2}\right) \left(\frac{1}{3}\right)^{k-1} [k^2 + k + 1] + c_1 3^{k-1}$$

$$= -\frac{1}{2}[k^2 + k + 1] + c_1 3^{k-1}, \quad 1 \le k \le K_2 \qquad (E.3)$$

$$= u_k^{(p)} + u_k^{(c)}$$

after using the following identity for summing the series of the first line leading to the above first term as the particular solution,

$$\Delta^{-1} \beta^k P_m(k) = \sum \beta^k P_m(k)$$

$$= \frac{\beta^k}{\beta - 1} \left[1 - \left(\frac{\beta}{\beta - 1}\right)\Delta + \left(\frac{\beta}{\beta - 1}\right)^2 \Delta^2 \right.$$

$$\left. - \left(\frac{\beta}{\beta - 1}\right)^3 \Delta^3 + \cdots \right] P_m(k), \quad \beta \ne 1 \qquad (E.4)$$

with $\beta = \frac{1}{3}$, $m = 2$, the degree of the polynomial $P_2(k) = k^2$, and where we see that all the terms involving Δ^3 and higher order operations on $P_2(k)$ vanish. This application of (E.4) is left for a simple exercise (see problem 2.) For the derivation of (E.4), see problem 5. This and many other operator identities were used in Section 3.3 in (3.41)-(3.56) and the exercises of the same section, where (E.4) is seen as a special case of (3.56).

Of course, this problem (E.1) with its constant coefficient and the UC nonhomogeneous term $r_k = k^2$ can be tried via solving its associated homogeneous equation,

$$u_{k+1} - 3u_k = (E - 3)u_k = 0$$

with $u_k^{(c)} = c_1 3^k$, $1 \le k \le K_2$. The particular solution can be found by the UC method, assuming the form $u_k^{(p)} = Ak^2 + Bk + C$, which we shall have for another exercise (see problem 3).

The Variation of Parameters Method for First Order Equations

The associated homogeneous first order equation with constant coefficient p of (3.121),

$$u_{k+1} - pu_k = (E - p)u_k = 0 \qquad (3.122)$$

can be easily solved, as we did in Section 3.3, and at the end of the above example, by letting $u_k = \lambda^k$, where we have the characteristic equation of (1.123) as $\lambda - p = 0$, $\lambda = p$. Hence the solution becomes $u_k = Ap^k$, where A is an arbitrary constant. In the case of the *variable coefficient p_k*,

$$u_{k+1} - p_k u_k = 0 \qquad (3.123)$$

we can generate the sequence u_{k+1} in terms of the preceding sequences u_k, u_{k-1}, $\cdots$, u_2, u_1 as follows:

$$u_{k+1} = p_k u_k,$$

$$u_k = p_{k-1} u_{k-1}$$

$$u_{k-1} = p_{k-2} u_{k-2}$$

$$\vdots$$

$$u_2 = p_1 u_1$$

Hence from all these relations we can write

$$u_{k+1} = p_k p_{k-1} p_{k-2} \cdots p_2 p_1 u_1, \qquad (3.124)$$

and if we take the initial value u_1 as an arbitrary constant A, we obtain the solution

$$u_{k+1} = A p_1 p_2 \cdots p_{k-1} p_k,$$

$$u_k = A p_1 p_2 \cdots p_{k-1} \qquad (3.125)$$

The method of *variation of parameters* for solving the general linear *nonhomogeneous* equation with variable coefficient (3.109)

$$u_{k+1} - p_k u_k = r_k \qquad (3.109)$$

centers on the main idea of letting the arbitrary constant A (of the solution of the variable coefficient homogeneous equation (3.105)) in (3.110) vary as a function $A(k)$ of k. This method happens to extend to higher order nonhomogeneous difference equations *provided that the complementary solution is known*, which we shall consider in the next Section 3.7 for second order difference equations.

So, if instead of the form in (3.125) we let

$$u_k = A(k) p_1 p_2 \cdots p_{k-1}, \qquad (3.126)$$

with its variable amplitude $A(k)$, in (3.109), we obtain

$$A(k+1) p_1 p_2 \cdots p_k - p_k A(k) p_1 p_2 \cdots p_{k-1} = r_k \qquad (3.109)$$

We then divide by $p_1 p_2 \cdots p_{k-1} p_k$, similar to what we did in (3.100), to generate the difference operation of Δ,

$$A(k+1) - A(k) = \Delta A(k) = \frac{r_k}{p_1 p_2 \cdots p_k}. \qquad (3.127)$$

This is a pivotal result of the method of variation of parameters, which also puts us at the same step (3.111) of the first *method of the summation factor*. So, if we operate on both sides of (3.127) by the summation operator Δ^{-1}, we obtain the variable amplitude $A(k)$,

$$A(k) = \Delta^{-1}\left(\frac{r_k}{p_1 p_2 \cdots p_k}\right), \qquad (3.128)$$

and upon using this $A(k)$ in the variable amplitude form (3.126) of the solution to (3.109), we obtain the final solution as

$$
\begin{aligned}
u_k &= p_1 p_2 \cdots p_{k-1} A(k) \\
&= p_1 p_2 \cdots p_{k-1} \Delta^{-1}\left(\frac{r_k}{p_1 p_2 \cdots p_k}\right) \\
&= p_1 p_2 \cdots p_{k-1} \sum_{j=1}^{k-1} \frac{r_j}{p_1 p_2 \cdots p_j} + c_1 p_1 p_2 \cdots p_{k-1}
\end{aligned}
\qquad (3.129)
$$

which is the same form as that of (3.117), that we obtained by the method of the summation factor.

Example 3.25 The Method of Variation of Parameters–A First Order Variable Coefficient Equation

Let us try to solve the first order nonhomogeneous difference equation with variable coefficient $p_k = k$,

$$u_{k+1} - k u_k = k \qquad (E.1)$$

With $r_k = k$, if we let

$$
\begin{aligned}
u_k &= p_1 p_2 \cdots p_{k-1} A(k) \\
&= 1 \cdot 2 \cdots (k-1) A(k) = (k-1)! A(k)
\end{aligned}
$$

and substitute in (E.1), we have

$$k! A(k+1) - k(k-1)! A(k) = k, \qquad (E.2)$$

Then, if we divide by $k!$, we obtain

$$A(k+1) - A(k) = \Delta A(k) = \frac{k}{k!} = \frac{1}{(k-1)!}, \qquad (E.3)$$

which is what we would have obtained had we used (3.128). So,

$$A(k) = \Delta^{-1}\left(\frac{1}{(k-1)!}\right) = \sum_{j=1}^{k-1} \frac{1}{(j-1)!} + c_1 \qquad (E.4)$$

and from (3.117), we have the solution

$$u_k = (k-1)! \sum_{j=1}^{k-1} \frac{1}{(j-1)!} + c_1(k-1)!, \quad 2 \le k \le K_2 \qquad (E.5)$$

If we have an *initial condition* such as

$$u_1 = 3 \qquad (E.6)$$

then, we may employ it in (E.5) to determine the arbitrary constant c_1, and hence the unique solution of (E.1), (E.6). However, the solution in (E.5) is not valid for $k = 1$, hence we must appeal to the difference equation (E.1), using $u_1 = 3$ to obtain u_2,

$$u_2 - u_1 = 1, \quad u_2 = u_1 + 1 = 3 + 1 = 4.$$

This u_2 is then used in (E.5) to determine c_1,

$$u_2 = 1! \cdot \frac{1}{0!} + c_1 = 1 + c_1 = 4, \quad c_1 = 3 \qquad (3.29)$$

Thus, the unique solution to the initial value problem (E.1), (E.6) is obtained from (E.5) with $c_1 = 3$ as

$$u_k = (k-1)! \sum_{j=1}^{k-1} \frac{1}{(j-1)!} + 3(k-1)!, \quad 2 \le k \le K_2 \qquad (E.7)$$

with $u_1 = 3$.

A simple but very interesting illustration that uses the method of this section, for solving first order variable coefficient equations, is that of Example 7.6 in Chapter 7. It models evaluating a family of integrals.

Exercises 3.6

1. Consider the first order equation with variable coefficients

$$u_{k+1} + ku_k = k^{(2)}. \qquad (E.1)$$

(a) Use the method of summation factor to find the general solution.

Hint: See Examples 3.23, 3.24.

(b) Use the variation of parameters method to solve (E.1).

Hint: See Example 3.25.

2. In Example 3.24, use the relation in (E.4) to verify the final result of the particular solution in (E.3). For the derivation of (E.4), see problem 5.

3. Use the UC method to solve the equation of Example 3.24 by first finding the particular solution.

4. Use (a) the method of summation factor and (b) the method of variation of parameters, as in problem 1, to solve the following first order equation with variable coefficient,

$$u_{k+1} - ku_k = (-1)^k(k-1)!. \qquad (E.1)$$

Hint: See Examples 3.23 and 3.25 for parts (a) and (b) respectively.

5. Use the identity (3.56),

$$\frac{1}{\Phi(E)}\beta^k P_m(k) = \frac{1}{\Phi(\beta E)}\beta^k P_m(k) \qquad (3.56)$$

to derive the identity (E.4) that was needed for Example 3.24, and illustrated in problem 2.

Hint: Write $\Delta^{-1} = \frac{1}{\Delta} = \frac{1}{E-1}$, whence $\Phi(E) = E - 1$, then expand $\frac{1}{\beta E+1} = \frac{1}{\beta-1} - \frac{1}{1+\frac{\beta}{\beta-1}\Delta}$ in an infinite series in powers of $\frac{\beta}{\beta-1}\Delta$.

Some nonlinear problems

In the whole book we covered only the solutions of linear difference equations. Here we present few nonlinear difference equations of very special forms that allow reaching a solution with the present methodes of solving linear problems.

6. In this exercise, we have a special class of first order *nonlinear* equations which is homogeneous in the sense that it can be reduced to a simpler equations in the quotient $v_k = \frac{u_{k+1}}{u_k}$. This may parallel what we have in the case of homogeneous first order differential equations.

(a) Consider the *nonlinear* equation in u_k

$$u_{k+1}^2 - 4u_{k+1}u_k - 5u_k^2 = 0. \qquad (E.1)$$

Show that this can be reduced to the following *simpler* equation in $v_k = \frac{u_{k+1}}{u_k}$,

$$v_k^2 - 4v_k - 5 = 0 \qquad (E.2)$$

$$= (v_k - 5)(v_k + 1) = 0$$

which as factorized above reduces to the *two linear* equations in u_k:

$$u_{k+1} - 5u_k = 0 \qquad (E.3)$$

or

$$u_{k+1} + u_k = 0 \qquad (E.4)$$

(b) Show that the two solutions are

$$u_k^{(1)} = c \cdot 5^k \quad \text{or} \quad u_k^{(2)} = c \cdot (-1)^k \qquad (E.5)$$

then verify your answer by simply substituting each of the two answers of (E.5) in (E.1).

7. Show that the *nonlinear* equation

$$u_{k+1}^2 + (3 - k)u_{k+1}u_k - 3ku_k^2 = 0$$

can be factorized into two linear equations, namely,

$$u_{k+1} + 3u_k = 0$$

or

$$u_{k+1} - ku_k = 0.$$

Solve these two equations to show the two solutions of (E.1) as

$$u_k = c \cdot (-3)^k \quad \text{or} \quad u_k = c \cdot (k - 1)!.$$

8. Consider the *nonlinear* first order equation in u_k

$$u_{k+1} - u_k^2 = 0.$$

(a) Write the equation as $u_{k+1} = u_k^2$, then take the logarithm of both sides to result in a *linear* equation in $v_k = \ln u_k$.

(b) solve the resulting linear equation in v_k to show that the solution is $v_k = \ln u_k = 2^{k-1} \ln u_1$, and hence $u_k = (u_1)^{2^k-1}$.

9. Consider the *nonlinear* second order equation in u_k,

$$u_{k+2} = \frac{u_{k+1}^2}{u_k}.$$

Use the method in problem 8, of taking the logarithm of both sides of the equation to reduce it to a linear second order equation in $v_k = \ln u_k$, then solve the latter equation to arrive at the final solution $u_k = e^{c_1 k + c_2}$, where c_1 and c_2 are arbitrary constants.

Hint: Note that the resulting second order linear difference equation in $v_k = \ln u_k$ can be written as $\Delta^2 v_k = 0$, *i.e.*, $\ln u_{k+2} - 2\ln u_{k+1} + u_k = 0 = \Delta^2(\ln u_k)$, whence $\ln u_k = c_1 k + c_2$.

10. We know from the familiar trigonometric identity $\cos 2x = 2\cos^2 x - 1$ that $f(x) = \cos x$ satisfies the nonlinear equation $f(2x) = 2f^2(x) - 1$. Use this idea for a change of variable that reduces the following nonlinear equation of first order in u_k,

$$u_{k+1} = 2u_k^2 - 1$$

to a linear one; then solve to show that the solution for the case $|u_1| \le 1$ is $u_k = \cos(2^{k-1} \cos^{-1} u_1)$.

Hint: Let $u_k = \cos \theta_k$, where $u_{k+1} = 2u_k^2 - 1$ becomes $\cos \theta_{k+1} = \cos 2\theta_k$, and by taking the arccosine of both sides we have the linear first order equation $\theta_{k+1} = 2\theta_k$ in θ_k, whose solution is simply $\theta_k = 2^{k-1}\theta_1$; $\theta_1 = \cos^{-1} u_1$.

11. Consider the *nonlinear* first order equation

$$u_{k+1} = 2u_k(1 - u_k).$$

Use the substitution $v_k = 1 - 2u_k$ to reduce this problem to the nonlinear form $v_{k+1} = v_k^2$ of problem 8, then use the same method to show that the solution is

$$u_k = \frac{1}{2}\left[1 - (1 - 2u_1)^{2^{k-1}}\right].$$

12. Similar to what we did as a substitution method in problem 10, use the trigonametric identity $\cos 3x = 4\cos^3 x - 3\cos x$ to find such substitution that reduces the *nonlinear* first order equation

$$u_{k+1} = 3u_k^3 - 3u_k, \quad |u_0| \le 1$$

to a linear equation, then solve the latter to show that the solution is $u_k = \cos(3^k \cos^{-1} u_0)$.

Hint: Let $u_k = \cos\theta_k$.

13. Consider the *nonlinear* equation of first order and its initial condition in (E.1)-(E.2)

$$u_{k+1} = A\sum_{l=0}^{k} u_l u_{k-l}, \qquad (E.1)$$

$$u_0 = 1, \qquad (E.2)$$

where the nonlinear sum on the right hand side is recognized as the *discrete self convolution product* $(u^*u)_k$ as defined in (4.42). We may also add that this self convolution product appears in the identity for the (generating) function $G(t) = \sum_{k=0}^{\infty} u_k t^k$ that we used for solving the linear problem 12 of Exercises 3.2 in Chapter 3

$$G^2(t) = \sum_{k=0}^{\infty} t^k \sum_{l=0}^{k} u_l u_{k-l} \qquad (E.3)$$

(a) Captalize on the form of (E.3) and operate on (E.1) to reduce it to the following (nonlinear) algebraic equation in $G(t)$,

$$\frac{1}{t}[G(t) - 1] = AG^2(t) \qquad (E.4)$$

Hint: Multiply both sides of (E.1) by t^k and sum from $k = 0$ to ∞, remebering to make the change of variable in the index $k + 1 = j$ because of u_{k+1} in (E.1).

(b) Solve for $G(t)$ in (E.4) then write the Taylor series expansion of the resulting $G(t)$ expression to have

$$G(t) = \sum_{k=0}^{\infty} \frac{(4A)^k \Gamma(k + \frac{1}{2})}{(k+1)! \Gamma(\frac{1}{2})} t^k \qquad (E.5)$$

whence the final solution to (E.1)-(E.2) is the coefficient in (E.5),

$$u_k = \frac{(4A)^k \Gamma(k + \frac{1}{2})}{(k+1)! \Gamma(\frac{1}{2})}$$

Hint: $G(t) = \frac{1}{2tA}\left[1 \pm \sqrt{1 - 4tA}\right]$, and we use the minus sign to have $u_0 = 1$ from using L'Hospital rule to have $G(0) = u_0 = 1$

3.7 Linear Second Order Difference Equations with Variable Coefficients

The Variation of Parameters and Other Methods

In the last section we discussed the method of variation of one parameter for solving the most general linear first order nonhomogeneous difference equation with variable coefficient p_k,

$$u_{k+1} - p_k u_k = r_k \equiv (E - p_k)u_k = r_k \qquad (3.109)$$

by letting

$$u_k = p_1 p_2 \cdots p_{k-1} A(k) \qquad (3.126)$$

with the *variable amplitude $A(k)$*.

When dealing with a linear *second* order nonhomogeneous equations with variable coefficients p_k and q_k,

$$u_{k+2} + p_k u_{k+1} + q_k u_k = r_k \qquad (3.130)$$

the above variation of parameters method, used for the first order equation (3.109) can be extended to solve the second order equation (3.131), provided that we know the two linearly independent solutions $u_k^{(1)}$ and $u_k^{(2)}$ of its associated homogeneous equation,

$$u_{k+2} + p_k u_{k+1} + q_k u_k = 0, \qquad (3.131)$$

or the complementary solution

$$u_k^{(c)} = c_1 u_k^{(1)} + c_2 u_k^{(2)} \qquad (3.132)$$

The *variation of parameters* form of a solution to (3.131), assumes the general form,

$$u_k = V_1(k) u_k^{(1)} + V_2(k) u_k^{(2)} \qquad (3.133)$$

with its two (unknown) variable amplitudes $V_1(k)$ and $V_2(k)$. The essential remaining part of this method is to give the process for finding these two variable amplitudes $V_1(k)$ and $V_2(k)$, hence the solution to (3.131) as u_k in (3.133).

The main question now is how to find the two linearly independent solutions $u_k^{(1)}$ and $u_k^{(2)}$ of the, in general, homogeneous *variable coefficient* second order equation (3.131). This is especially important since all we did up until now for second order equations was limited to constant coefficient ones, which we discussed, primarily, in Section 3.3. Hence, at this point, if we are to illustrate the method of variation of parameters, it would be for only nonhomogeneous, constant coefficient, second order equations

$$u_{k+2} + p u_{k+1} + q u_k = r_k \qquad (3.134)$$

with p and q as constants. But such second order equations with constant coefficients were covered in Section 3.3 by the use of the

UC method, except that such a method was limited to the nonhomogeneous terms r_k of (3.116) being in a special class of sequences, namely, the UC sequences. So, if we are to use the variation of parameters method for constant coefficient equations, the only possible advantage we may have is to cover larger class of sequences for the nonhomogeneous term r_k in (3.116) being not necessarily UC sequences.

The Method of Reduction of Order

So, if we are to search for solving homogeneous equations with variable coefficients, we find that our first exposure to any such variable coefficient equations was that of the first order only, which we had already discussed and illustrated in Section 3.6. This means that there is a glimpse of hope in solving the second order homogeneous equation (3.132), if we have a method that reduces the second order homogeneous equation to a first order equation, whence our above method of the reduction of order can cover such an equation with its variable coefficients. It turns out that there is such a method, namely, the same *reduction of order* method, which depends on a lesser condition, that is just to know one of the two solutions $u_k^{(1)}$ or $u_k^{(2)}$ of the second order homogeneous equation (3.132). More specifically, this method states that if we know one solution of (3.132), say $u_k^{(1)}$, then we can use it with a variable amplitude as

$$u_k = V(k)u_k^{(1)} = V_k u_k^{(1)} \tag{3.135}$$

in (3.132), where it will reduce (3.113) to a first order equation in the unknown amplitude $V(k)$, hence the name *reduction of order*. Then, with $V(k)$ in (3.136), we have the second linearly independent solution (or the general solution) of (3.136) as we shall discuss and illustrate next.

If we substitute $u_k = V_k u_k^{(1)}$ in (3.132), we have

$$V_{k+2}u_{k+2}^{(1)} + p_k V_{k+1}u_{k+1}^{(1)} + q_k V_k u_k^{(1)} = r_k \tag{3.136}$$

What we should do next is to have the unknown sequence V_k under the first order operator Δ, which means a *first order* equation in V_k.

This is possible if we can introduce $V_{k+2} - V_{k+1} = \Delta V_{k+1}$ and $V_{k+1} - V_k = \Delta V_k$ instead of just V_{k+2} and V_{k+1} of (3.136), respectively. We will have this if we consider (3.132), which is, of course, valid for its first solution $u_k^{(1)}$,

$$u_{k+2}^{(1)} + p_k u_{k+1}^{(1)} + q_k u_k^{(1)} = 0 \qquad (3.137)$$

then multiply it by V_{k+1},

$$V_{k+1} u_{k+2}^{(1)} + p_k V_{k+1} u_{k+1}^{(1)} + q_k V_{k+1} u_k^{(1)} = 0$$

and, finally, subtract it from (3.136) to have

$$(V_{k+2} - V_{k+1}) u_{k+2}^{(1)} + p_k V_{k+1} u_{k+1}^{(1)} - p_k V_{k+1} u_{k+1}^{(1)}$$

$$-q_k (V_{k+1} - V_k) u_k^{(1)} = 0,$$

$$u_{k+2}^{(1)} \Delta V_{k+1} - q_k u_k^{(1)} \Delta V_k = 0 \qquad (3.138)$$

as a first order, homogeneous equation in ΔV_k with variable coefficient $p_k = \frac{q_k u_k^{(1)}}{u_{k+2}^{(1)}}$ as in (3.120). This equation (3.138) should be easily solved after letting $z_k = \Delta V_k$, where it becomes

$$u_{k+2}^{(1)} z_{k+1} - q_k u_k^{(1)} z_k = 0 \qquad (3.139)$$

which can be solved for z_k by the *reduction of order* method discussed in Section 3.6 for first order equations. The next step is to solve for V_k from

$$\Delta V_k = z_k \qquad (3.140)$$

which can be obtained by the simple summation *operation* of Δ^{-1},

$$V_k = \Delta^{-1} z_k = \sum_{j=1}^{k-1} z_j + c \qquad (3.141)$$

However this is not the end of the story, since there is still a catch for this method, which is knowing one solution $u_k^{(1)}$ of the homogeneous equation (3.132). Of course, there is always the inspection with some luck! but we must have a method of finding this $u_k^{(1)}$, even if it is a limited one. For example, in the case of the homogeneous equation with variable coefficients

$$u_{k+2} - (k+2)u_{k+1} + ku_k = 0 \qquad (3.142)$$

we can easily verify that $u_k = (k-1)!$ is one of its two solutions, since

$$(k+1)! - (k+2)k! + k(k-1)! = (k+1)! - (k+1)k! - k! + k!$$
$$= (k+1)! - (k+1)! - k! + k! = 0.$$

As another example, it is also easy to verify that $u_k^{(1)} = k - 1$ is one of the solutions of the homogeneous equation

$$u_{k+2} - \frac{2k+1}{k}u_{k+1} + \frac{k}{k-1}u_k = 0, \qquad (3.143)$$

since

$$(k+1) - \frac{2k+1}{k}k + \frac{k}{k-1}(k-1) = k+1 - (2k+1) + k = 0.$$

Reduction of Order with Factorized Operators

Now, aside from a good try, how could we come up with such one solution of (3.132)? There is one approach, which reduces the solving of the second order homogeneous equation (3.132) to solving *two* first order equations in succession, but this needs a *special form* for the difference operator, $E^2 + p_k E + q_k$ of the equation,

$$(E^2 + p_k E + q_k)u_k = 0, \qquad (3.144)$$

namely, to be *factorizable*, i.e.,

$$(E^2 + p_k E + q_k) = (E - A_k)(E - B_k) \qquad (3.145)$$

as we shall illustrate in the following example.

Example 3.26   A Factorized Difference Operator

Consider the difference operator in (3.142), it is factorizable, since we can write it as

$$u_{k+2} - (k+2)u_{k+1} + ku_k = [E^2 - E(k+1) + k]u_k$$
$$= (E-1)(E-k)u_k, \qquad (E.1)$$

This can be verified, after noting how the operator E in the factor $(E - 1)$ acts on the $-k$ in the factor $(E - k)$ to make it $-(k + 1)$,

$$
\begin{aligned}
(E - 1)(E - k)u_k &= (E - 1)(u_{k+1} - ku_k) \\
&= E(u_{k+1} - ku_k) - 1 \cdot (u_{k+1} - ku_k) \\
&= u_{k+2} - (k + 1)u_{k+1} - u_{k+1} + ku_k \\
&= u_{k+2} - (k + 2)u_{k+1} + ku_k.
\end{aligned}
$$

We should note that the middle term $E(k+1)$ in $E^2 - E(k+1) + k$ of (E.1) is different from $(k + 1)E$ due to the *noncommutivity* of the operator E with the *variable* coefficient $(k + 1)$. The factorization with the latter choice gives $E^2 - (k + 1)E + k = (E - 1)(E - k)$, which is not the same, as we shall illustrate in the next example.

Example 3.27   The Noncommutivity of the Variable Coefficient
                    Factors of the Difference Operator

The sensitivity of the action of E on k as $Ek = k + 1$ in the above process of verification in Example 3.26 should signal to us that the above two factors $(E - 1)$ and $(E - k)$ with the variable term k in the factor $(E - k)$, are not commutative, since

$$
\begin{aligned}
(E - k)(E - 1)u_k &= (E - k)(u_{k+1} - u_k) \\
&= E(u_{k+1} - u_k) - k(u_{k+1} - u_k) \\
&= u_{k+2} - u_{k+1} - ku_{k+1} + ku_k \\
&= u_{k+2} - (k + 1)u_{k+1} + ku_k
\end{aligned}
\tag{E.1}
$$

and

$$
(E - 1)(E - k)u_k = u_{k+2} - (k + 2)u_{k+1} + ku_k.
\tag{E.2}
$$

the latter product $(E - 1)(E - k)$ in (E.2) is, obviously, not equal to $(E - 1)(E - k)$ of (E.1). This should illustrate that the two above different difference equations (E.1) and (E.2) have their own *unique* factorization with due note to the noncommutivity of the difference operator and the variable coefficient involved.

The noncommutivity of the two factors of the difference operator with variable coefficients, as illustrated in the above example, can be easily shown for the general case $(E - A_k)(E - B_k)$ with the variable sequences A_k and B_k. This is in contrast to the special case $(E - A)(E - B)$ with constants A and B, which is commutative, and which we have used often in Section 3.3.

It turns out, as we shall see shortly, that if we have this factorization property for the difference operator of a difference equation, it will also amount to a method of reduction of order. The result is that the second order equation will be reduced to a first order one in a new variable w_k, $w_k = \Delta u_k$. Then this last first order equation is solved for u_k by a simple direct summation of the operator Δ^{-1}, to obtain the desired solution $u_k = \Delta^{-1} w_k$, which we will illustrate next in Example 3.28. We will also illustrate in Example 3.29 the first more general reduction of order method, with its final result in (3.135)-(3.141), however for this method, we have to depend on us being handed the first solution $u_k^{(1)}$ of (3.132) by some means! One such possibility that may be resorted to represents, again, one of the most general method of solutions, namely, the *series solution method*, which assumes the following form for the solution

$$u_k = \sum_{j=-\infty}^{\infty} c_j k^{(j)} \tag{3.146}$$

as a $k^{(j)}$–*factorial polynomial* infinite series, and which we shall mention very briefly at the end of this section, and attempt to illustrate by an exercise with detailed supporting hints.

Example 3.28 The Reduction of Order Method with a Factorized Difference Operator

Consider the second order nonhomogeneous difference equation with variable coefficients,

$$u_{k+2} - (k + 2)u_{k+1} + ku_k = k$$

$$= [E^2 - (k + 2)E + k]u_k = k \tag{E.1}$$

whose homogeneous case is (3.142), and where we have already verified that its difference operator is factorizable as

$$[E^2 - (k+2)E + k] = [E^2 - E(k+1) + k]$$
$$= (E-1)(E-k), \tag{E.2}$$

There, we also noted and showed in Example 3.27 that such factors of (E.2), with variable coefficients, are not commutative, i.e., $(E - 1)(E - k) \neq (E - k)(E - 1)$. With this factorization of the operator in (E.2), the difference equation (E.1) can be written in the following form that is most susceptible to the reduction of its order process,

$$(E - 1)(E - k)u_k = k \tag{E.3}$$

As we discussed above, we let $w_k = (E - k)u_k$, whence (E.3) becomes a first order nonhomogeneous equation in this new (intermediate) variable w_k,

$$(E - 1)w_k = k$$
$$= \Delta w_k = k \tag{E.4}$$

which, as we discussed in the preceding part on first order equations, can be solved directly via the Δ^{-1} summation operation,

$$w_k = \Delta^{-1}k = \Delta^{-1}k^{(1)}$$
$$= \frac{k^{(2)}}{2} + c_1 = \frac{1}{2}k(k-1) + c_1 \tag{E.5}$$

Here we used the pair

$$\Delta^{-1}k^{(n)} = \frac{k^{(n+1)}}{n+1},$$

where $k^{(n)}$ is the factorial polynomial

$$k^{(n)} = k(k-1)\cdots(k-n+1); \quad k \equiv k^{(1)}.$$

With this solution for $w_k = \frac{1}{2}k(k-1) + c_1$, we can immediately use the simple relation of w_k to the desired solution u_k,

$$(E - k)u_k = w_k = \frac{1}{2}k(k-1) + c_1,$$

$$u_{k+1} - ku_k = \frac{1}{2}k(k-1) + c_1 \qquad\qquad (E.6)$$

as a first order, nonhomogeneous equation with variable coefficient, which we have already covered in detail in Section 3.6, and where we illustrated it for the same variable coefficient operator $(E - k)$ in Example 3.25, but with nonhomogeneous term k instead of $\frac{1}{2}k(k-1) + c_1$ of the present case in (E.6).

So, in parallel to what we did in Example 3.25, if we compare (E.6) to the general first order equation (3.109), we have $p_k = k$ and $r_k = \frac{1}{2}k(k-1) + c_1$, whence we can obtain the solution of (E.6) as a special case of the solution of (3.109) in (3.116)-(3.117),

$$u_k = p_1 p_2 \cdots p_{k-1} \Delta^{-1} \left(\frac{r_k}{p_1 p_2 \cdots p_k} \right)$$

$$= p_1 p_2 p_{k-1} \sum_{j=1}^{k-1} \frac{r_j}{p_1 p_2 \cdots p_j} + c_2 p_1 p_2 \cdots p_{k-1}, \qquad (3.117)$$

as

$$u_k = 1 \cdot 2 \cdots (k-1) \sum_{j=1}^{k-1} \frac{\frac{1}{2}j(j-1) + c_1}{1 \cdot 2 \cdots j} + c_2 \cdot 1 \cdot 2 \cdots (k-1)$$

$$u_k = (k-1)! \sum_{j=1}^{k-1} \frac{\frac{1}{2}j(j-1) + c_1}{j!} + c_2(k-1)! \qquad (E.7)$$

Here, we used one arbitrary constant c_1 for the first summation operator Δ^{-1} in (E.5), and a different second arbitrary constant c_2 for the second Δ^{-1} operation of (3.117a) that was necessary for solving (E.6). These two arbitrary constants c_1 and c_2, as we may expect, belong to the two terms of the complementary solution $u_k^{(c)}$ of (E.1). Hence, in (E.7) we have the general solution of the second order equation (E.1),

$$u_k = \frac{(k-1)!}{2} \sum_{j=1}^{k-1} \frac{j(j-1)}{j!} + c_1(k-1)! \sum_{j=1}^{k-1} \frac{1}{j!} + c_2(k-1)!$$

$$= u_k^{(p)} + c_1(k-1)! \sum_{j=1}^{k-1} \frac{1}{j!} + c_2(k-1)!$$

$$= u_k^{(p)} + u_k^{(c)}$$

$$(E.8)$$

where, as expressed above, the first term, without any arbitrary constant, represents the particular solution of the (nonhomogeneous) equation (E.1), while the sum of the last two terms with their arbitrary constants c_1 and c_2 represents the complementary solution of (E.1), i.e., the general solution $u_k^{(c)}$ to the associated homogeneous equation

$$u_{k+2} - (k+2)u_{k+1} + ku_k = 0 \qquad (E.9)$$

of (E.1).

The general solution of (E.1) in (E.8) is, of course, not unique, because of the two arbitrary constants c_1 and c_2 in (E.8). So, in order to have a unique solution for (E.1), we must have two initial conditions on u_k to determine c_1 and c_2. If we assume the following two initial conditions

$$u_1 = 1 \qquad (E.10)$$

$$u_2 = 3, \qquad (E.11)$$

we note that we cannot use the first initial condition with $k = 1$, since the middle term in (E.8) is not defined for $k = 1$ because the index j starts at $j = 1$ and stops at $j = k-1$. So, as we did in Example 3.25, we should appeal to the difference equation (E.1) with $u_1 = 1$ and $u_2 = 3$ to generate u_3, then use u_2 and u_3 in (E.8) to determine c_1 and c_2. For $k = 1$ in (E.1), we have

$$u_3 - 3u_2 + u_1 = 1,$$

$$u_3 = 3u_2 - u_1 + 1 = 9 - 1 + 1 = 9. \qquad (E.12)$$

Now, we substitute $u_2 = 3$ and $u_3 = 9$ in (E.8) to have the two linear equations in c_1 and c_2,

$$u_2 = 3 = \frac{1!}{2} \sum_{j=1}^{1} \frac{j(j-1)}{j!} + c_1 \cdot 1 \cdot \sum_{j=1}^{1} \frac{1}{j!} + c_2 \cdot 1,$$

$$3 = 0 + c_1 + c_2, \quad c_2 = 3 - c_1,$$

$$u_3 = 9 = \frac{2!}{2} \sum_{j=1}^{2} \frac{j(j-1)}{j!} + 2c_1 \sum_{j=1}^{2} \frac{1}{j!} + c_2 \cdot 2!,$$

$$9 = 1[0+1] + 2c_1[1 + \frac{1}{2}] + 2c_2 = 1 + 3c_1 + 2c_2$$

$$= 1 + 3c_1 + 6 - 2c_1, c_1 = 2; \quad c_2 = 1.$$

Hence, with $c_1 = 2$ and $c_2 = 1$, the unique solution to the initial value problem (E.1), (E.10)-(E.11) becomes,

$$u_k = \frac{(k-1)!}{2} \sum_{j=1}^{k-1} \frac{j(j-1)}{j!} + 2(k-1)! \sum_{j=1}^{k-1} \frac{1}{j!} + (k-1)!;$$

$$3 \leq k \leq K_2$$
$$(E.13)$$

and

$$u_1 = 1, \quad u_2 = 3, \quad u_3 = 9. \qquad (E.14)$$

Next, we will illustrate the first method of reducing the order of the second order homogeneous equation with variable coefficients (3.132) to the first order equation in ΔV_k or $z_k = \Delta V_k$,

$$u_{k+2}^{(1)} \Delta V_{k+1} - q_k u_k^{(1)} \Delta V_k = 0 \qquad (3.147)$$

or with $z_k = \Delta V_k$, we have

$$u_{k+2}^{(1)} z_{k+1} - q_k u_k^{(1)} z_k = 0 \qquad (3.148)$$

where V_k is the variable amplitude, of one of the two solutions $u_k^{(1)}$ of the homogeneous equation (3.131), that gives the second solution $u_k = V_k u_k^{(1)}$ of (3.132). We must stress again that this method of reduction of order does depend entirely on knowing one of the solutions, say $u_k^{(1)}$, of the associated second order homogeneous equation (3.131). We must also remind, that the sought general solution

$$u_k^{(c)} = c_1 u_k^{(1)} + c_2 u_k^{(2)} \qquad (3.149)$$

(of this homogeneous equation (3.131)) is being prepared for the main purpose of solving the second order nonhomogeneous equation with variable coefficients (3.130), and where $u_k^{(1)}$ and $u_k^{(2)}$ are predetermined in the development of the *variation of parameters* method that assumes the form of solution for (3.131) as

$$u_k = V_1(k) u_k^{(1)} + V_2(k) u_k^{(2)}, \qquad (3.133)$$

which we shall discuss in detail after the following example of illustrating finding $u_k^{(2)}$, given that we know $u_k^{(1)}$ of the homogeneous equation (3.131).

Example 3.29   The Reduction of Order Method After Knowing
                    One of the Two Solutions

Consider the homogeneous second order difference equation with variable coefficients,

$$u_{k+2} - k(k+1) u_k = 0$$

$$= [E^2 - k(k+1)] u_k = 0 \qquad (E.1)$$

We first note that the difference operator here is factorizable, since

$$[E^2 - k(k+1)] = E \cdot [E - (k-1)k]$$
$$= E^2 - (k-1+1)(k+1)$$
$$= E^2 \quad k(k+1).$$

This means that we can solve (E.1) by the more direct factorization of Example 3.28. However, we would like to illustrate the other method of using variation of parameters with $u_k = V_k u_k^{(1)}$ to have its reduced first order form (3.139)-(3.140) in $z_k = \Delta V_k$. Again, this method needs knowing $u_k^{(1)}$, so we either assume having it, or try inspection, for the case of this simple equation (E.1) to show that $u_k^{(1)} = (k-1)!$ is indeed a solution, since

$$u_{k+2} - k(k+1)u_k = (k+1)! - k(k+1)(k-1)!$$
$$= (k+1)! - (k+1)! = 0.$$

Of course, this assumption of having one of the two solutions of the homogeneous equation cannot be counted on for general homogeneous equations. In the present case, we may say that $u_k^{(1)} = (k-1)!$ was found as a part of the solution of the reduction of order via the factorization process of

$$[E^2 - k(k+1)] = E[E - (k-1)k].$$

So, if we let $u_k = V_k u_k^{(1)} = (k-1)!V_k$, then according to (3.138), the second order equation (E.1) in u_k reduces to the following first order equation in V_k,

$$u_{k+2}^{(1)}\Delta V_{k+1} - q_k u_k^{(1)}\Delta V_k = u_{k+2}^{(1)}\Delta V_{k+1} - k(k+1)u_k^{(1)}\Delta V_k \quad (E.2)$$

and with $z_k = \Delta V_k$, we have, as was done in (3.139), the first order equation in z_k,

$$u_{k+2}^{(1)}z_{k+1} - q_k u_k^{(1)}z_k = u_{k+2}^{(1)}z_{k+1} + k(k+1)u_k^{(1)}z_k = 0 \quad (E.3)$$

which we shall solve for z_k, then find $V_k = \Delta^{-1}z_k$. First, we write (E.3) in the form to suit the general first order equation (3.109), where we divide by $u_{k+2}^{(1)} \neq 0$ to have

$$z_{k+1} + (k+1)k\frac{u_k^{(1)}}{u_{k+2}^{(1)}}z_k = 0 \qquad (E.4)$$

with $p_k = -k(k+1)\frac{u_k^{(1)}}{u_{k+2}^{(1)}}$, and $r_k = 0$ in (3.109), and with $u_k^{(1)} = (k-1)!$. So, with

$$p_k = -\frac{k(k+1)(k-1)!}{(k+1)!} = -1,$$

(E.4) as a homogeneous equation has, according to (3.117), the solution

$$z_k = c_1 p_1 p_2 \cdots p_{k-1} = (-1)^{k-1} c_1 \qquad (E.5)$$

Now, we return to $z_k = \Delta V_k$ as in (3.140)

$$\Delta V_k = V_{k+1} - V_k = z_k = (-1)^{k-1} c_1$$

$$V_k = \Delta^{-1} z_k$$

and its solution, according to (3.116), with $p_k = 1$ and $r_k = (-1)^{k-1} c_1$, is

$$
\begin{aligned}
V_k &= p_1 p_2 \cdots p_{k-1} \left[\sum_{j=1}^{k-1} \frac{(-1)^{j-1} c_1}{p_1 p_2 \cdots p_j} + c_2 \right] \\
&= c_1 \sum_{j=1}^{k-1} (-1)^{j-1} + (-1)^{k-1} c_2 = (-1)^{k-1} c_1 \sum_{j=1}^{k-1} -1 + (-1)^{k-1} c_2 \\
&= c_1 (-1)^k + c_2 (-1)^{k-1}, \quad c_1 = 0 \text{ when } k \text{ is odd} \\
&= c_1 (-1)^k + (-1)^k c_3 \\
&= c_4 (-1)^k
\end{aligned}
$$

where $c_3 = -c_2$, $c_1 + c_3 = c_4$. So, the second solution of (E.1) becomes

$$u_k^{(2)} = V_k u_k^{(1)} = (-1)^k u_k^{(1)} = (-1)^k (k-1)!$$

The Reduction of Order Using the Casoratian

In the following example, we solve the above homogeneous problem of Example 3.29 with the help of the following property of the Casoratian $C(k)$ of its two linearly independent solutions $u_k^{(1)}$ and $u_k^{(2)}$,

$$C(k) = C_k = u_k^{(1)} u_{k+1}^{(2)} - u_{k+1}^{(1)} u_k^{(2)} \qquad (3.150)$$

where we can show here that $C(k)$ satisfies the following first order equation

$$C_{k+1} - q_k C_k = 0 \qquad (3.151)$$

where q_k is as in (3.137), or (3.131). This equation can be proved easily, when we substitute C_k from (3.150) in (3.151).

$$C_{k+1} - q_k C_k = u_{k+1}^{(1)} u_{k+2}^{(2)} - u_{k+2}^{(1)} u_{k+1}^{(2)}$$
$$- q_k u_k^{(1)} u_{k+1}^{(2)} + q_k u_{k+1}^{(1)} u_k^{(2)},$$

then we add and subtract $p_k u_{k+1}^{(2)} u_{k+1}^{(1)}$ and regroup to have

$$C_{k+1} - q_k C_k = [u_{k+2}^{(2)} + p_k u_{k+1}^{(2)} + q_k u_k^{(2)}] u_{k+1}^{(1)}$$
$$- [u_{k+2}^{(1)} + p_k u_{k+1}^{(1)} + q_k u_k^{(1)}] u_{k+1}^{(2)}$$
$$= [0] u_{k+1}^{(1)} - [0] u_k^{(2)} = 0$$

The terms in the above two parentheses vanish because $u_k^{(2)}$ and $u_k^{(1)}$ are given as solutions of the homogeneous equation (3.132),

$$u_{k+2} + p_k u_{k+1} + q_k u_k = 0 \tag{3.132}$$

Now, knowing $u_k^{(1)}$, and in order to involve a first order equation in $u_k^{(2)}$, we may note that, using the definition of $C(k)$ in (3.150), we have

$$\frac{C(k)}{u_k^{(1)} u_{k+1}^{(1)}} = \frac{u_k^{(1)} u_{k+1}^{(2)} - u_{k+1}^{(1)} u_k^{(2)}}{u_k^{(1)} u_{k+1}^{(1)}} = \Delta \frac{u_k^{(2)}}{u_k^{(1)}} \tag{3.152}$$

after using the formula $\Delta \left(\frac{u_k}{v_k} \right)$ in (1.19) for differencing a quotient of sequences.

From (3.152), we have

$$\frac{u_k^{(2)}}{u_k^{(1)}} = \Delta^{-1} \frac{C(k)}{u_k^{(1)} u_{k+1}^{(1)}}, \tag{3.153}$$

So, from this result we can find $u_k^{(2)}$ after solving the simple first order homogeneous equation (3.151) in $C(k)$, which according to (3.117) (for the homogeneous case) we have its solution as,

$$C(k) = A q_1 q_2 \cdots q_{k-1} \tag{3.154}$$

So the final solution $u_k = u_k^{(2)}$ from (3.153) and this result (3.154) becomes

$$
\begin{aligned}
u_k &= u_k^{(1)} \Delta^{-1} \left[\frac{A q_1 q_2 \cdots q_{k-1}}{u_k^{(1)} u_{k+1}^{(1)}} \right] \\
&= u_k^{(1)} \left[A \sum_{j=1}^{k-1} \frac{q_1 q_2 \cdots q_{j-1}}{u_j^{(1)} u_{j+1}^{(1)}} + B \right]
\end{aligned}
\tag{3.155}
$$

where both A and B are arbitrary constants. Indeed with $B \neq 0$, the above solution in (3.153) is the general solution, where $u_k^{(2)}$ is the term involving the summation with its arbitrary coefficient A. If we leave $B \neq 0$, then u_k in (3.153) represents the general solution $u_k^{(c)} = A u_k^{(2)} + B u_k^{(1)}$ of (3.132).

Example 3.30 The Reduction of Order Method Via the Casoratian Determinant

We will illustrate the above method, of using the Casoratian of the two solutions $u_k^{(1)}$ and $u_k^{(2)}$ of the homogeneous second order equation, to solve the same problem of Example 3.29,

$$
u_{k+2} - k(k+1) u_k = 0 \tag{E.1}
$$

where we assume knowing one solution, namely, $u_k^{(1)} = (k-1)!$. Here we have $q_k = -k(k+1)$, and if we prepare

$$
\begin{aligned}
q_1 q_2 \cdots q_{k-1} &= (-1)(2)(-2)(3)(-3)(4) \cdots (-1)(k-1)^2 k \\
&= (-1)^{k-1} 2^2 \cdot 3^2 \cdot 4^2 \cdots (k-1)^2 k \\
&= (-1)^{k-1} [(k-1)!]^2 \cdot k
\end{aligned}
$$

with $u_k^{(1)} = (k-1)!$ and $u_{k+1}^{(1)} = k!$ for (3.153), we obtain

$$
\begin{aligned}
u_k &= (k-1)! \left[A \sum_{j=1}^{k-1} \frac{(-1)^{j-1}[(j-1)!]^2 j}{(j-1)!(j-1)! j} + B \right] \\
&= (k-1)! \left[A \sum_{j=1}^{k-1} (-1)^{j-1} + B \right] \\
&= (k-1)! A (-1)^k + B(k-1)! \\
&= A(-1)^k (k-1)! + B(k-1)! \\
&= A u_k^{(2)} + B u_k^{(1)}
\end{aligned}
$$

where we have now the second linearly independent solution $u_k^{(2)}$ in the above first term as

$$u_k^{(2)} = (-1)^k(k-1)!.$$

The Variation of Parameters Method–Continuation

Now we return to the second order nonhomogeneous equation with variable coefficients (3.130)

$$u_{k+2} + p_k u_{k+1} + q_k u_k = r_k \qquad (3.130)$$

where, as we started it in (3.133), we write its *variation of parameters* form of solution as

$$u_k = V_1(k)u_k^{(1)} + V_2(k)u_k^{(2)} \qquad (3.133)$$

where $u_k^{(1)}$ and $u_k^{(2)}$, the solutions of the associated homogeneous equation (3.131),

$$u_{k+2} + p_k u_{k+1} + q_k u_k = 0 \qquad (3.131)$$

are given, and we are to find a method for determining the variable amplitudes $V_1(k)$ and $V_2(k)$ in (3.133) to have a particular (or even the general) solution for the nonhomogeneous equation (3.130).

Let us emphasize again that all the preceding discussions and illustrations of the method, of reduction of order (or different variations!) in this section, were aimed at finding these two linearly independent solutions $u_k^{(1)}$ and $u_k^{(2)}$ of (3.131) for the general case of the *variable coefficients* p_k and q_k in (3.131). We may also recall that the spirit of the method of reduction of order is that after becoming one of the two solutions $u_k^{(1)}$ of (3.132), we can then give it a variable amplitude $V(k)$, which resulted in the second order equation having reduced to a first order one in $z_k = \Delta V_k$, in $V(k)$. The solution to the latter is much more certain and is easier to find than that of the second order one, thus we end up finding the second solution $u_k^{(2)}$ after knowing $u_k^{(1)}$. The spirit of the method of variation of parameters for solving second order equations is in this same direction, where,

here, it uses two unknown amplitudes $V_1(k)$ and $V_2(k)$ as in (3.133), then it imposes a constraint that reduces the nonhomogeneous second order equation in u_k to *two simultaneous first order* equations in the first order difference $\Delta V_1(k)$ and $\Delta V_2(k)$. Such constraint, as we may anticipate, will be one that prevents the effect of the second order differencing of equation (3.131) on the two unknows $V_1(k)$ and $V_1(k)$ in its proposed variation of parameters solution of (3.132). This can be shown easily, when we first write u_{k+1} for u_k in (3.133),

$$
\begin{aligned}
u_{k+1} &= V_1(k+1)u^{(1)}_{k+1} + V_2(k+1)u^{(2)}_{k+1} \\
&= V_1(k)u^{(1)}_{k+1} + \Delta V_1 u^{(1)}_{k+1} + V_2(k)u^{(2)}_{k+1} + \Delta V_2 u^{(2)}_{k+1} \\
&= V_1(k)u^{(1)}_{k+1} + V_2(k)u^{(2)}_{k+1} + u^{(1)}_{k+1}\Delta V_1 + u^{(2)}_{k+1}\Delta V_2
\end{aligned}
$$

(3.156)

where we used $V_i(k+1) = V_i(k) + \Delta V_i, i = 1, 2$. Here we note the appearance of the first order difference operators in $\Delta V_1(k)$ and $\Delta V_2(k)$. Now, we also need u_{k+2} for (3.131), which, when used in (3.154) above, will produce second order differencing $\Delta^2 V_1(k)$ and $\Delta^2 V_2^{(k)}$ on $V_1(k)$ and $V_2(k)$ that are not wanted if we are to look for first order difference equations in $V_1(k)$ and $V_2(k)$. A way to do just this, is to impose *the constraint that the sum of the last two terms in (3.154) above, which involve ΔV_1 and ΔV_2, vanishes,*

$$
u^{(1)}_{k+1}\Delta V_1 + u^{(2)}_{k+1}\Delta V_2 = 0.
$$

(3.157)

Hence, we obtain in this equation (3.155), the first of the two sought linear first order difference equations in the two unknown variable amplitudes $V_1(k)$ and $V_2(k)$. With this assumption, u_{k+1} of the nominated solution becomes a much simpler one, and free of the differencing of Δ on the two unknowns $V_1(k)$ and $V_2(k)$,

$$
u_{k+1} = V_1(k)u^{(1)}_{k+1} + V_2(k)u^{(2)}_{k+1}.
$$

(3.158)

The second sought first order equation in $V_1(k)$ and $V_2(k)$ is obtained, when we let the solution of (3.133), with the above adjustment in (3.158), satisfy the nonhomogeneous equation (3.130), namely, if we use u_k of (3.113), the above (adjusted) u_{k+1} in (3.158) and its following shifted value

$$
u_{k+2} = V_1(k+1)u^{(1)}_{k+2} + V_2(k+1)u^{(2)}_{k+2}
$$

(3.159)

in the original equation (3.130)

$$u_{k+2} + p_k u_{k+1} + q_k u_k = r_k, \qquad (3.130)$$

In the process after these substitutions, we must note that, *by definition*,

$$u_{k+2}^{(1)} + p_k u_{k+1}^{(1)} + q_k u_k^{(1)} = 0 \qquad (3.160)$$

and

$$u_{k+2}^{(2)} + p_k u_{k+1}^{(2)} + q_k u_k^{(2)} = 0 \qquad (3.161)$$

since $u_k^{(1)}$ and $u_k^{(2)}$ are assumed to be the two linearly independent solutions of the homogeneous equation(3.131). So, if we substitute u_k of (3.133), u_{k+1} of (3.158) and u_{k+2} of (3.159) in (3.130), we have

$$V_1(k+1)u_{k+2}^{(1)} + V_2(k+1)u_{k+2}^{(2)} + p_k[V_1(k)u_{k+1}^{(1)}$$

$$+V_2(k)u_{k+1}^{(2)}] + q_k[V_1(k)u_k^{(1)} + V_2(k)u_k^{(2)}] = r_k$$

$$= V_1(k)u_{k+2}^{(1)} + u_{k+2}^{(1)}\Delta V_1 + V_2(k)u_{k+2}^{(2)}$$

$$+u_{k+2}^{(2)}\Delta V_2 + p_k V_1(k)u_{k+1}^{(1)} + p_k V_2(k)u_{k+1}^{(2)}$$

$$+q_k V_1(k)u_k^{(1)} + q_k V_2(k)u_k^{(2)} = r_k$$

$$= u_{k+2}^{(1)}\Delta V_1 + u_{k+2}^{(2)}\Delta V_2 +$$

$$+V_1(k)[u_{k+2}^{(1)} + p_k u_{k+1}^{(1)} + q_k u_k^{(1)}]$$

$$+V_2(k)[u_{k+2}^{(2)} + p_k u_{k+1}^{(2)} + q_k u_k^{(2)}] = r_k$$

$$= u_{k+2}^{(1)}\Delta V_1 + u_{k+2}^{(2)}\Delta V_2 = r_k \qquad (3.162)$$

as the second (sought) linear first order equation in $V_1(k)$ and $V_2(k)$. In the above steps we used $V_i(k+1) = V_i(k) + \Delta V_i$, $i = 1, 2$, rearranged terms, and used (3.160),(3.161) for the vanishing of the expressions in the two above parenthesises of the equation leading to the final equation (3.162).

In equations (3.157) and (3.162) we have now the two *first order* difference equations

$$u_{k+1}^{(1)} \Delta V_1 + u_{k+1}^{(2)} \Delta V_2 = 0 \tag{3.157}$$

$$u_{k+2}^{(1)} \Delta V_1 + u_{k+2}^{(2)} \Delta V_2 = r_k \tag{3.162}$$

as simultaneous equations in $\Delta V_1(k)$ and $\Delta V_2(k)$. Their solution is guaranteed, since their *Casoratian*

$$C(k) = u_{k+1}^{(1)} u_{k+2}^{(2)} - u_{k+1}^{(2)} u_{k+2}^{(1)} \neq 0,$$

because the two solutions $u_k^{(1)}$ and $u_k^{(2)}$ of the homogeneous equation (3.131) are assumed to be given as linearly independent.

Before we start illustrating this important method of variation of parameters for solving nonhomogeneous linear equations with variable coefficients, we note that this method can be easily extended to nth order linear equations with variable coefficients.

In our case of second order nonhomogeneous equations, we again have to face this method's main requirement of having the two linearly independent solutions $u_k^{(1)}$ and $u_k^{(2)}$ of the associated homogeneous equations. The illustration of this method, thus becomes much easier if we consider nonhomogeneous second order equations with constant coefficients, since obtaining $u_k^{(1)}$ and $u_k^{(2)}$ becomes an easy task as was discussed in Section 3.3. In order not to limit our illustrations here to constant coefficients, which is very tempting! We have labored enough in the preceding part of this section for solving the homogeneous equation with *variable coefficients*. Such attempts involved few variations on the method *of the reduction of order*, which we illustrated in Examples 3.28, 3.29 and 3.30. So, with this preparation, we will illustrate in the next example, the method of variation of parameters for solving a nonhomogeneous second order equation with variable coefficients, where to keep the details to minimum, we shall consider the same homogeneous equation with variable coefficients that we used in Examples 3.29 and 3.30.

Example 3.31   Variation of Parameters Method for a Variable Coefficient Equation

Consider the second order nonhomogeneous equation with variable coefficient,

$$u_{k+2} - k(k+1)u_k = 2(k+1)!, \qquad (E.1)$$

where we have already solved its associated homogeneous equation,

$$u_{k+2} - k(k+1)u_k = 0 \qquad (E.2)$$

with the methods of reduction of order in Examples 3.29 and 3.30, and found its two linearly independent solution to be $u_k^{(1)} = (k-1)!$ and $u_k^{(2)} = (-1)^k(k-1)!$.

To find the particular solution (or the general solution) of (E.1) via the present method of variation of parameters using the solution form in (3.132),

$$\begin{aligned} u_k &= V_1(k)u_k^{(1)} + V_2(k)u_k^{(2)} \\ &= V_1(k) \cdot (k-1)! + V_2(k) \cdot (-1)^k(k-1)!, \end{aligned} \qquad (E.3)$$

we appeal to the two simultaneous linear equations, (3.157) and (3.162) in the differences $\Delta V_1(k)$ and $\Delta V_2(k)$ of the two unknown amplitudes $V_1(k)$ and $V_2(k)$,

$$u_{k+1}^{(1)}\Delta V_1 + u_{k+1}^{(2)}\Delta V_2 = 0, \qquad (3.157)$$

and

$$u_{k+2}^{(1)}\Delta V_1 + u_{k+2}^{(2)}\Delta V_2 = r_k \qquad (3.162)$$

to have

$$k!\Delta V_1 + (-1)^{k+1}k!\Delta V_2 = 0, \qquad (E.4)$$

and

$$(k+1)!\Delta V_1 + (-1)^{k+2}(k+1)!\Delta V_2 = 2(k+1)!. \qquad (E.5)$$

To find ΔV_1, we just multiply the first equation $(E.4)$ by $(k+1)$ then add to $(E.5)$ to have

$$2(k+1)!\Delta V_1 = 2(k+1)!, \quad \Delta V_1 = 1. \qquad (E.6)$$

Hence

$$V_1(k) = \Delta^{-1}(1) = \sum_{j=1}^{k-1} 1 + c_1 = (k-1) + c_1. \qquad (E.7)$$

We may note again here that keeping the arbitrary constant c_1 will generate one of the solutions of the homogeneous equation that we already know. So, there is no loss if c_1 is dropped, since we are looking for the particular solution of (E.1), whose initial construction have already been started with the above term $(k-1)$, which, as expected, is without an arbitrary constant. Now, to find the second variable amplitude $V_2(k)$, we use $\Delta V_1 = 1$ of (E.6) in (E.4) to have

$$\Delta V_2 = -(-1)^{k+1}\Delta V_1 = (-1)^k \Delta V_1 = (-1)^k.$$

Hence

$$\Delta V_2 = -(-1)^{k+1}\Delta V_1 = (-)^k \Delta V_1 = (-1)^k$$

$$V_2 = \Delta^{-1}(-1)^k = \sum_{j=1}^{k-1}(-1)^j + c_2,$$

$$V_2 = \begin{cases} -1 + c_2, & k \text{ even} \\ 0 + c_2, & k \text{ odd.} \end{cases} \qquad (E.8)$$

With these determined variable amplitudes $V_1(k)$ and $V_2(k)$ in (E.7), (E.8), substituted in (E.3), we have the final general solution of (E.1) as

$$
\begin{aligned}
u_k &= (k-1)!V_1(k) + (-1)^k(k-1)!V_2(k) \\
&= (k-1)![(k-1)+c_1] + (-1)^k(k-1)!\begin{cases} -1+c_2, & k \text{ even} \\ 0+c_2, & k \text{ odd} \end{cases} \\
&= (k-1)!(k-1) + \begin{cases} (-1)^{k+1}(k-1)!, & k \text{ even} \\ 0, & k \text{ odd} \end{cases} \\
&\qquad\qquad + c_1(k-1)! + c_2(-1)^k(k-1)! \\[1em]
&= u_k^{(p)} + c_1(k-1)! + c_2(-1)^k(k-1)! \\
&= u_k^{(p)} + u_k^{(c)}
\end{aligned}
\qquad (E.9)
$$

where, the first part (without an arbitrary constant) represents the *particular solution* of (E.1), while the sum of the two parts with arbitrary constants c_1 and c_2, clearly, represent the complementary solution of (E.1).

It is a simple matter to verify that the first part of (E.9) is indeed
a solution of (E.1). For example, if we substitute it in (E.1) for the
case of even k, we have

$$
\begin{aligned}
u_{k+2}^{(p)} \quad - \quad & k(k+1)u_k^{(p)} = (k+1)!(k+1) - (k+1)! \\
& -k(k+1)[(k-1)!(k-1) - (k-1)!] \\
= \quad & (k+1)![k+1-1] - k(k+1)(k-1)![k-1-1] \\
= \quad & (k+1)! \cdot k - (k+1)! \cdot k + 2(k+1)! = 2(k+1)!
\end{aligned}
$$

which $r_k = 2(k+1)!$ in (E.1).

We may note that the two solutions of the variable coefficient
equation (E.1) in the above example, namely, $u_k^{(1)} = (k-1)!$ and
$u_k^{(2)} = (-1)^k(k-1)!$ represent as special case with very simple closed
forms, which facilitated the illustration of the above example. In
general, solutions of the homogeneous equations are with more complicated forms such as the second term we had in the answer (E.8)
of Example 3.28 for the homogeneous equation

$$u_{k+2} - (k+2)u_{k+1} + ku_k = 0$$

in (E.1) of that example. This is the main reason for finding that
most of the illustrations, for the method of variation of parameters,
are done, for nonhomogeneous equations with constant coefficients,
which is what we shall do in the next example.

Example 3.32   Variation of Parameter for a Constant Coefficient
                    Equation

Consider the nonhomogeneous second order equation

$$u_{k+2} - 5u_{k+1} + 6u_k = k^2$$

$$= [E^2 - 5E + 6]u_k = k^2 \tag{E.1}$$

we note that it's associated homogeneous equation

$$u_{k+2} - 5u_{k+1} + 6u_k = 0 \tag{E.2}$$

is with constant coefficients. Hence according to the method of solution, discussed in Section 3.3, it is an easy matter to find its two

linearly independent solutions as $u_k^{(1)} = 2^k$ and $u_k^{(2)} = 3^k$ after, simply, consulting the characteristic equation of (E.2)

$$\lambda^2 - 5\lambda + 6 = 0, \quad (\lambda - 2)(\lambda - 3) = 0,$$

$$\lambda_1 = 2, \quad \lambda_2 = 3 \qquad (E.3)$$

with its two distinct real roots $\lambda_1 = 2$ and $\lambda_2 = 3$, which guarantees us their corresponding two linearly independent solutions $u_k^{(1)} = 2^k$ and $u_k^{(2)} = 3^k$. To use the method of variations of parameter for solving the nonhomogeneous equation (E.1), we use the form in (3.115),

$$u_k = V_1(k)u_k^{(1)} + V_2(k)u_k^{(2)} = 2^k V_1(k) + 3^k V_2(k) \qquad (E.4)$$

then appeal to the two simultaneous linear equations(3.138) and (3.143) in $\Delta V_1(k)$ and $\Delta V_2(k)$ with $r_k = k^2$ of (E.1) to have

$$2^{k+1}\Delta V_1 + 3^{k+1}\Delta V_2 = 0 \qquad (E.5)$$

and

$$2^{k+2}\Delta V_1 + 3^{k+2}\Delta V_2 = k^2 \qquad (E.6)$$

To find ΔV_1, we multiply (E.5) by 3, then subtract the resulting equation from (E.6) to have

$$2^{k+2}\Delta V_1 - 3 \cdot 2^{k+1}\Delta V_1 = k^2,$$

$$(1 - \frac{3}{2})2^{k+2}\Delta V_1 = k^2, \quad \Delta V_1 = -\frac{k^2}{2^{k+1}} \qquad (E.7)$$

So,

$$V_1(k) = -\Delta^{-1}\frac{k^2}{2^{k+1}}. \qquad (E.8)$$

In evaluating this summation operation, we have to depend on the following identity, or the other identity used in (E.4) of Example 3.25:

$$F(E)\beta^k = \beta^k F(\beta E). \qquad (E.9)$$

So, we will write Δ^{-1} in terms of E as

$$\Delta^{-1} = \frac{1}{\Delta} = \frac{1}{E-1},$$

then apply this $\frac{1}{E-1}$ in(E.8) with the use of the property (E.9) to have

$$V_1(k) = -\frac{1}{E-1}\left(\frac{1}{2}\right)^k \cdot \frac{k^2}{2}$$

$$= -\left(\frac{1}{2}\right)^k \frac{1}{\frac{1}{2}E-1}\left(\frac{k^2}{2}\right) \qquad (E.10)$$

$$= \left(\frac{1}{2}\right)^k \frac{1}{2-E}(k^2) + c_1$$

Now, we can expand $\frac{1}{2-E} = \frac{1}{1-\Delta}$ in powers of Δ, and prepare $k^2 = k^{(2)} + k^{(1)}$ as sum of factorial polynomials, to be ready for the action, of the powers of the Δ operator in the resulting expansion of $\frac{1}{2-E} = \frac{1}{1-\Delta}$,

$$V_1(k) = \left(\frac{1}{2}\right)^k \frac{1}{1-\Delta}(k^{(2)} + k^{(1)}) + c_1$$

$$= \left(\frac{1}{2}\right)^k [1 + \Delta + \Delta^2 + \cdots](k^{(2)} + k^{(1)}) + c_1 \qquad (E.11)$$

But since $\Delta k^{(2)} = 2k^{(1)}$, $\Delta k^{(1)} = 1$, $\Delta^2 k^{(2)} = 2$, with the rest of the results of the higher order Δ being zeros, we have

$$V_1(k) = \left(\frac{1}{2}\right)^k [k^{(2)} + k^{(1)} + 2k^{(1)} + 1 + 2] + c_1$$

$$= \left(\frac{1}{2}\right)^k [k^2 + 2k + 3] + c_1 \qquad (E.12)$$

after using $k^{(2)} = k(k-1)$ and $k^{(1)} = k$. Now, we return to (E.5) to find

$$\Delta V_2 = \left(\frac{2}{3}\right)^k \Delta V_1 = -\left(\frac{2}{3}\right)^k \cdot \frac{k^2}{2^{k+1}} \qquad (E.13)$$

where we solve for $V_2(k)$,

$$V_2(k) = -\Delta^{-1}\left(\frac{2}{3}\right)^k \cdot \frac{k^2}{2^{k+1}}$$

$$= \frac{1}{2}\frac{1}{1-E}\left(\frac{1}{3}\right)^k \cdot k^2, \qquad (E.14)$$

which can be found, with parallel steps to that of (E.10)-(E.12), as

$$V_2(k) = -\frac{1}{2} \left(\frac{1}{3}\right)^k (k^2 + k + 1) + c_2. \qquad (E.15)$$

So, with $V_1(k)$ from (E.11) and $V_2(k)$ in (E.15), the general solution (E.4) of (E.1) becomes

$$
\begin{aligned}
u_k &= 2^k V_1(k) + 3^k V_2(k) \\
&= 2^k \left[\left(\frac{1}{2}\right)^k (k^2 + 2k + 3) + c_1 \right] \\
&\quad + 3^k \left[-\frac{1}{2} \left(\frac{1}{3}\right)^k (k^2 + k + 1) + c_2 \right] \\
&= k^2 + 2k + 3 - \frac{1}{2}k^2 - \frac{1}{2}k - \frac{1}{2} + c_1 \cdot 2^k + c_2 \cdot 3^k \qquad (E.16) \\
&= \frac{1}{2}k^2 + \frac{3}{2}k + \frac{5}{2} + c_1 2^k + c_2 3^k \\
&= u_k^{(p)} + c_1 2^k + c_2 3^k \\
&= u_k^{(p)} + u_k^{(c)},
\end{aligned}
$$

where the first part, without the arbitrary constant makes the particular solution $u_k^{(p)} = \frac{1}{2}k^2 + \frac{3}{2}k + \frac{5}{2}$ of (E.1), which can be verified very easily upon substituting it in (E.1).

Other Methods

As we mentioned earlier, there are other methods for solving second order equations with variable (polynomial) or constant coefficients, a very important one of which is that of the *factorial polynomial infinite series solutions*. In summary, the method assumes an infinite factorial polynomial series (3.146) for the solution,

$$u_k = \sum_{j=-\infty}^{\infty} c_j k^{(j)}, \qquad (3.146)$$

This solution is plugged in the homogeneous equation, then the terms of similar factorial polynomial $k^{(n)}$ are grouped and equated to zero.

This, as done in the power series solutions of differential equations, will result in recurrence relations among the coefficients c_j to be determined. It also means that such method will result in a difference equation to be solved for the coefficients c_j of (3.147). We shall leave the illustration of this method for an exercise with detailed supporting hints (see problem 7, also problems 10-12).

Exercises 3.7

1. Verify that $u_k^{(1)} = 2^k$ is one of the two solutions of the associated homogeneous equation of

$$u_{k+2} - 5u_{k+1} + 6u_k = k^2, \qquad (E.1)$$

   then use the method of reduction of order to solve (E.1), which is the same equation that was solved by the variation of parameters in Example 3.32.

2. (a) Verify that $u_k^{(1)} = k - 1$ is one of the two solutions of the equation

$$u_{k+2} - \frac{2k+1}{k}u_{k+1} + \frac{k}{k-1}u_k = 0.$$

   (b) Use the method of reduction of order to find $u_k^{(2)}$, the second linearly independent solution.

3. Consider the equation

$$u_{k+2} - u_{k+1} - k^2 u_k = (E^2 - E - k^2)u_k = 0. \qquad (E.1)$$

   (a) Write it in a factorized form.
      Hint: In factorizing the operator $E^2 - E - k^2$, watch for the action of E in the first factor on the variable term in the second factor.
   (b) Verify that $u_k^{(1)} = (k-1)!$ is one of the solutions of (E.1), then use reduction of order for the second solution.
   (c) Use another method of reduction of order to solve (E.1)
      Hint: Use the factorization approach that was illustrated in Examples 3.28, 3.29.

4. (a) Verify that $u_k^{(1)} = 2^k$ and $u_k^{(2)} = 1$ are the two linearly independent solutions of the associated homogeneous equation of

$$u_{k+2} - 3u_{k+1} + 2u_k = 3k^2 + 4^k. \qquad (E.1)$$

    (b) Use the method of variations of parameters to solve (E.1).

5. Solve the following equation by the method of variation of parameters

$$u_{k+2} - 2u_{k+1} + u_k = 3 + k + 4^k.$$

    Hint: Note that we have two repeated roots for the characteristic equation.

6. (a) Solve the following equation by the method of variation of parameters.

$$u_{k+2} - u_k + 1 + u_k = \frac{1}{k!}.$$

    (b) This is a second order equation with constant coefficients, what is the simplest method, or methods to be used for solving it.

    Hint: See that $r_k = \frac{1}{k!}$ is not in the UC form considered in Section 3.3 for constant coefficients equations.

7. As for linear differential equations, even with some polynomial variable coefficients, an *infinite series* expansion in terms of the *factorial polynomial* $k^{(j)}$ (that is the compatible polynomial with the difference operator Δ as oppose to the power x^j for the differential operator),

$$u_k = \sum_{j=0}^{\infty} c_j k^{(j)}, \qquad (E.1)$$

may be assumed for the solution of a linear homogeneous difference equation. Then the terms of the same factorial polynomial $k^{(m)}$ are grouped, and their coefficients are equated to zero. This results in the recurrence relations (new difference equations) in the coefficients c_j of the assumed series solution

(E.1) for u_k. We will illustrate this method in the form of this and a few more exercises. For example, consider the homogeneous second order difference equation with variable (first degree polynomial) coefficients

$$(k+1)u_{k+2} - (3k+2)u_{k+1} + (2k-1)u_k = 0. \qquad (E.2)$$

(a) With the series solution (E.1) being written in terms of the factorial polynomial $k^{(j)}$, we know that

$$\Delta k^{(j)} = jk^{(j-1)}, \qquad (E.3)$$

hence we should write (E.2) in terms of the difference operator Δ instead of E. Show that (E.2) reduces to the following

$$(k+1)\Delta^2 u_k - k\Delta u_k - 2u_k = 0. \qquad (E.4)$$

(b) Plug in the series solution (E.1) in (E.4), use (E.3), then write the limits of the series in each of the three terms to show that (E.4) becomes

$$\sum_{j=0}^{\infty} \left[(j+2)(j+1)^2 c_{j+2} - (j+2)c_j \right] k^{(j)} = 0 \qquad (E.5)$$

to have the recurrence relation in the coefficient c_j, of the desired solution u_k in (E.1), as

$$(j+1)^2 c_{j+2} - c_j = 0, \quad j = 0, 1, 2, \cdots. \qquad (E.6)$$

(c) Instead of solving (E.6) with its variable coefficient $(j+1)^2$, solve it recursively to find c_{2j} and c_{2j+1} in terms of the initial values c_0 and c_1, respectively, as

$$c_2 = \frac{c_0}{1^2}, \quad c_4 = \frac{c_2}{3^2} = \frac{c_0}{1^2 \cdot 3^2}, \quad \cdots,$$

$$c_{2n} = \frac{c_0}{1^2 \cdot 2^2 \cdots (2n-1)^2},$$

$$c_3 = \frac{c_1}{2^2}, \quad c_5 = \frac{c_3}{4^2} = \frac{c_0}{2^2 \cdot 4^2}, \quad \cdots,$$

$$c_{2n+1} = \frac{c_1}{2^2 \cdot 4^2 \cdots (2n)^2}. \qquad (E.7)$$

for the final (general) solution u_k in (E.1) to become

$$u_k = c_0 \sum_{j=0}^{\infty} \frac{k^{(2j)}}{1^2 \cdot 3^2 \cdots (2j-1)^2} + c_1 \sum_{j=0}^{\infty} \frac{k^{(2j+1)}}{2^2 \cdot 4^2 \cdots (2j)^2}. \qquad (E.8)$$

(d) Verify that u_k in (E.8) is the general solution of (E.2).

8. Use another method to solve the difference equation (E.6) of problem 7, and compare the answer with that of (E.7).

   Hint: Consult the methods of this section.

9. Use the method of this section to solve the difference equation,

$$u_{k+1} - 2u_k = 0,$$

   and compare your answer with that of (E.8) in part (c) in the same problem.

10. (a) Use the power (factorial polynomial) series method of problem 7 to solve the first order homogeneous variable coefficient equation,

$$u_{k+1} - k u_k = 0$$

   that is associated with Examples 3.23, 3.25 of Section 3.6.

   (b) Compare answer of part (a) with that found in (E.5) of Example 3.25 in Section 3.6.

11. (a) Use the power (factorial polynomial) series method of problem 7 to solve the difference equation

$$u_{k+2} - u_{k+1} + u_k = 0$$

   of problem 7 in Exercises 3.2, and compare the two answers.

   (b) Compare the answer in part (a) with that of the generating function method in problem 13 of Section 3.2.

(c) Use the power (factorial polynomial) series method of problem 7 to solve the difference equation of problem 3, that was solved by the method of the generating function.

(d) Compare the two answers.

12. Use the power (factorial polynomial) series method to solve the second order homogeneous variable coefficient equation (E.1) of Example 3.30.

3.8 Convergence, Equilibrium, and Stability of Linear Equations

A. Convergence and the Roots of the Characteristic Equation

This section may be introduced by saying that the methods discussed in the preceding sections for obtaining, even a unique solution, are all *formal*! The question we raise here is whether such solution is *useful* under all limiting situations. This includes the auxiliary conditions as the initial conditions, the form of the constant coefficients of the equation, and the limiting case of the independent variable $t_k \equiv k$ for the sought useful solution $u(t_k) \equiv u_k$ as k approached ∞. An example is when t_k is time, then a solution of a difference equation in the form $u_k = 2^k$ is not useful, since after a very long time (as $k \to \infty$) the solution diverges to ∞. On the other hand the solution $u_k = \left(\frac{1}{3}\right)^k$ is useful, since it is always bounded and it converges to 0 as $k \to \infty$. We may recall that the 2 in 2^k and the $\frac{1}{3}$ in $\left(\frac{1}{3}\right)^k$ were zeros of the characteristic equation of a linear homogeneous equation, as we discussed it in detail in Section 3.2. Hence we should expect the good behavior of the solution to depend on the form of *the zeroes* of the characteristic equation associated with the linear homogeneous equation. We shall soon give a statement of an important theorem to that effect. But, first let us examine a couple of our examples of Section 3.2 for their usefulness.

Consider the homogeneous equation of Example 3.8,

$$u_{k+2} - 6u_{k+1} + 8u_k = 0 \tag{3.163}$$

with its two linearly independent solutions

$$u_k^{(1)} = 2^k \text{ and } u_k^{(2)} = 4^k$$

corresponding to the two distinct roots of the characteristic equation $\lambda^2 - 6\lambda + 8 = (\lambda - 2)(\lambda - 4) = 0$, which are $\lambda_1 = 2$ and $\lambda_2 = 4$, respectively. We note that both of these solutions (on their own) diverge as $k \to \infty$. Furthermore, the general solution of the equation $u_k = c_1 2^k + c_2 4^k$, even though it depends on c_1 and c_2, which are usually determined by the initial conditions $u_0 = A$ and $u_1 = B$, still will not give a convergent solution regardless of the values A and B. The exception are those values A and B that render $c_1 = c_2 = 0$, which results in $u_k \equiv 0$. Such a trivial solution is not useful since we know in advance that it is a solution of any homogeneous equation. The other example is

$$u_{k+2} - 3u_{k+1} + 2u_k = 0 \qquad (3.164)$$

with the two linearly independent solutions $u_k^{(1)} = 1$ and $u_k^{(2)} = 2^k$. Here there is a hope in the first (bounded) solution. So for the general solution $u_k = c_1 + c_2 2^k$ we may have the appropriate initial conditions that keep c_1 and render $c_2 = 0$. This may be accomplished by having equal values $u_0 = u_1 = A \neq 0$, whence

$$u_0 = A = c_1 + c_2,$$

$$u_1 = A = c_1 + 2c_2,$$

and upon subtracting the first equation from the second, we have $c_2 = 0$, which results in $c_1 = A$, and the final (useful) solution becomes $u_k = A$, a constant, but nevertheless, bounded.
The general solution to the equation

$$2u_{k+2} + 3u_{k+1} - 2u_k = 0 \qquad (3.165)$$

is

$$u_k = c_1 \left(\frac{1}{2}\right)^k + c_2(-2)^k \qquad (3.166)$$

where the first solution $u_k^{(1)} = \left(\frac{1}{2}\right)^k$ converges to 0 as $k \to \infty$, while the second one $u_k^{(2)} = (-2)^k$ diverges in an oscillating way. So, only some initial conditions that make $c_2 = 0$, result in a final useful (bounded) solution for (3.165). It can be verified that the initial conditions $u_0 = 2$ and $u_1 = 1$ give $c_2 = 0$, $c_1 = 2$, where we end up with the bounded solution $u_k = 2(\frac{1}{2})^k = 2^{-k+1}$.

Now we state a theorem concerning the convergence of the solution of the linear homogeneous second order equation with constant coefficients

$$u_{k+2} + a_1 u_{k+1} + a_2 u_k = 0, \qquad (3.167)$$

which depends mainly on the magnitudes $|\lambda_1|$ and $|\lambda_2|$ of the zeroes λ_1, λ_2 of its characteristic equation

$$\lambda^2 + a_1 \lambda + a_2 = 0 \qquad (3.168)$$

First we define the number

$$\rho = max(|\lambda_1|, |\lambda_2|) \qquad (3.169)$$

which is the larger of the (positive) numbers $|\lambda_1|$ and $|\lambda_2|$ if they are different, and either $|\lambda_1|$ or $|\lambda_2|$ if they are equal. For a complex root $\lambda_1 = r_1(\cos\alpha_1 + i\sin\alpha_1)$, we use its modulus for $|\lambda_1| = (r_1^2 \cos^2\alpha_1 + r_1^2 \sin^2\alpha_1)^{\frac{1}{2}} = (r_1^2(\cos^2\alpha_1 + \sin^2\alpha_1))^{\frac{1}{2}} = r_1$.

Theorem 3.6   Convergence for Solutions of Linear, Homogeneous, Constant Coefficient Equations

For λ_1 and λ_2 as the two roots of the characteristic equation (3.168) of the linear, homogeneous, second order, constant coefficient equation (3.167), consider the number $\rho = max(|\lambda_1|, |\lambda_2|)$ as introduced in (3.169). Then $\rho < 1$ is *necessary* and *sufficient* condition for the solution u_k of (3.167) to converge to zero regardless of the initial values $u_0 = A$ and $u_1 = B$. We ommit the proof here, and continue the illustration of this theorem with the following example and the exercises.

Example 3.33   Convergence of the Sequence Solution

Clearly, the above theorem applies to the following difference equation

$$4u_{k+2} + u_k = 0, \qquad (E.1)$$

which is linear, homogeneous with constant coefficients. The characteristic equation is

$$4\lambda^2 + 1 = (2\lambda + i)(2\lambda - i) \qquad (E.2)$$

whose (complex) roots are $\lambda_1 = \frac{1}{2}i$ and $\lambda_2 = -\frac{1}{2}i$, and the general solution of (E.1), as we saw in Section 3.2, can be written as

$$u_k = A(\frac{1}{2})^k \cos\left(\frac{k\pi}{2} + B\right) \qquad (E.3)$$

This is, obviously, a bounded solution, which converges to zero as $k \to \infty$, primarily because of the decaying amplitude $(\frac{1}{2})^k$. This is supported by the above Theorem 3.6, since $|\lambda_1| = |\frac{1}{2}i| = \frac{1}{2} = |\lambda_2| = |-\frac{1}{2}i| = \frac{1}{2} < 1$, and

$$\rho = max(|\lambda_1|, |\lambda_2|) = max(\frac{1}{2}, \frac{1}{2}) = \frac{1}{2} < 1, \qquad (E.4)$$

as the necessary and sufficient condition of the theorem for the solution of equation (E.1) to converge to zero.

We may note that for the nonhomogeneous equation

$$u_{k+2} + a_1 u_{k+1} + a_2 u_k = r_k$$

with its general solution

$$u_k = u_k^{(c)} + u_k^{(p)} = c_1 u_k^{(1)} + c_2 u_k^{(2)} + u_k^{(p)}$$

where $u_k^{(c)}$ is the general solution of its associated homogeneous equation (3.167), even if $u_k^{(c)}$ is convergent, we may still have an unbounded solution due to the particular solution $u_k^{(p)}$.

B. Stability of the Equilibrium Solution to Linear Equations

In the last part A of this section, we showed how the convergence of the solution of the linear, second order, homogeneous, constant coefficient equation (3.167)

$$u_{k+2} + a_1 u_{k+1} + a_2 u_k = 0 \qquad (3.167)$$

depends entirely on the two roots λ_1 and λ_2 of its characteristic equation

$$\lambda^2 + a_1\lambda + a_2 = 0 \qquad (3.168)$$

and that it is independent of the initial conditions $u_0 = A$ and $u_1 = B$ as stated very clearly in Theorem 3.6. Here, we will consider the nonhomogeneous case (3.170) with a constant nonhomogeneous term $r_k = r$,

$$u_{k+2} + a_1 u_{k+1} + a_2 u_k = r \qquad (3.170)$$

and present another theorem that gives three conditions on the (constant) coefficients a_1 and a_2 for the *equilibrium* solution $u_k^{(e)}$ of (3.170) (when it materializes!) to be *stable*. Our discussion, of course, is limited to *linear* equations, and we will try to be brief.

First we shall explain what we mean by an equilibrium solution $u_k^{(e)}$ of (3.170). We may get a better understanding of the equilibrium concept if we think of the above second order difference equation as the difference analog (or an approximation to) a second order differential equation in $u(t)$,

$$m\frac{d^2u}{dt^2} + b\frac{du}{dt} + ku = F \qquad (3.171)$$

where E^2 and E in (1.170) relate in some approximation way to the second and the first order derivatives of the first two terms in (3.171), respectively. In (3.171) we see a viscous force term $b\frac{du}{dt}$, which we know that it acts as a dampening influence, ku as the Hooke's law force, and F as a constant driving force corresponding to r in (3.170). From our experience, we expect the solution of (3.171), in general, to be oscillatory, but with a decaying amplitude due to the viscous term $b\frac{du}{dt}$. Hence, after a long time the solution of (3.171) will ultimately decay to zero, which is what we term an *equilibrium state* $u^{(e)} = \lim_{t\to\infty} u(t) = 0$.

In general, we have an equilibrium solution when the latter settles to a *constant value* $u^{(e)} = c$. So, for the difference equation, if it settles for a solution $u^{(e)} = c = $ constant as k becomes very large, then this solution $u^{(e)}$ is called an equilibrium solution of (3.170). Since with this equilibrium state, we have $u_{k+2}^{(e)} = u_{k+1}^{(e)} = u_k^{(e)} = $

$u^{(e)}$, then from (3.170) we find that such an equilibrium solution $u^{(e)}$ satisfies the following equation

$$u^{(e)} + a_1 u^{(e)} + a_2 u^{(e)} = r \qquad (3.172)$$

where

$$u^{(e)} = \frac{r}{1 + a_1 + a_2}, \quad 1 + a_1 + a_2 \neq 0 \qquad (3.173)$$

provided that $1 + a_1 + a_2 \neq 0$.

We may remark that if any two consecutive sequences, say u_{k+1} and u_k of (3.170) are equal to the equilibrium solution $u^{(e)}$, then all the succeeding sequences take the equilibrium value $u^{(e)}$ as seen from substituting $u_k = u_{k+1} = u^{(e)}$ in (1.170), where u_{k+2} becomes $u^{(e)}$, since

$$u_{k+2} + a_1 u^{(e)} + a_2 u^{(e)} = r$$

implies that u_{k+2} is a constant, and for it to be a solution, it must take the equilibrium value $u^{(e)}$.

Another remark is that the simple condition $1 + a_1 + a_2 \neq 0$ in (3.173) gives us the first glimpse of the dependence of a, possibly good behaving, solution of (3.170) on its coefficients a_1 and a_2. This is in the sense that this condition is a necessary condition for an equilibrium solution of (3.173). In addition, a condition in this direction, namely, $1 + a_1 + a_2 > 0$ will be among three conditions in the following Theorem 3.7 that will provide us with *necessary* and *sufficient* conditions for such an equilibrium solution $u^{(e)}$ of (3.170) to be *stable*. The simplest way of defining what we mean here by a stable solution of an equation such as (3.170) is that its *output* u_k should not experience a very large change, when the *input*, in this case what we supply as the two initial conditions, changes by a small value. In other words, there should be a *continuous dependence*, of the change in the output on that of the input. A very conservative way of saying this is if the solution does not depend on the initial conditions $u_0 = A$ and $u_1 = B$. With this, an equilibrium solution $u^{(e)}$ of (3.170) is called stable, or that equation (3.170) is stable if every solution of it converges to that equilibrium value $u^{(e)}$, where this convergence is independent of the given initial conditions u_0 and u_1. We shall state the following Theorem 3.7, which, as mentioned earlier, gives three conditions on the coefficients a_1 and a_2 of (3.170) for its equilibrium solution $u^{(e)}$ to be stable in the sense we defined above. We shall ommit the proof, and be satisfied with it's illustration here and in the exercises.

It turned out that the necessary and sufficient condition $\rho = max(|\lambda_1|, |\lambda_2|) < 1$, for the convergence of the solution of the homogeneous equation (3.167), is also what is needed for the stability of the equilibrium solution $u^{(e)}$ of (3.170) as given in (3.173). However, the theorem that ties the stability of the equilibrium solution to the coefficients a_1 and a_2 of (3.170) is the following, whose proof, as expected, depends on the above statement or theorem.

Theorem 3.7   The Dependence of the Stability on the Coefficients of the Equation

For the difference equation (3.170), the following three conditions

$$1 + a_1 + a_2 > 0$$

$$1 - a_1 + a_2 > 0 \qquad\qquad (3.174)$$

and

$$1 - a_2 > 0$$

are *necessary* and *sufficient* for the stability of the equilibrium solution $u_k^{(e)}$ of (3.170).

These three conditions of (3.174) are not so easy to come by, which is understood, since we are speaking of an equilibrium solution first, then we want that solution to be stable. For example, consider the difference equation

$$u_{k+2} - 3u_{k+1} + 2u_k = 1 \qquad\qquad (3.175)$$

which is a special case of (3.170) with $a_1 = -3$, $a_2 = 1$ and $r = 1$. The best way we can tell about whether or not it has an equilibrium solution, is to look at its general solution (that is found in problem 3 of this section's exercises), which is derived with the UC method as

$$u_k = c_1 + c_2 2^k - k$$

If we apply the initial conditions

$$u_0 = 1 = c_1 + c_2, \qquad\qquad (3.176)$$

$$u_1 = -1 = c_1 + 2c_2 - 1, \tag{3.177}$$

we have $c_1 = 3$ and $c_2 = -2$, and the solution to the initial value problem (3.175), (3.176)-(3.177) is

$$u_k = 3 - 2^{k+1} - k \tag{3.178}$$

which does not have an equilibrium state, since it diverges as $k \to \infty$.

Let us consider another special case of (3.170),

$$u_{k+2} - u_{k+1} + \frac{1}{2}u_k = 1 \tag{3.179}$$

whose characteristic equation

$$\lambda^2 - \lambda + \frac{1}{2} = 0, \tag{3.180}$$

$$2\lambda^2 - 2\lambda + 1 = 0$$

has two complex roots $\frac{1}{2} \mp i\frac{1}{2} = \frac{1}{\sqrt{2}}e^{\mp i\frac{\pi}{4}}$ with the complementary solution of (3.179) written as

$$
\begin{aligned}
u_k^{(c)} &= \left(\frac{1}{\sqrt{2}}\right)^k [c_1 \cos \frac{\pi}{4}k + c_2 \sin \frac{\pi}{4}k] \\
&= 2^{-\frac{k}{2}} [c_1 \cos \frac{\pi}{4}k + c_2 \sin \frac{\pi}{4}k]
\end{aligned}
\tag{3.181}
$$

This is a decaying oscillatory solution that dies out to zero as $k \to \infty$. We still have to deal with the general solution of (3.179), which involves finding a particular solution.

The particular solution of the nonhomogeneous equation (3.179) can be obtained easily via the UC method, where letting $u_k^{(p)} = A$ in (3.179) determines A to be 2,

$$A - A + \frac{1}{2}A = 1, \quad A = 2 \tag{3.182}$$

Hence, the general solution to (3.179) is

$$u_k = 2^{-\frac{k}{2}} [c_1 \cos \frac{\pi k}{4} + c_2 \sin \frac{\pi k}{4}] + 2,$$

and it will reach an equilibrium state of $u^{(e)} = 2$ as $k \to \infty$. So the equilibrium solution of (3.179) is $u^{(e)} = 2$. Since in this special example, we have a constant particular solution, the initial conditions $u_0 = A$ and $u_1 = B$ will only affect $u_k^{(c)}$, which is dying out anyway. Thus, and as it appears, this equilibrium solution is stable. What remains now is to show that Theorem 3.7 supports our observation. This means that we would like to see that the necessary and sufficient conditions on the coefficients of our equation (3.179) as specified in (3.174) of Theorem 3.7 are satisfied to guarantee this stability. With $a_1 = -1$ and $a_2 = \frac{1}{2}$ in (3.179), the three conditions of Theorem 3.7: $1 + a_1 + a_2 > 0$, $1 - a_1 + a_2 > 0$ and $1 - a_2 > 0$ are satisfied, since

$$1 + a_1 + a_2 = 1 - 1 + \frac{1}{2} = \frac{1}{2} > 0,$$

$$1 - a_1 + a_2 = 1 + 1 + \frac{1}{2} = \frac{5}{2} > 0,$$

and

$$1 - a_2 = 1 - \frac{1}{2} = \frac{1}{2} > 0.$$

Exercises 3.8

1. Consider the nonhomogeneous difference equation

$$u_{k+2} + u_k = \sin \frac{k\pi}{2} \qquad (E.1)$$

with its general solution

$$u_k = u_k^{(c)} + u_k^{(p)} = c_1 \cos \frac{k\pi}{2} + c_2 \sin \frac{k\pi}{2} - \frac{1}{2} k \sin \frac{k\pi}{2}$$

$$= A \cos \left(\frac{k\pi}{2} + B \right), \qquad (E.2)$$

where $\tan B = -\frac{c_2}{c_1}$, $c_1 = A \cos B$; $c_2 = -A \sin B$.

(a) Verify Theorem 3.6 as it predicts the divergence of the complementary solution in (E.2), which of course we can see that $u_k^{(c)}$ diverges in a finitely oscillating way as $k \to \infty$.

Hint: Find the two complex roots $\lambda_1 = i\sqrt{2}$ and $\lambda_2 = -i\sqrt{2}$ and evaluate their moduli $|\lambda_1| = \sqrt{2}$ and $|\lambda_2| = \sqrt{2}$.

(b) Verify that u_k of (E.2) is indeed the general solution of (E.1).

2. Consider the nonhomogeneous equation

$$8u_{k+2} - 6u_{k+1} + u_k = 2^k \qquad (E.1)$$

with its general solution

$$u_k = u_k^{(c)} + u_k^{(p)} = c_1\left(\frac{1}{2}\right)^k + c_2\left(\frac{1}{4}\right)^k + \frac{2^k}{21} \qquad (E.2)$$

(a) As it appears in $u_k^{(c)}$ of (E.2), which converges to zero as $k \to \infty$, illustrate Theorem 3.6 to support that, i.e., show that the zeroes λ_1 and λ_2, of the characteristic equation of the homogeneous equation

$$8u_{k+2} - 6u_{k+1} + u_k = 0, \qquad (E.3)$$

do satisfy the condition in Theorem 3.16.

(b) Show that even though the $u_k^{(c)}$ converges to zero, the general solution in (E.2) diverges.

3. Consider the nonhomogeneous difference equation

$$u_{k+2} - 3u_{k+1} + 2u_k = 1 \qquad (E.1)$$

and its homogeneous case

$$u_{k+2} - 3u_{k+1} + 2u_k = 0 \qquad (E.2)$$

(a) Show that the general solution of (E.1) is

$$u_k = u_k^{(c)} + u_k^{(p)} = c_1 + c_2 2^k - k \qquad (E.3)$$

(b) Use Theorem 3.6 to show that $u_k^{(c)}$, as seen in (E.3), indeed diverges.

4. For equation (3.179)

$$u_{k+2} - u_{k+1} + \frac{1}{2}u_k = 1, \qquad (E.1)$$

use theorem 3.6 to predict that the solution of its homogeneous case converges to zero.

5. Consider the slightly different equation than that of problem 4,

$$u_{k+2} - 0.8u_{k+1} + 0.5u_k = 1 \qquad (E.1)$$

(a) Use Theorem 3.6 to prove the convergence of its complementary solution $u_k^{(c)}$.

(b) Solve (E.1) to determine whether it has a stable equilibrium solution?

(c) Use Theorem 3.7 to validate your observation of the solution obtained in part (b).

6. Consider the nonhomogeneous equation

$$u_{k+2} - 2.4u_{k+1} + 1.6u_k = 1$$

Use the steps in problems 5 and 4 to get some idea about the behavior of its solution, i.e., convergence, stability, etc.

Chapter 4

DISCRETE FOURIER TRANSFORMS

4.1 Continuous to Discrete

In Section 2.3 of Chapter 2, the notion of an *integral transform* was introduced in a general way. The emphasis there was on the use of such transforms, like the Laplace and Fourier transforms, in reducing differential operators to algebraic operators, hence the "operational calculus" method of solving differential equations. A special case in which the transform variable λ takes on only certain discrete values was then considered. This leads to the idea of a *finite transform* and the associated *orthogonal expansion* of a function. In Section 2.4 we took one final step, by considering the situation in which both the transform variable λ and the original independent variable are restricted to discrete sets of values. This leads to the idea of *discrete transform*, where, again we had our emphasis and illustrations in Section 2.4 on the use of the discrete Fourier transforms in reducing difference operators to algebraic operators, hence the "operational sum calculus" method of solving difference equations. The latter concept represents an added feature of this book.

It is now time to dispense with general discussion and turn to the detailed discussion of the discrete Fourier transform, which we need for facilitating such discrete transforms method. We will begin by a very brief presentation of the Fourier integral transform and its associated finite transform that is necessary to show their relation, or parallel to the main subject of this chapter, namely, the discrete Fourier transform.

We begin by simply giving the Fourier *integral* transform pair that relates a given function $G(f)$ to its transform $g(t)$ as defined in (2.52),

$$g(t) = \int_{-\infty}^{\infty} G(f)e^{i2\pi ft}\,df, \quad -\infty < t < \infty \tag{4.1}$$

where $i = \sqrt{-1}$. In keeping with the notation of Section 2.3, the kernel of this transform is $K(f,t) = e^{i2\pi ft}$. The transform variable, t, often represents time, in which case f represents frequency with dimension of Hertz (cycles / second). The *inverse* transform is given by a symmetric relation.

$$G(f) = \int_{-\infty}^{\infty} g(t)e^{-i2\pi ft}\,dt, \quad -\infty < t < \infty. \tag{4.2}$$

The use of f (instead of ν) for frequency, that may be confused with the usual use of f for a function, is a tradition in electrical engineering literature. There, the fast Fourier transform (FFT) was developed as the most efficient algorithm of computing the discrete Fourier transform (DFT). thus most references on these subjects, including the software of the FFT, use this notation. The same goes for the limits of the frequency $0 < f < \frac{1}{T}$ instead of the desirable $0 < f < f_0$, or most appropriately $0 < \nu < \nu_0$.

Consider now the case in which the given function $G(f)$ is *periodic* with period $\frac{1}{T}$; that is, $G(f + \frac{1}{T}) = G(f)$. We may work on any interval of length $\frac{1}{T}$. Let us choose $0 < f < \frac{1}{T}$. On this interval, the function $G(f)$ has a *Fourier series* expansion of the form

$$G(f) = \sum_{k=-\infty}^{\infty} g(t_k)e^{-i2\pi ft_k}, \quad 0 < f < \frac{1}{T}. \tag{4.3}$$

This representation is a special case of (2.65). The *Fourier coefficients* $\{g(t_k)\}$ are given by the *finite* Fourier transform

$$g(t_k) = T\int_0^{\frac{1}{T}} G(f)e^{i2\pi ft_k}\,df \tag{4.4}$$

which corresponds to (2.64) of Section 2.3 with its finite interval $I' = (0, \frac{1}{T})$. Notice that the transform variable t now takes on only the discrete values $\{t_k\}$. It is also possible to show that the kernel $K(f, t_k)$ satisfies the orthogonality relation (2.62) on the interval $(0, \frac{1}{T})$.

We now move to introduce the discrete Fourier transform in which not only the transform variable t is discrete, but also the original independent variable f. Again we will consider the given function $G(f)$ to be periodic with period $\frac{1}{T}$ and will consider $G(f)$ on the interval $0 < f < \frac{1}{T}$. We must first establish the grid points in the variable f and t. On the interval $0 < f < \frac{1}{T}$, we will choose N equally spaced grid points $\{f_n\}$ separated by a distance $\Delta f = \frac{1}{NT}$ as shown in Fig. 4.1.

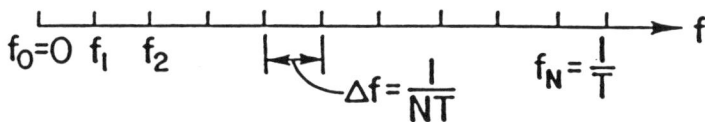

Figure 4.1: Equally spaced grid points at $f_n = n \cdot \frac{1}{NT}$

For the variable t, we must determine the interval on which we should work. Looking at the kernel $e^{i2\pi ft}$, we see that as a function of t, this has a period of $\frac{1}{f}$. The frequencies f are now restricted to the discrete set $\{f_n = \frac{n}{NT}\}$. The longest period of the kernel in the variable t is therefore $\frac{1}{f_1} = \frac{1}{\Delta f} = NT$. (Choosing $f = f_0 = 0$ reduces the kernel to a constant function whose period is infinite). Therefore we choose the interval $0 < t < NT$ and partition it by a set of equally spaced grid points $\{t_k\}$ separated by a distance $\Delta t = \frac{NT}{N} = T$ as shown in Fig. 4.2.

We now have the framework in which to develop the discrete Fourier transform. To obtain the discrete transform of $G(f)$, we return to its finite transform (4.4) and replace the integral over the interval $0 < f < \frac{1}{T}$ by a sum over the grid points $\{f_n\}$. This transition will be reflected in the notation by replacing G by $\tilde{G}$ and g by

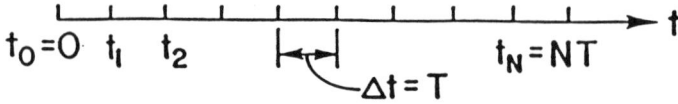

Figure 4.2: The partitioning of the interval $(0, NT)$ by N points

$\tilde{g}$. We then have

$$\tilde{g}(t_k) = T \sum_{n=0}^{N-1} \tilde{G}(f_n) e^{i2\pi f_n t_k} \, \Delta f.$$

Inserting the values of the grid points, $f_n = n\Delta f = \frac{n}{NT}$ and $t_k = k\Delta t = kT$, we have the discrete transform $\tilde{g}(t_k)$ of $\tilde{G}(f_n)$

$$\mathcal{F}\{\tilde{G}\} = \tilde{g}(t_k) = \frac{1}{N} \sum_{n=0}^{N-1} \tilde{G}(f_n) e^{\frac{i2\pi nk}{N}}, \quad 0 \leq k \leq N-1. \qquad (4.5)$$

We may also return to the (infinite) expansion (4.3) and replace it by a finite sum. The result is the *inverse* discrete Fourier transform $\tilde{G}(f_n)$ of $\tilde{g}(t_k)$

$$\mathcal{F}^{-1}\{\tilde{g}\} = \tilde{G}(f_n) = \sum_{k=0}^{N-1} \tilde{g}(t_k) e^{-i2\pi f_n t_k}$$

or

$$\mathcal{F}^{-1}\{\tilde{g}\} = \tilde{G}(f_n) = \sum_{k=0}^{N-1} \tilde{g}(t_k) e^{\frac{-i2\pi nk}{N}}, \quad 0 \leq n \leq N-1. \qquad (4.6)$$

Several things should be mentioned at this point. In deriving (4.5) and (4.6), we have argued by analogy and replaced infinite or continuous operations (integration and infinite summation) by finite sums. The result is a pair of relations between the functions $\tilde{G}$ and $\tilde{g}$ which are defined on their own sets of grid points $\{f_n\}$ and $\{t_k\}$, respectively. In this process, we can be sure that we have made *some approximations* and that $\tilde{G}$ and $\tilde{g}$ are not Fourier transforms of each other in the sense of (4.1) and (4.2). Our job now is to show that $\tilde{G}$ and $\tilde{g}$ are, nevertheless, exact discrete Fourier transforms of each other and that this transform pair inherits many of the properties of the Fourier integral transform pair (4.1), (4.2) and the finite transform pair (4.3), (4.4). Then, as we have already explained in Section 2.4, we shall use the discrete Fourier transform to solve difference equations with boundary conditions just as the Fourier integral (or continuous) transforms (4.1), (4.2) and the finite transforms (4.3), (4.4) of Section 2.3 are used to solve differential equations with boundary conditions. Finally it should be mentioned that there are many ways to arrive at the discrete Fourier transform pair (4.5), (4.6). Some of these alternate approaches may be found in the author's book on the subject of integral and discrete transforms and in a number of other references, which are listed at the end of the book under "Operational Calculus" or "Fast Fourier Transforms." Our intent in this section has been to introduce the discrete Fourier transform in a brief, but believable way which does not necessarily constitute an exact derivation. With this tool in hand, it is now time to become familiar with it and learn its main use in this book, namely, to solve difference equations with boundary conditions as we have already discussed in Section 2.4. We stress again that the discrete Fourier transforms $\tilde{g}$, and $\tilde{G}$ are only approximations (may be very rough) of the Fourier transforms g, and G. As we did in Section 2.4, since we will be working mainly with the discrete transforms in the rest of this book, we shall drop the $\sim$ and use $g(t_k)$ and $G(t_k)$ or merely g_k and G_k for such discrete transform pairs. Also, for the operators $\tilde{\mathcal{F}}$ and $\tilde{\mathcal{F}}^{-1}$ we shall write $\mathcal{F}$ and $\mathcal{F}^{-1}$ without the fear of confusing them with those operators used for the exponential Fourier transform and its inverse in Section 2.3.

4.2 Properties of the Discrete Fourier Transform (DFT)

The principal result of the preceding section may be summarized as follows. If a given function G is sampled at N equally spaced points

$\{f_n\}$ on an interval $0 < f < \frac{1}{T}$, then the *discrete Fourier transform* of G is given by

$$\mathcal{F}\{G_n\} \equiv g_k = g(t_k) = \frac{1}{N} \sum_{n=0}^{N-1} G(f_n) e^{\frac{i2\pi nk}{N}}, \quad 0 \le k \le N - 1. \quad (4.5)$$

where the set $\{t_k\}$ contains equally spaced grid points on the interval $0 \le t \le NT$. Furthermore, an *inverse* transform has been proposed (to be proved soon) which is given by

$$\mathcal{F}^{-1}\{g_k\} \equiv G_n = G(f_n) = \sum_{k=0}^{N-1} g(t_k) e^{\frac{-i2\pi nk}{N}}, \quad 0 \le n \le N - 1. \quad (4.6)$$

We begin our investigation of the discrete Fourier transform (henceforth DFT) by noting some *periodic properties* of (4.5), (4.6). Since $e^{\frac{i2\pi n(k+N)}{N}} = e^{\frac{i2\pi nk}{N}} \cdot e^{i2\pi n} = e^{\frac{i2\pi nk}{N}}$ and also $e^{\frac{i2\pi(n+N)k}{N}} = e^{\frac{i2\pi nk}{N}}$, it is clear that the kernel of the DFT is periodic in both n and k with period N. This means that the sequences $\{G(f_n)\}$ and $\{g(t_k)\}$ defined by (4.5), (4.6) are also periodic in n and k with period N; that is $g(t_{k+N}) = g(t_k)$ and $G(f_{n+N}) = G(f_n)$, as was shown in (2.102). The fact that the limits of summation in the DFT pair are not symmetric about zero is primarily a matter of convention and convenience. Because of the periodicity just mentioned, these pairs may be defined with limits of summation n_0 and $n_0 + N - 1$, where n_0 is any integer. To obtain symmetric limits of summation, we may take $n_0 = -M$ and an odd N, $N = 2M + 1$. Recalling that $t_k = kT$ and $f_n = \frac{n}{NT}$, we have the symmetric transform pair

$$G(\frac{n}{(2M+1)T}) = \sum_{k=-M}^{M} g(kT) e^{\frac{-i2\pi nk}{2M+1}}, \quad -M \le n \le M \quad (4.7)$$

and

$$g(kT) = \frac{1}{(2M+1)} \sum_{k=-M}^{M} G(\frac{n}{(2M+1)T}) e^{\frac{i2\pi nk}{2M+1}}, -M \le k \le M \quad (4.8)$$

In our coming work, we may choose either of the pairs (4.5), (4.6) or (4.7), (4.8) in a way that is most suitable to the problem at hand.

The case of even $N = 2M$, and a variety of other forms of the discrete (complex exponential) Fourier transform that are used in other books, and most importantly in the different routines of computing the fast Fourier transform (FFT), are the subject of problems 4 and 5 in the Exercises.

It is time to verify that the relation given in (4.5), (4.6) does in fact constitute a transform pair. Toward this end, it will be useful to investigate the (discrete) orthogonality property of the kernel of the DFT. The kernel of the DFT is the set $\{e^{i2\pi f_n t_k}\}$ or equivalently $\{e^{\frac{i2\pi nk}{N}}\}$. Proving orthogonality amounts to showing that the sum

$$\sum_{n=0}^{N-1} e^{i2\pi f_n t_k} \cdot e^{-i2\pi f_n t_l} = \sum_{n=0}^{N-1} e^{\frac{i2\pi nk}{N}} \cdot e^{-\frac{i2\pi nl}{N}}$$

is zero unless $k = l$, in which case the sums have a non-zero value (which is one for an orthonormal kernel). This was established in (2.97) of Chapter 2, but we repeat it here for emphasis and for its extreme importance in verifying that $g(t_k)$ in (4.5) and $G(f_n)$ in (4.6) are exact discrete Fourier transforms of each other. Therefore, consider the sum

$$\sum_{n=0}^{N-1} e^{\frac{i2\pi nk}{N}} \cdot e^{-\frac{i2\pi nl}{N}} = \sum_{n=0}^{N-1} e^{\frac{i2\pi n(k-l)}{N}}.$$

We recognize that this sum may be written as a geometric series and then summed as in (2.96)–(2.97) of Section 2.4. This gives us

$$\sum_{n=0}^{N-1} e^{\frac{i2\pi n(k-l)}{N}} = \sum_{n=0}^{N-1} \left\{ e^{\frac{i2\pi(k-l)}{N}} \right\}^n = \frac{1 - \left\{ e^{\frac{i2\pi(k-l)}{N}} \right\}^N}{1 - e^{\frac{i2\pi(k-l)}{N}}}$$

$$= \frac{1 - e^{i2\pi(k-l)}}{1 - e^{\frac{i2\pi(k-l)}{N}}} = 0, \quad \text{if } k \neq l.$$

If $k \neq l$, the numerator is zero, while the denominator is non-zero. However, if $k = l$ the last expression is indeterminate and it is easiest to return to the original sum and note that

$$\sum_{n=0}^{N-1} e^{\frac{i2\pi n(k-l)}{N}} = \sum_{n=0}^{N-1} 1 = N, \quad \text{if } k = l.$$

Thus, we have shown the (discrete) orthogonality of the DFT kernel which may be summarized as

$$\sum_{n=0}^{N-1} e^{\frac{i2\pi n(k-l)}{N}} = N\delta_{k,l} \tag{4.9}$$

where the *Kronecker delta* $\delta_{k,l}$ is defined by

$$\delta_{k,l} = \begin{cases} 0 & \text{if } k \neq l \\ 1 & \text{if } k = l. \end{cases} \tag{4.10}$$

With this orthogonality result, we may now show that G and g as given in (4.5), (4.6) are exact discrete transforms of each other. To do this, we begin with the sum in (4.6) and substitute the expression for g as given in (4.5)

$$\begin{aligned}
\sum_{k=0}^{N-1} g(t_k) e^{-\frac{i2\pi nk}{N}} &= \sum_{k=0}^{N-1} \left\{ \frac{1}{N} \sum_{l=0}^{N-1} G(f_l) e^{\frac{i2\pi lk}{N}} \right\} e^{\frac{-i2\pi nk}{N}} \\
&= \frac{1}{N} \sum_{l=0}^{N-1} G(f_l) \sum_{k=0}^{N-1} e^{\frac{i2\pi k(l-n)}{N}} \\
&= \frac{1}{N} \sum_{l=0}^{N-1} G(f_l) \delta_{l,n} \cdot N \\
&= G(f_n).
\end{aligned}$$

after interchanging the order of the two summations in the second step, and using the discrete orthogonality property (4.9) in the final step. Expression (4.6) has been verified using the definition of g given in (4.5). It is also possible to begin with (4.5) and verify it using the definition of G given in (4.6). In either case, the orthogonality property has a fundamental role in these relations.

Having verified that (4.5), (4.6) constitute a discrete transform pair, we may now compute the DFT of some specific sequences. Before doing so, it will be useful to simplify the notation. We will denote $G(f_n)$ and $g(t_k)$ by G_n and g_k, respectively. In addition, the quantity $e^{\frac{i2\pi}{N}}$ is the *nth root of unity*, since

$$\left(e^{\frac{i2\pi}{N}} \right)^N = e^{i2\pi} = \cos 2\pi + i \sin 2\pi = 1,$$

and is conventionally denoted ω (in some books W is used). With these changes, the DFT pair (4.5), (4.6) appears as

$$g_k = \frac{1}{N} \sum_{n=0}^{N-1} G_n \omega^{nk}, \quad 0 \le k \le N-1 \qquad (4.11)$$

$$G_n = \sum_{n=0}^{N-1} g_k \omega^{-nk}, \quad 0 \le n \le N-1 \qquad (4.12)$$

We will now look at some examples.

Example 4.1 The Discrete Fourier Transform of $\{G_n\} = \{n\}$

Consider the sequence $\{G_n\}_{n=0}^{N-1} = \{n\}_{n=0}^{N-1}$. We wish to compute the DFT of $\{G_n\}$ using (4.11). Letting $\mathcal{F}$ represent the operation of taking the DFT we have

$$g_k = \mathcal{F}\{G_n\} = \mathcal{F}\{n\} = \frac{1}{N} \sum_{n=0}^{N-1} n \omega^{nk}.$$

We are left with a sum on the right hand side which (hopefully) should suggest summation by parts. Furthermore, this sum might also be recognized from the result of Example 2.4 in Section 2.2, namely that

$$\sum_{n=0}^{N-1} n a^n = \frac{a}{(a-1)^2}[(N-1)a^N - Na^{N-1} + 1] \qquad (4.13)$$

with $a = \omega^k$, then we have

$$\mathcal{F}\{G_n\} = \frac{1}{N} \frac{\omega^k}{(\omega^k-1)^2}[(N-1)\omega^{kN} - N\omega^{k(N-1)} + 1]$$

This may be simplified considerably by noting that

$$\omega^{kN} = \left(e^{\frac{i2\pi k}{N}}\right)^N = e^{i2\pi k} = \cos 2\pi k + i \sin 2\pi k = 1.$$

Therefore,

$$
\begin{aligned}
g_k &= \mathcal{F}\{G_n\} = \frac{1}{N} \frac{\omega^k}{(\omega^k-1)^2}[(N-1) - N\omega^{-k} + 1] \\
&= \frac{\omega^k}{(\omega^k-1)^2}(1 - \omega^{-k}) = \frac{1}{\omega^k - 1}.
\end{aligned}
$$

This result holds for $1 \leq k \leq N - 1$, but clearly does not make sense for $k = 0$. For $k = 0$, we return to (4.13) and set $k = 0$ giving

$$g_0 = \frac{1}{N} \sum_{n=0}^{N-1} n = \frac{1}{N} \frac{N(N-1)}{2} = \frac{N-1}{2}$$

Therefore,

$$g_k = \mathcal{F}\{n\} = \begin{cases} \frac{1}{\omega^k - 1} & 1 \leq k \leq N - 1 \\ \frac{N-1}{2} & k = 0. \end{cases} \tag{4.14}$$

Example 4.2   The Discrete Fourier Transform of $\{G_n\} = \{1\}$

The constant sequence $\{G_n\} = \{1\}$ for $0 \leq n \leq N - 1$ has a DFT given by

$$\begin{aligned} g_k &= \mathcal{F}\{1\} = \frac{1}{N} \sum_{n=0}^{N-1} 1 \cdot \omega^{nk} = \frac{1}{N} \sum_{n=0}^{N-1} (\omega^k)^n \\ &= \frac{1}{N} \frac{1 - \omega^{kN}}{1 - \omega^k} \end{aligned}$$

after recognizing the last sum as a geometric series. As long as $k \neq 0$, the numerator of this last expression is zero (since $\omega^{kN} = 1$) and the denominator is non-zero. If $k = 0$, we return to the original sum to find

$$g_0 = \frac{1}{N} \sum_{n=0}^{N-1} 1 \cdot \omega^{n \cdot 0} = \frac{1}{N} \cdot N = 1.$$

Therefore,

$$g_k = \mathcal{F}\{1\} = \begin{cases} 0 & \text{if } k \neq 0 \\ 1 & \text{if } k = 0. \end{cases} \equiv \delta_{k,0} \tag{4.15}$$

Example 4.3   The Discrete Fourier Transform of $\{G_n\} = \{\cos \frac{2\pi \alpha n}{N}\}$

Consider the function $G(x) = \cos \alpha x$, where $0 \leq \alpha \leq N - 1$ is an integer, sampled at the grid points $\{\frac{2\pi n}{N}\}_{n=0}^{N-1}$ on the interval $0 \leq x \leq$

2π. The sampled values produce a sequence $\{G_n\} = \{\cos\frac{2\pi\alpha n}{N}\}_{n=0}^{N-1}$ for which we will find the DFT. This calculation is simplified by using the Euler identity $\cos\theta = \frac{1}{2}(e^{i\theta} + e^{-i\theta})$. Then we have

$$g_k = \mathcal{F}\{\cos\frac{2\pi\alpha n}{N}\} = \frac{1}{N}\sum_{n=0}^{N-1}(\cos\frac{2\pi\alpha n}{N})\omega^{nk}$$

or

$$\begin{aligned}
g_k &= \frac{1}{N}\sum_{n=0}^{N-1}\frac{1}{2}(\omega^{\alpha n} + \omega^{-\alpha n})\,\omega^{nk}, \quad \omega \equiv e^{\frac{i2\pi}{N}} \\
&= \frac{1}{2N}(N\delta_{-\alpha,k} + N\delta_{\alpha,k}) \\
&= \frac{1}{2}\delta_{-\alpha,k} + \frac{1}{2}\delta_{\alpha,k} \qquad\qquad (4.16)
\end{aligned}$$

after using the orthogonality property for the two sums.

To interpret the result, recall the periodicity of the discrete Fourier transform g_k. We found earlier that $g_k = g_{k+N}$, from which it also follows that $g_{-k} = g_{N-k}$, when we look at the periodic extension of the sequence $\{g_k\}_{k=0}^{N-1}$ to the left of its basic period $0 < k \le N - 1$. In the above example, if α is an integer satisfying $0 < \alpha < N$, then $g_\alpha = \frac{1}{2}$ and $g_{-\alpha} = g_{N-\alpha} = \frac{1}{2}$. If $\alpha = 0$, we recover the result of Example 4.2.

Before getting too involved with calculation and generating transform pairs, it might be well to try to understand what these transforms actually mean. In the previous three example, we were given a sequence $\{G_n\}$ and then computed its DFT $\{g_k\}$ using (4 11) This process is called the *forward transform* or *analysis* of the sequence $\{G_n\}$. In order to gain more familiarity with these operations, let us now take the transform $\{g_k\}$ found in the previous three examples and try to reconstruct the original sequence $\{G_n\}$ using (4.12). This process is called the *inverse* or *backward transform* or *synthesis*. Here are the inverse transforms of our three examples.

Example 4.4 Computing the Inverse Discrete Fourier Transform

For $\{G_n\} = \{n\}$ we found that

$$
g_k = \begin{cases} \frac{1}{\omega^k - 1} & 1 \le k \le N - 1. \\[2ex] \frac{N-1}{2} & k = 0 \end{cases}
$$

Now applying (4.12) to recover $\{G_n\}$ we have

$$
G_n = \sum_{k=0}^{N-1} g_k \omega^{-nk} = \frac{N-1}{2} + \sum_{k=1}^{N-1} \frac{\omega^{-nk}}{\omega^k - 1}.
$$

Unfortunately, this last sum is rather difficult to evaluate for arbitrary values of n. But we can choose special values of n. For example with $n = 0$, we have

$$
\begin{aligned}
G_0 &= \frac{N-1}{2} + \sum_{k=0}^{N-1} \frac{1}{\omega^k - 1} \\[2ex]
&= \frac{N-1}{2} + \sum_{k=1}^{N-1} \frac{1}{(\cos k\theta - 1) + i \sin k\theta}
\end{aligned}
$$

where we have let $\theta = \frac{2\pi}{N}$ and used $\omega^k = e^{ik\theta} = \cos k\theta + i \sin k\theta$. From here, if we multiply and divide inside the sum by $(\cos k\theta - 1) - i \sin k\theta$, it follows that

$$
\begin{aligned}
G_0 &= \frac{N-1}{2} + \sum_{k=1}^{N-1} \frac{(\cos k\theta - 1) - i \sin k\theta}{2(1 - \cos k\theta)} \\[2ex]
&= \frac{N-1}{2} - \frac{1}{2} \sum_{k=0}^{N-1} 1 - \frac{i}{2} \sum_{k=1}^{N-1} \frac{\sin k\theta}{1 - \cos k\theta} \\[2ex]
&= \frac{N-1}{2} - \frac{N-1}{2} - \frac{i}{2} \sum_{k=1}^{N-1} \cot \frac{k\theta}{2} \\[2ex]
&= 0.
\end{aligned}
$$

where the last sum vanishes by the antisymmetry of $\cot \frac{k\theta}{2}$ about $\frac{k\theta}{2} = \frac{\pi}{2}$, where $k = \frac{N}{2}$. We see that we have recovered the $n = 0$

term of the original sequence. A similar calculation (see exercise 2) also gives $G_{\frac{N}{2}} = \frac{N}{2}$.

Example 4.5 The Inverse DFT of $\{\delta_{k,0}\}_{k=0}^{N-1}$

In the second example above we found the transform of the constant sequence $\{G_n\}_{n=0}^{N-1} = \{1\}_{k=0}^{N-1}$ to be $\{g_k\}_{k=0}^{N-1} = \{\delta_{k,0}\}_{k=0}^{N-1}$. Using (4.12) to recover $\{G_n\}$, we have

$$G_n = \sum_{k=0}^{N-1} g_k \omega^{-nk} = \sum_{k=0}^{N-1} \delta_{k,0}\omega^{-nk} = \omega^0 = 1$$

which holds for $0 \le n \le N - 1$.

Example 4.6 Another Inverse DFT

In the third example we considered the sequence $\{\cos\frac{2\pi\alpha n}{N}\}_{n=0}^{N-1}$ where $0 \le \alpha \le N - 1$ is an integer. The transform sequence was found to be $\{g_k\}_{k=0}^{N-1} = \{\frac{1}{2}(\delta_{\alpha,k} + \delta_{\alpha,N-k})\}_{k=0}^{N-1}$. Once again using (4.12) to compute the inverse transform, we have

$$
\begin{aligned}
G_n &= \sum_{k=0}^{N-1} g_k \omega^{-nk} = \frac{1}{2}\sum_{k=0}^{N-1}(\delta_{\alpha,k} + \delta_{\alpha,N-k})\omega^{-nk} \\
&= \frac{1}{2}(\omega^{-n\alpha} + \omega^{-n(N-\alpha)}) \\
&= \frac{1}{2}(\omega^{n\alpha} + \omega^{-n\alpha}) \\
&= \frac{1}{2}(e^{\frac{i2\pi n\alpha}{N}} + e^{\frac{-i2\pi n\alpha}{N}}) \\
&= \cos\frac{2\pi n\alpha}{N}
\end{aligned}
$$

after recalling that $\omega^{nN} = 1$.

Exercises 4.2

1. Compute the DFT of the following sequences

(a) $\{G_n\} = \{\delta_{n,p}\}$ where $0 \le p \le N$

(b) $\{G_n\} = \{n^{(2)}\}$

(c) $\{G_n\} = \{n^{(3)}\}$ in terms of $\mathcal{F}\{n^{(2)}\}$

(d) $\{G_n\} = \{n^{(p)}\}$ in terms of $\mathcal{F}\{n^{(p-1)}\}$

2. In example 4.4 of this section, it was shown that

$$g_k = \mathcal{F}\{n\} = \begin{cases} \frac{1}{\omega^k - 1} & 1 \le k \le N - 1 \\ \frac{N-1}{2} & k = 0. \end{cases}$$

Show that $G_{\frac{N}{2}} = \frac{N}{2}$ using the inverse transform.

3. Compute the DFT of

(a) the step function

$$G_n = \begin{cases} 0 & \text{for } 0 \le n < \frac{N}{2} \\ 1 & \text{for } \frac{N}{2} \le n \le N. \end{cases}$$

(b) the triangular pulse

$$G_n = \begin{cases} n & \text{for } 0 \le n \le \frac{N}{2} \\ N - n & \text{for } \frac{N}{2} \le n \le N. \end{cases}$$

4. (a) For our basic definition of the DFT and its inverse in (4.11)-(4.12) or (4.5)-(4.6), show that the following definition is equivalent to (4.11)-(4.12) aside from the multiplicative constant $\frac{1}{N}$ being used here with the sum of the inverse DFT in (E.2) instead of the sum of our DFT in (4.11).

$$g_k = \sum_{n=0}^{N-1} G_n \omega^{nk}, \qquad 0 \le k \le N - 1 \qquad (E.1)$$

$$G_n = \frac{1}{N} \sum_{k=0}^{N-1} g_k \omega^{-nk}, \qquad 0 \le n \le N - 1 \qquad (E.2)$$

(b) For the case of the positive *odd* integer $N = 2M + 1$ of (4.7)-(4.8), show that these sums with their limits in terms of the odd positive integer N become

$$g_k = \frac{1}{N} \sum_{n=-\frac{N-1}{2}}^{\frac{N-1}{2}} G_n \omega^{nk}, \quad -\frac{N-1}{2} \leq k \leq \frac{N-1}{2} \quad \omega \equiv e^{\frac{2\pi i}{N}}$$

$$G_n = \sum_{k=-\frac{N-1}{2}}^{\frac{N-1}{2}} g_k \omega^{-nk}, \quad -\frac{N-1}{2} \leq n \leq \frac{N-1}{2}$$

(c) For the case of N an *even* integer, show that (4.11)-(4.12) can be written as

$$g_k = \frac{1}{N} \sum_{n=-\frac{N}{2}+1}^{\frac{N}{2}} G_n \omega^{nk}, \quad -\frac{N}{2}+1 \leq k \leq \frac{N}{2}, \quad \omega \equiv e^{\frac{2\pi i}{N}}$$

$$G_n = \sum_{k=-\frac{N}{2}+1}^{\frac{N}{2}} g_k \omega^{-nk}, \quad -\frac{N}{2}+1 \leq n \leq \frac{N}{2}$$

5. The following variety of equivalent forms of the DFT and its inverse pairs are used with the main routines of computing the fast Fourier transform. Establish this equivalence by showing for each pair that the DFT and its inverse are discrete Fourier transforms of each other, as was done for the pair (4.5)-(4.6) with the help of the discrete orthogonality property (4.9).

(a) The DFT pair in part (a) of problem 4.

(b) The DFT pair in part (b) of problem 4.

(c) The DFT pair in part (c) of problem 4.

(d)

$$g_k = \frac{1}{\sqrt{N}} \sum_{n=0}^{N-1} G_n \omega^{-nk}, \quad 0 \leq k \leq N-1$$

$$G_n = \frac{1}{\sqrt{N}} \sum_{k=0}^{N-1} g_k \omega^{nk}, \quad 0 \leq n \leq N-1$$

(e)

$$g_k = \frac{1}{N} \sum_{n=1}^{N} G_n e^{-\frac{i2\pi(n-1)(k-1)}{N}}, \qquad 1 \le k \le N$$

$$G_n = \sum_{k=1}^{N} g_k e^{\frac{i2\pi(n-1)(k-1)}{N}}, \qquad 1 \le n \le N$$

(f)

$$g_k = \frac{1}{\sqrt{N}} \sum_{n=1}^{N} G_n e^{\frac{i2\pi(n-1)(k-1)}{N}}, \qquad 1 \le k \le N$$

$$G_n = \frac{1}{\sqrt{N}} = \sum_{k=1}^{N} g_k e^{\frac{i2\pi(n-1)(k-1)}{N}}, \qquad 1 \le n \le N$$

(g)

$$g_k = \sum_{n=0}^{N-1} G_{n+1} \omega^{-nk}, \qquad 0 \le k \le N-1,$$

$$G_{n+1} = \frac{1}{N} \sum_{k=0}^{N-1} g_k \omega^{nk}, \qquad 0 \le n \le N-1$$

where we note here that the input G_n is indexed $G_1, G_2, \ldots, G_N$ which is different from that of the output $g_0, g_1, \ldots, g_{N-1}$

(h)

$$g_{k+1} = \frac{1}{N} \sum_{n=0}^{N-1} G_{n+1} \omega^{nk}, \qquad 0 \le k \le N-1$$

$$G_{n+1} = \sum_{k=0}^{N-1} g_{k+1} \omega^{-nk}, \qquad 0 \le k \le N-1$$

and we note that both input and output sequences start with the first terms g_1 and G_1 instead of the usual zeroth terms g_0 and G_0.

6. Show that the first term g_0 in the output of the discrete Fourier transform in (4.5) is the average of the input sequence $\{G_n\}_{n=0}^{N-1}$, which is parallel to what we also have from the zeroth Fourier cosine coefficient a_0 of the Fourier cosine series in (2.84).

Hint: From (2.84) we have

$$a_0 = \frac{1}{\pi}g_c(0) = \frac{1}{\pi}\int_0^\pi G(x)dx$$

which is the average value of $G(x)$ over the interval of definition $(0,\pi)$.

4.3 Sine – Cosine (DST & DCT) Form of the DFT

Some additional understanding of the DFT may be gained by presenting it in another, entirely equivalent form. Consider the synthesis step of the transform (4.12)

$$G_n = \sum_{n=0}^{N-1} g_k \omega^{-nk}, \ \ 0 \le n \le N-1 \qquad (4.12)$$

Matters will be simplified somewhat if N is assumed to be *even*. Therefore, let $N = 2M$. Rewriting (4.12) with $N = 2M$ gives

$$
\begin{aligned}
G_n &= \sum_{k=0}^{2M-1} g_k e^{-\frac{i\pi nk}{M}} \\
&= g_0 + \sum_{k=1}^{M-1} g_k e^{-\frac{i\pi nk}{M}} + g_M(-1)^n + \sum_{k=M+1}^{2M-1} g_k e^{-\frac{i\pi nk}{M}}.
\end{aligned}
$$

In the last sum, define a new index by letting $l = 2M - k$ and recall the periodicity property $e^{-\frac{i\pi n(2M-l)}{M}} = e^{-i2\pi} \cdot e^{\frac{i\pi nl}{M}} = e^{\frac{i\pi nl}{M}}$

$$
\begin{aligned}
G_n &= g_0 + \sum_{k=1}^{M-1} g_k e^{-\frac{i\pi nk}{M}} + g_M(-1)^n + \sum_{l=1}^{M-1} g_{2M-l} e^{\frac{i\pi nl}{M}} \\
&= g_0 + \sum_{k=1}^{M-1} g_k e^{-\frac{i\pi nk}{M}} + g_M(-1)^n + \sum_{k=1}^{M-1} g_{2M-k} e^{\frac{i\pi nk}{M}} \\
&= g_0 + \sum_{k=1}^{M-1} \left[(g_k + g_{2M-k}) \cos\frac{\pi nk}{M} \right.
\end{aligned}
$$

$$+ i(g_{2M-k} - g_k) \sin \frac{\pi nk}{M} \Bigg] + g_M(-1)^n$$

$$\equiv \frac{a_0}{2} + \sum_{k=1}^{M-1} \left[a_k \cos \frac{\pi nk}{M} + b_k \sin \frac{\pi nk}{M} \right]$$

$$+ \frac{a_M(-1)^n}{2}, \quad 0 \le k \le N - 1. \qquad (4.17)$$

In the final line, we have defined the *cosine coefficients*, a_k, and the *sine coefficients*, b_k using expression (4.11) for g_k to find that

$$a_k = g_k + g_{2M-k} = \frac{2}{N} \sum_{n=0}^{N-1} G_n \cos \frac{2\pi nk}{N}, \quad 0 \le k \le \frac{N}{2} = M \qquad (4.18)$$

$$b_k = i(g_{2M-k} - g_k) = \frac{2}{N} \sum_{n=0}^{N-1} G_n \sin \frac{2\pi nk}{N}, \quad 0 \le k \le \frac{N}{2} - 1 = M - 1$$
$$(4.19)$$

The representation (4.17) is referred to as the *sine–cosine form of the DFT*. Given the complex coefficients $\{g_k\}$ of a sequence $\{G_n\}$, it is always possible to find the sine and cosine coefficients $\{a_k\}$ and $\{b_k\}$ through the relations

$$a_0 = 2g_0, \quad a_M = 2g_M$$

$$a_k = g_k + g_{2M-k}, \quad b_k = i(g_{2M-k} - g_k), \quad 1 \le k \le M - 1 \qquad (4.20)$$

Notice for a sequence $\{G_n\}_{n=0}^{N-1}$, there are N complex coefficients or alternatively a total of N sine and cosine coefficients.

In the case that the sequence $\{G_n\}$ is *real*, then from (4.18) we see that the $\{a_k\}$ and $\{b_k\}$ are also real. When $\{G_n\}$ is real we have the further symmetry that $g_{-k} = g_{2M-k} = \overline{g_k}$ which allows the cosine and sine coefficients to be written

$$a_k = g_k + g_{2M-k} = g_k + \overline{g_k} = 2Re\{g_k\} \qquad (4.21)$$

$$b_k = i(g_{2M-k} - g_k) = i(\overline{g_k} - g_k) = 2Im\{g_k\} \qquad (4.22)$$

We may also deduce some symmetries of the cosine and sine coefficients in extended ranges of the indices. For example, when $\{G_n\}$ is real,

$$a_{-k} = 2Re\{g_{-k}\} = 2Re\{\overline{g_k}\} = 2Re\{g_k\} = a_k \qquad (4.23)$$

Therefore since $a_k = a_{k+N}$, we also have

$$a_{N-k} = a_k, \quad \text{for} \quad 0 \le k \le \frac{N}{2}. \qquad (4.24)$$

In a similar way for the sine coefficients,

$$b_{-k} = 2Im\{g_{-k}\} = 2Im\{\overline{g_k}\} = -2Im\{g_k\} = -b_k, \qquad (4.25)$$

also

$$b_{N-k} = b_{-k} = -b_k, \quad \text{for} \quad 0 < k < \frac{N}{2}. \qquad (4.26)$$

from which it follows that $b_0 = b_{\frac{N}{2}} = 0$.

We close the section with a few words of interpretation. Given a sequence $\{G_n\}$, we have seen that it may be expressed as a linear combination of either complex exponentials (4.12) or sines and cosines (4.17). In either case, the coefficients in the linear combination may be found by (4.11) and (4.21), (4.22) respectively. These representations for $\{G_n\}$ indicate how the sequence $\{G_n\}$ may be constructed (or synthesized) from periodic functions of various frequencies. The coefficients g_k or (a_k and b_k) multiply the kth harmonic $e^{\frac{2i\pi nk}{N}}$ (or $\sin \frac{2\pi nk}{N}$ and $\cos \frac{2\pi nk}{N}$) which has k complete periods on the interval $0 \le n \le N$ with a period of $\frac{N}{k}$. The coefficient g_k or (a_k and b_k) indicates how much the kth harmonic is weighted in the complete representation (see Figs. 4.3, 4.4). Any sequence of N points can be represented by using roughly the first $\frac{N}{2}$ harmonics. In our examples above, the sequence $\{G_n\} = \{n\}$ requires all of the harmonics on the interval $[0, N]$ as shown in Fig. 4.3, while the sequences $\{1\}$ and $\{\cos \frac{2\pi\alpha n}{N}\}$ are themselves exact harmonics and require only $k = 0$ and $k = \alpha$ harmonics respectively for their representation, as shown in Fig. 4.5.

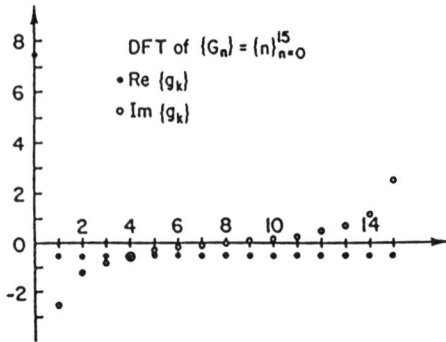

Figure 4.3: DFT of $\{G_n\} = \{n\}_{n=0}^{15}$, $Re\,\{g_k\}$, $Im\,\{g_k\}$

(Most of the figures here are from the author's book, "Integral and Discrete Transforms with Applications and Error Analysis", courtesy of Marcel Dekker, Inc.)

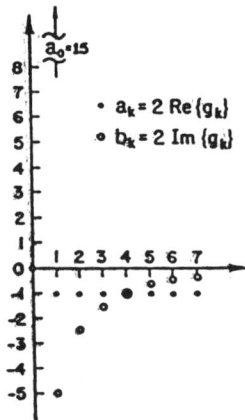

Figure 4.4: Sine and cosine coefficients $\{a_k\}$ and $\{b_k\}$ for $\{G_n\} = \{n\}_{n=0}^{15}$

Often the analysis or decomposition of a sequence $\{G_n\}$ into its harmonics is presented by giving its *spectrum*. This is a graph of the harmonic number k with the absolute value of the kth coefficient $|g_k| = \frac{1}{2}\sqrt{a_k^2 + b_k^2}$. The spectrums of $\{G_n\} = \{1\}$ and $\{G_n\} = \{\cos\frac{2\pi\alpha n}{N}\}$ are given in Fig. 4.5, while those and $\{G_n\} = \{n\}$ of $\{G_n\} = \{\sin\frac{2\pi\alpha n}{N}\}$, $\alpha = 4$ are given in Fig. 4.6.

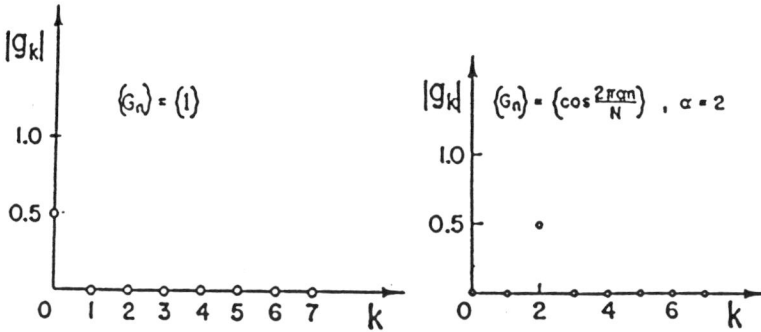

Figure 4.5: Spectrum $|g_k| = \frac{1}{2}(a_k^2 + b_k^2)^{\frac{1}{2}}$ for two 16 point sequences $\{G_n\} = \{1\}$ and $\{G_n\} = \{\cos\frac{2\pi\alpha n}{N}\}$, $\alpha = 2$

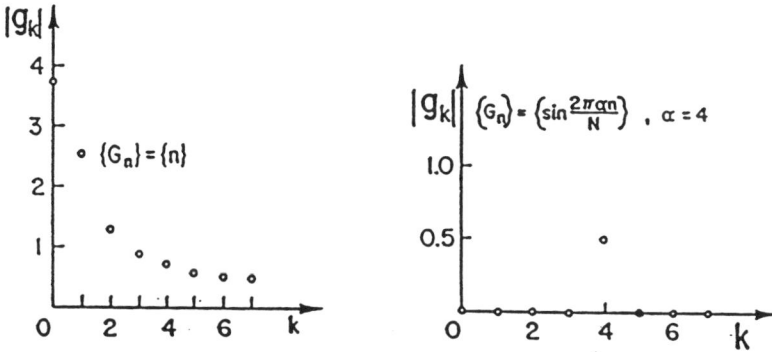

Figure 4.6: Spectrum $|g_k| = \frac{1}{2}(a_k^2 + b_k^2)^{\frac{1}{2}}$ for two 16 point sequences and $\{G_n\} = \{n\}$ $\{G_n\} = \{\sin\frac{2\pi\alpha n}{N}\}$, $\alpha = 4$

Exercises 4.3

1. Find the sine and cosine coefficients of the N-point sequences.

   (a) $\{G_n\} = \{n\}$

   (b) -$\{G_n\} = \{1\}$

   (c) $\{G_n\} = \{n^{(2)}\}$

   in two ways:

   > i direct summation using (4.18)

   > ii from the complex coefficients using (4.20) or (4.21), (4.22).

2. It is possible to derive the sine-cosine form of the DFT from scratch without using the complex form. In order to do this, it is necessary to have the orthogonality properties of the sequences $\{\cos \frac{2\pi\alpha k}{N}\}$ and $\{\sin \frac{2\pi\beta k}{N}\}$. Prove the following discrete orthogonality relations.

   i

   $$\sum_{k=0}^{N-1} \cos \frac{2\pi\alpha k}{N} \cos \frac{2\pi\beta k}{N} = \begin{cases} 0 & \alpha \neq \beta \\ \frac{N}{2} & \alpha = \beta \neq 0,\ N \\ N & \alpha = \beta = 0,\ N, \end{cases}$$

   Hint: Use the identity $\cos A \cos B = \frac{1}{2}[\cos(A+B)+\cos(A-B)]$

   ii

   $$\sum_{k=0}^{N-1} \sin \frac{2\pi\alpha k}{N} \sin \frac{2\pi\beta k}{N} = \begin{cases} 0 & \alpha \neq \beta \\ \frac{N}{2} & \alpha = \beta \neq 0,\ N \\ N & \alpha = \beta = 0,\ N, \end{cases}$$

   iii

   $$\sum_{k=0}^{N-1} \cos \frac{2\pi\alpha k}{N} \sin \frac{2\pi\beta k}{N} = 0.$$

4.4   Solution of a Difference Equation by the DFT–Traffic Network

Before moving on to study further properties of the DFT and to compute additional DFT pairs, it might be helpful to illustrate with more details the traffic network problem, where we shall model it as a difference equation with boundary conditions. The detailed mathematical modeling of a number of interesting problems as *boundary value problems* associated with difference equations, are found in Chapter 7. In Chapter 6, there is another practical problem in combinatories, which is modeled as an *initial value problem* associated with difference equations.

Example 4.7   Traffic Flow in Equilibrium

The following example may be somewhat idealized, but it does give an indication of the type of problem that can give rise to a difference equation. Consider a traffic network which may be represented as a closed ring as shown in Fig. 4.7.

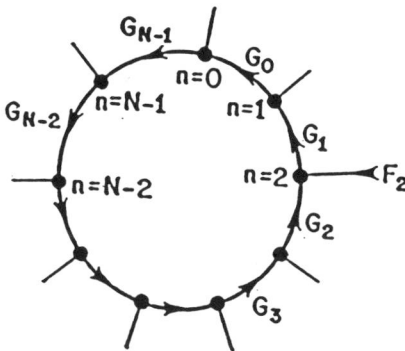

Figure 4.7: A traffic network with N intersections

The nodes (dots) represent intersections. The arcs (lines) represent streets. There are N nodes numbered from $n = 0$ to $n = N - 1$ where we agree that $n = N$ and $n = 0$ denote the same intersection. We will associate a variable G_n with the arc connecting node n and node $(n+1)$. We will also choose a direction or positive sense to each arc and mark it by an arrow. This choice is arbitrary and the figure indicates one particularly simple choice. The variable G_n represents the number of vehicles passing along the nth arc per unit of time and is positive when the traffic flow is in the direction of the arrow on that arc. We will also assume that each intersection may be either an entrance or an exit for traffic to or from the loop. We will denote the traffic flow into or out of intersection n by F_n where a positive F_n is taken to mean flow into the intersection (entrance) and a negative F_n is taken to mean flow out of the intersection (exit). The goal is to model this traffic network and to find the flow rates, G_n, along the streets when the entire traffic pattern reaches an equilibrium or steady state.

In order to model this system, we must relate the flow along one arc to the flow along neighboring arcs and through neighboring intersections. A few observations make this possible. The key words in the statement of the problem are "equilibrium" and "steady state". A steady state traffic pattern depends upon two conditions being met.

    i The total traffic flow into the loop from entrances must balance the total traffic flow out of the loop along exits. A more concise way of stating this is

$$\sum_{n=0}^{N-1} F_n = 0.$$

    ii At each intersection, traffic flow in must equal traffic flow out.

It should be clear that if either of these conditions is not satisfied, then there is the possibility of traffic accumulation somewhere in the loop or perhaps vanishing from the loop. These two conditions amount to a *conservation of traffic law*. The first condition restricts the data in the problem. If the data does not meet the condition, then there is no point in looking for a steady state solution. The second condition allows us to find the equations that govern the flows, G_n.

Consider intersection n where $0 \leq n \leq N - 1$. Using the sign and arrow conventions we have established, the flow into intersection n

is $F_n + G_n$. The flow out of intersection n is G_{n-1}. Condition (2) above can now be expressed as

$$G_n - G_{n-1} + F_n = 0$$

which holds for intersections $0 \leq n \leq N - 1$. To complete the statement of the problem, we include the *periodic boundary condition* which states that $G_0 = G_N$. The complete (boundary value) problem now appears as

$$G_n - G_{n-1} = -F_n, \quad 0 \leq n \leq N - 1 \qquad (4.27)$$

$$G_0 = G_N \qquad (4.28)$$

which consists of a first order, constant coefficient, non-homogeneous difference equation (4.27) and a (typical) periodic boundary condition (4.28).

This problem (4.27)–(4.28) was already considered in Example 2.11, without the above derivation, for the purpose of illustrating the direct use of the DFT in transforming the difference equation (4.27) to an algebraic equation, and at the same time involving the periodic boundary condition (4.28), which is what we termed "the operational sum calculus method." As we did in Example 2.13 of Chapter 2 for the DST, we will also illustrate the (indirect) method of using the inverse discrete Fourier transform G_n of (4.6) as a *series solution* to be plugged in the difference equation, then the coefficients of the resulting series, on both sides of the difference equation (4.27) are equated to determine g_k, the discrete Fourier transform of the unknown G_n, hence we obtain G_n via (4.6). The operational sum method of the DFT that we used in Example 2.11 will be repeated here for completeness and comparison. It starts by taking the discrete Fourier transform of the difference equation (4.27), that reduces it to an algebraic equation in the discrete Fourier transform g_k, which is, of course, a much easier equation to solve. As was mentioned earlier, we have already used this direct DFT transform method for solving the boundary value problem (4.27) in Example 2.11 of Section 2.4, but only to the extent of illustrating the algebraization process of this method. We, then, left illustrating the complete details of arriving at the final solution of the present traffic problem, with a specific non-homogeneous term $F_n = V(-1)^n = Ve^{i\pi n}$ in (4.27), for this section and its following example.

Example 4.8   A Particular Pattern of Traffic Flow in Equilibrium

Let us now consider a specific traffic network in which the streets which connect to the loop are alternately entrances and exits, each carrying V vehicles per unit time. The sequence $\{F_n\}$ which describes this pattern is $F_n = V(-1)^n = Ve^{i\pi n}$. Note that in order to satisfy condition (i) above we must also assume that N is *even*.

(a) The DFT Method as a Series Solution

The job now is to find a sequence of flow rates $\{G_n\}$ which satisfies (4.27), (4.28) with this particular choice of $\{F_n\}$. The periodic boundary condition suggests that we express the solution $\{G_n\}$ in terms of the DFT since, as we have seen, a sequence so defined has the necessary periodicity. Therefore, we will look for a solution sequence of the form

$$G_n = \sum_{k=0}^{N-1} g_k \omega^{-nk}, \ \ 0 \le n \le N - 1 \qquad (4.12)$$

where the g_k's must now be determined. A necessary step at this very stage, of the DFT series solution method, is to express the right hand side sequence $\{F_n\}$ of (4.27) in terms of a DFT. We may interject here that this step is not needed at this stage for the direct DFT transform method (or the "operational sum calculus method"), and may not be needed in the last step of this method if the discrete Fourier transform g_k can be inverted analytically to find the solution of the difference equation G_n. If this analytical inversion is not possible, then the worst that can happen is to have $f_k = \mathcal{F}\{F_n\}$ as part of the expression for g_k. Then a final DFT (numerical) inverse is computed to find the final solution G_n with the help, of course, of the FFT (Fast Fourier Transform) algorithm. Another advantage to this second operational (or transform) method is that for F_n, even if it is given as a numerical data, its transform f_k can wait until the end to be included in the expression for g_k. This is in the hope that G_n, the inverse of the g_k, can be expressed in terms of some operation on F_n, hence avoiding the computation of f_k entirely. This often happens in the transforms methods after we learn most of their efficient tools, which we shall continue their study after this example.

To express the non-homogeneous term F_n on the right hand side of (4.27) as a DFT series, we write

$$F_n = \sum_{k=0}^{N-1} f_k \omega^{-nk} = \sum_{k=0}^{N-1} f_k e^{-\frac{i2\pi kn}{N}}$$

and recalling that $F_n = V e^{i\pi n}$, it is evident that

$$f_k = \begin{cases} 0 & \text{if } k \neq \frac{N}{2} \\ \\ V & \text{if } k = \frac{N}{2}. \end{cases} \tag{4.29}$$

In this case, F_n is a multiple of the $\frac{N}{2}$ harmonic, so all coefficients in the DFT are zero except $f_{\frac{N}{2}}$. We may now substitute for G_n and F_n in (4.27), (4.28) giving

$$\sum_{k=0}^{N-1} g_k \omega^{-nk} - \sum_{k=0}^{N-1} g_k \omega^{-(n-1)k} = -\sum_{k=0}^{N-1} f_k \omega^{-nk} \tag{4.30}$$

Combining sums yields

$$\sum_{k=0}^{N-1} [g_k(1 - \omega^k) + f_k] \omega^{-nk} = 0, \quad \text{for } 0 \leq n \leq N - 1.$$

This is an essential step, to the present method of the DFT as a series solution, which amounts to equating the coefficients of the series on both sides of equation (4.30), a procedure that is used in all series solution methods including that often used for solving differential equations, which we have already illustrated in Example 1.5 of Section 1.1.

This sum can vanish for all $0 \leq n \leq N - 1$ only if each term of the sum vanishes independently. That is, we must have

$$g_k(1 - \omega^k) + f_k = 0, \quad 0 \leq k \leq N - 1,$$

$$g_k = \frac{-f_k}{1 - \omega^k}, \quad \text{for } 0 \le k \le N - 1. \tag{4.31}$$

Recalling the expression for f_k from (4.29), we have

$$g_k = \begin{cases} 0 & \text{if } k \ne \frac{N}{2} \\ -\frac{V}{1-\omega^k} & \text{if } k = \frac{N}{2}. \end{cases} \tag{4.32}$$

Notice that the coefficients $\{g_k\}$ of the solution were found simply by solving an *algebraic equation*, a clear advantage for the present use of the DFT in solving difference equations. To recover the solution $\{G_n\}$, it remains only to do the synthesis or inverse DFT. First note that

$$g_{\frac{N}{2}} = -\frac{V}{1 - \omega^{\frac{N}{2}}} = -\frac{V}{1 - e^{i\pi}} = -\frac{V}{2}.$$

Therefore,

$$\begin{aligned} G_n &= \sum_{k=0}^{N-1} g_k \omega^{-nk} \\ &= g_{\frac{N}{2}} \omega^{-\frac{nN}{2}} = -\frac{V}{2} e^{-i\pi n} = -\frac{V}{2}(-1)^n. \end{aligned} \tag{4.33}$$

since the only non-vanishing g_k is $g_{\frac{N}{2}}$ as seen in (4.32). This says that the flow along each street has a volume of $\frac{V}{2}$ and is along the arrow (counter clockwise) on odd numbered arcs and opposes the arrow on even numbered arcs. The solution for a traffic circle with $N = 4$ intersections is shown in Fig. 4.8.

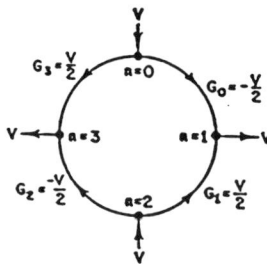

Figure 4.8: A traffic circle with $N = 4$ intersections

This example illustrates the simplest pattern of inflow and out-flow to the loop. More interesting and perhaps more realistic traffic patterns can be handled by the same method. These problems will be investigated in the exercises (see problem 1).

(b) The Direct Operational Sum Calculus Method of the DFT.

Even though we have started illustrating this method for this same problem of traffic flow in Example 2.11 for the sake of showing the main idea of this method in algebraizing the difference equation (4.27) (and at the same time involving the boundary condition (4.28)), it can only help to repeat these few steps and finish the problem to its final solution. In this way we will have a clear idea about the difference between using the DFT as a series in the first series solution method of part (a), and the present operational method, where the DFT as a transform is used to algebraize the "difference operator" E^{-1} in the term $G_{n-1} = E^{-1}G_n$ of the difference equation (4.27).

If we take the discrete Fourier transform of both sides of the difference equation (4.27), using the operational pair (2.113),

$$\mathcal{F}\{G_{n-1}\} = \frac{\omega^k}{N}[G_N - G_0] + \omega^k g_k, \qquad (2.113)$$

we have:

$$\mathcal{F}\{G_n - G_{n-1}\} = -\mathcal{F}\{F_n\} = g_k - \frac{\omega^k}{N}[G_N - G_0] - \omega^k g_k = -f_k, \quad (4.34)$$

where $f_k = \mathcal{F}\{F_n\}$, which we shall leave it the general f_k until the last step of this method. Now, since we are given the periodic boundary condition (4.28) of G_n being periodic with period N, i.e., $G_N = G_0$, it causes the boundary condition term $G_N - G_0$ in (4.34) to vanish, and we have,

$$g_k - \omega^k g_k = (1 - \omega^k)g_k = -f_k,$$
$$g_k = \frac{-f_k}{1 - \omega^k}, \qquad (4.35)$$

which is the same result (4.31) we obtained above via the first method of the DFT as a series solution method. However, as we explained

before starting this example, the present operational method of using the DFT transform did not require writing the non-homogeneous
term F_n of the difference equation (4.27) in a DFT series form as
done in (4.30). This, as we mentioned earlier, is an advantage of the
present operational or transform method. As was mentioned at the
end of Example 2.11, we may also add that another advantage of this
direct DFT method, especially via the use of its more general operational pair (4.33), is that it will take care of a jump $J = G_N - G_0 \neq 0$
between the values of the end points $n = N, n = 0$, unlike that of
the typical periodic condition (4.28), which involves no jump, i.e.,
$J = G_N - G_0 = 0$ in (4.28). The general case of periodicity with
the above jump cannot be addressed directly with the series solution method of part (a). In the present example we do have periodic boundary condition without a jump, hence this direct transform
method will proceed in the same way of (4.32) for finding the inverse
DFT of g_k in (4.35) with the particular f_k as found in (4.29).

We should stress here that the success of this direct DFT transform method as an operational method depends a great deal on developing a good number of DFT operational transform pairs as well as
sequence pairs, which is the case for all operational calculus methods
such as the Laplace and Fourier integral transforms methods, where
they have extensive tables of their pairs. Until now we developed
only very few DFT pairs that are essential for our first example of
illustrating the use of the DFT in a series solution method and in
the direct transform, or operational sum calculus method. The DFT
operational pairs included the ones in (2.107) and (2.108) to illustrate how the DFT algebraizes $EG_n = G_{n+1}$ and $E^2 G_n = G_{n+2}$,
respectively. The DFT sequence pairs were also limited to very few
including the DFT transforms of the sequences $\{G_n\}_{n=0}^{N-1} = \{1\}_{n=0}^{N-1}$,
$\{n\}_{n=0}^{N-1}$ and $\{\cos \frac{2\pi a n}{N}\}_{n=1}^{N-1}$ that we developed in Examples 4.1, 4.2,
and 4.3, respectively. This is besides the sequence pair of (4.29) in
this example, few in Exercises 4.2, and few more for the discrete
sine (DST) and cosine (DCT) Fourier transforms in Section 4.3. Of
course, in order to make a viable method of this direct DFT transform, we must have more extensive tables of the DFT pairs, especially
in regard to the operational pairs that facilitate the mechanics of the
method. The DFT sequence pairs are not as necessary as those

for the Laplace transform, thanks to the FFT algorithm that can compute very fast, a comment that became a reality only after this algorithm was developed in 1965. Nevertheless, more DFT sequences pairs are welcome for facilitating all kinds of analytical computations for the direct DFT transform method. In passing we may mention that in the case of the z–transform, as a direct transform method for initial value problems that we shall discuss in Chapter 6, we must have extensive tables of this transform pairs, since, unfortunately, we do not have a fast algorithm like the FFT of the discrete Fourier transform. In the next section we will develop more DFT pairs, where the derivation of the operational pairs, as expected, will depend on the remaining basic properties of the DFT, most of which parallel those of Fourier and Laplace transforms.

As we mentioned very briefly at the end of Section 3.5 for partial difference equations with boundary conditions, and presented in some detail at the end of Section 2.4, the DFT can be extended to transform sequences of *two* and more variables. With such extensions, both of the above series solution method as well as the direct operational sum method can be extended to cover partial difference equations in two or more dimensions. Hence the need for the tables of the operational as well as the sequence pairs of the *two dimensional* discrete Fourier transforms.

Exercises 4.4

1. Assume N is a multiple of 3. Find the steady state traffic flow through the streets of a traffic loop when the intersections are alternately an entrance admitting V vehicles per unit time, and 2 exits allowing $\frac{V}{2}$ vehicles per unit time to leave. Notice that the sequence describing flow into and out of the loop is $\{\cos\frac{2\pi n}{3}\}$. Show that a steady state traffic pattern exists in this case.

2. What assumptions are required to insure that an N intersection loop has a steady traffic flow when the entrance/exit traffic at intersection n is given by $\{\sin\frac{\pi n}{3}\}$? Find the flow of traffic through the streets of this network.

3. Consider the difference equation with the periodic boundary condition

$$G_{n+1} + G_n = n \,, \qquad (E.1)$$

$$G_N = G_0 \,. \qquad (E.2)$$

(a) Find the Discrete Fourier Transform of (E.1) as an algebraic equation in $g_k = \mathcal{F}\{G_n\}$.

Hint: Consider the DFT pairs (2.107) and (4.14).

(b) Find g_k of part (a), then its inverse DFT as the solution G_n of (E.1)–(E.2). Note that you may have to do the inverse DFT of the resulting g_k numerically.

(c) Attempt to use the UC method of Section 3.3 to solve (E.1)–(E.2) and compare with the answer in part (b).

(d) Use the Discrete Fourier Transform in a series solution method, as illustrated in part (a) of Example 4.8, to solve (E.1)–(E.2), then compare the answer with those of parts (a)–(b) of the DFT.

(e) Use the operational sum method.

Hint: Use (2.107).

4. Use the DFT in a series solution method as we did in part (a) of Example 4.8 to solve the second order recurrence relation (E.1) in G_n with its two auxiliary conditions (E.2)–(E.3):

$$G_{n+1} - G_n - G_{n-1} = 0 \,, \qquad (E.1)$$

$$G_0 = 0 \,, \qquad (E.2)$$

$$G_1 = 1 \qquad (E.3)$$

to show that the solution is:

$$G_n = \frac{1}{\sqrt{5}} \left(\frac{1+\sqrt{5}}{2}\right)^n - \frac{1}{\sqrt{5}} \left(\frac{1-\sqrt{5}}{2}\right)^n \,, \quad n \geq 0 \,.$$

Hint: The resulting algebraic equation for using the DFT series solution is quadratic, where you expect two distinct zeros $\lambda_1 = \frac{1+\sqrt{5}}{2}$ and $\lambda_2 = \frac{1-\sqrt{5}}{2}$. So let $G_n = c_1 \lambda_1^k + c_2 \lambda_2^k$ and find the arbitrary constants c_1 and c_2 from using the two initial conditions (E.2), (E.3).

5. Consider the second order homogeneous difference equation with its (periodic) boundary condition that involves a jump $J = G_N - G_0 = 1$

$$G_{n+1} - G_n - G_{n-1} = 0, \qquad (E.1)$$

Find the DFT of (E.1), where the periodicity condition that involves a jump between the values at the end points is satisfied.

Hint: Use the operational sum method of the DFT via its operational pairs (2.107), (2.113).

4.5 Further Properties of the DFT for its Operational Sum Calculus Method

In the work that follows, there is a continual choice of notation. The DFT may be written in the form of expression (4.5), (4.6) in which the time and frequency grid points are indicated. Alternatively, we may work with expressions (4.11), (4.12) in which the time and frequency variables are replaced by integers $0 \leq k \leq N - 1$ and $0 \leq n \leq N - 1$. The choice is one of convenience and is not important mathematically (see problems 4 and 5 of Exercises 4.2 for the varied notations of the DFT). Often the nature of the problem at hand will suggest one notation as opposed to the other. Because expression (4.11), (4.12) are somewhat simpler, we will use them in the following section. To present them again, we have the following DFT pair which relates the sequences $\{G_n\}_{n=0}^{N-1}$ and $\{g_k\}_{k=0}^{N-1}$. We will also introduce the operator notation $\mathcal{F}$ for the DFT and $\mathcal{F}^{-1}$ for the inverse DFT.

$$g_k = \mathcal{F}\{G_n\} = \frac{1}{N} \sum_{n=0}^{N-1} G_n \omega^{nk}, \quad 0 \leq k \leq N - 1 \qquad (4.11)$$

$$G_n = \mathcal{F}^{-1}\{g_k\} = \sum_{k=0}^{N-1} g_k \omega^{-nk}, \quad 0 \leq n \leq N - 1 \qquad (4.12)$$

where $\omega = e^{\frac{i2\pi}{N}}$. We will now investigate some additional properties of this transform pair.

In contrast to what we did in Section 2.4 for the operational sum calculus of the DFT's, where the more general periodicity with jump discontinuities $G_N - G_0 \neq 0$ at the end points are considered as in (2.107), we shall limit the present very basic discussion to the case of no jump discontinuities at the end points.

Periodicity

We have already mentioned the periodicity of the sequence $\{G_n\}$ and $\{g_k\}$ which are related by the DFT pair. To summarize, we have for any integer r

$$
\begin{aligned}
G_{n+rN} &= \sum_{k=0}^{N-1} g_k \omega^{-(n+rN)k} = \sum_{k=0}^{N-1} g_k \omega^{-rNk} \omega^{-nk} \\
&= \sum_{k=0}^{N-1} g_k \omega^{-nk} = G_n.
\end{aligned}
\tag{4.33}
$$

This, and the corresponding property $g_{k+rN} = g_k$, results from the periodicity of the kernel ω^{nk}; that is,

$$
\begin{aligned}
\omega^{(n+rN)k} &= e^{\frac{i2\pi}{N}(n+rN)k} = e^{\frac{i2\pi nk}{N}} \cdot e^{i2\pi rk} \\
&= e^{\frac{i2\pi nk}{N}} = \omega^{nk}.
\end{aligned}
$$

Linearity

The DFT is a *linear operator* in that (i) the transform of the sum of two sequences is the sum of the transforms of the individual sequences and (ii), the transform of a constant multiple of a sequence is that constant times the transform of the sequence. Stated more compactly,

i

$$
\mathcal{F}\{G_n + H_n\} = \mathcal{F}\{G_n\} + \mathcal{F}\{H_n\} = \{g_k\} + \{h_k\}.
\tag{4.36}
$$

ii

$$
\mathcal{F}\{\alpha G_n\} = \alpha \mathcal{F}\{G_n\} = \alpha\{g_k\}, \quad \text{for} \ \ \alpha \in R.
\tag{4.37}
$$

These properties follow from the corresponding linearity properties of the finite sums.

Symmetry

The symmetry property expresses the result of applying the DFT to a sequence twice. Because of the symmetry of (4.11), (4.12), the effect of applying the DFT twice is to recover the original sequence within a minor modification. A short calculation using (4.11) gives

$$\mathcal{F}\{\mathcal{F}\{G_n\}\} \;=\; \mathcal{F}\{g_k\} = \frac{1}{N} \sum_{k=0}^{N-1} g_k \omega^{nk}$$

$$= \frac{1}{N} G_{-n} \tag{4.38}$$

A similar property (see exercises) holds for the inverse transform, namely, $\mathcal{F}^{-1}\{\mathcal{F}^{-1}\{g_k\}\} = N g_{-k}$.

Shift in n (frequency) and the Operational Sum Calculus Method

The following two shift properties facilitate the solution of difference equations. They are also helpful in finding new transform pairs. Consider the DFT operating on a shifted sequence $\{G_{n-l}\}$ where l is some integer.

$$\mathcal{F}\{G_{n-l}\} \;=\; \frac{1}{N} \sum_{n=0}^{N-1} G_{n-l} \omega^{nk}$$

$$= \frac{1}{N} \sum_{m=-l}^{N-1-l} G_m \omega^{(m+l)k}$$

$$= \frac{\omega^{lk}}{N} \sum_{m=-l}^{N-1-l} G_m \omega^{mk} \tag{4.39}$$

$$= \frac{\omega^{lk}}{N} \sum_{n=0}^{N-1} G_n \omega^{nk}$$

$$= \omega^{lk} g_k = e^{\frac{i2\pi lk}{N}} g_k$$

after letting $m = n - l$ in the second line and using the periodicity of G_n and ω^{nk} in the third line.

This property was used implicitly in solving the difference equation of Example 4.8(a) in the previous section where, for example, we found that $\mathcal{F}\{G_{n-1}\} = \omega^k g_k$, for the periodic G_n (without jump discontinuity at the end points). (See (2.107) also).

We may add here that this property (4.39) of the DFT represents the backbone of our attempt at using the DFT to transform a difference equation in G_n to an *algebraic* equation in the transformed sequence g_k. For example, in $\mathcal{F}\{G_{n-1}\} = \omega^{-k} g_k$, we see how the difference operation $EG_n = G_{n+1}$ is being transformed via the DFT to the algebraic multiplication $(\omega^{-k} g_k)$ of the transformed sequence g_k by ω^k. As we mentioned in the traffic flow problem of Example 4.8, we term the above direct method of transforming the difference equation (as well as satisfying the boundary conditions) to an algebraic equation, as the operational sum calculus method, which we introduced in Section 2.4.

Shift in k (time) and the Operational Sum Calculus Method

In a very similar way, it is left as an exercise to show that

$$\mathcal{F}\{e^{\frac{-i2\pi ln}{N}} G_n\} = g_{k-l}. \tag{4.40}$$

This pair and symmetry with that of (4.39) may also represent the action of the inverse DFT in algebraizing the difference g_{k-l} as we transform it back to the n-space. However, if we are transforming to the k-space, then (4.40) would represent a drawback for the DFT as it transforms the algebraic term $e^{-\frac{j2\pi kn}{N}} G_n$ with its *variable* coefficient $e^{-\frac{j2\pi kn}{N}}$ to the more complicated *"differenced"* term g_{k-l}.

Discrete Convolution Product

The *discrete convolution product* of two sequences $\{G_n\}$ and $\{H_n\}$ is given by

$$(G * H)_n = \frac{1}{N} \sum_{l=0}^{N-1} G_l H_{n-l}. \tag{4.41}$$

The corresponding convolution product in the time (k) domain of two sequences $\{g_k\}$ and $\{h_k\}$ is

$$(g * h)_k = \sum_{l=0}^{N-1} g_l h_{k-l}. \tag{4.42}$$

With these two convolution products defined in this manner, the following convolution theorems hold.

$$\mathcal{F}^{-1}\{g_k h_k\} = (G * H)_n \tag{4.43}$$

and

$$\mathcal{F}\{G_n H_n\} = (g * h)_k. \tag{4.44}$$

We will give the proof of the first of these results.

$$
\begin{aligned}
\mathcal{F}^{-1}\{g_k h_k\} &= \sum_{k=0}^{N-1} g_k h_k \omega^{-nk} \\
&= \sum_{k=0}^{N-1} \left\{ \frac{1}{N} \sum_{l=0}^{N-1} G_l \omega^{lk} \right\} h_k \omega^{-nk} \\
&= \frac{1}{N} \sum_{l=0}^{N-1} G_l \sum_{k=0}^{N-1} h_k \omega^{-k(n-l)} \\
&= \frac{1}{N} \sum_{l=0}^{N-1} G_l H_{n-l} \\
&= (G * H)_n,
\end{aligned}
$$

where we used the definition of g_k inside the sum of the second line, interchanged the two sums in the third line, then used the shift property (4.39) to have the fourth line.

The second convolution theorem is proved in a similar way. These results can be very useful in transforming and inverting sequences, especially in cases where a direct calculation would be difficult. For example, consider the problem of computing $\mathcal{F}^{-1}\{g_k^2\}$ where $g_k = \mathcal{F}\{G_n\}$. A direct calculation using the definition (4.12) would be quite involved. However, if we use the convolution theorem (4.41) we have

$$\mathcal{F}^{-1}\{g_k^2\} = (G * G)_n = \frac{1}{N} \sum_{l=0}^{N-1} l(n-l)$$

$$= \frac{n}{N} \sum_{l=0}^{N-1} l - \frac{1}{N} \sum_{l=0}^{N-1} l^2$$

$$= \frac{n(N-1)}{2} - \frac{(N-1)(2N-1)}{6}.$$

It is not difficult to show that the discrete convolution products are *commutative*, i.e.,

$$G * H = H * G \quad \text{and} \quad g * h = h * g.$$

Also, setting $n = 0$ in (4.41) produces the discrete version of *Parseval's equality*,

$$\sum_{k=0}^{N-1} g_k \overline{h_k} = \frac{1}{N} \sum_{l=0}^{N-1} G_l \overline{H_l} \tag{4.45}$$

where $-$ denotes complex conjugation, and the fact that $\mathcal{F}\{\overline{H_{-n}}\} = \{\overline{h_k}\}$ has been used.

Discrete Fourier Transform Pairs

We now take the opportunity to summarize in Table 4.1 the properties of the DFT which have been studied in the previous pages. In addition, several of the most commonly used transform pairs are included for future use. Those transform pairs which have not already been derived make excellent exercises. The notation $\{g_k\} = \mathcal{F}\{G_n\}$ and $\{h_k\} = \mathcal{F}\{H_n\}$ is used in the following table.

Compared to this brief Table 4.1, a more detailed table of such DFT pairs is found in reference 8 among the operational calculus references at the end of this book.

	Sequence	Transform	
		(a) Operational pairs	
1.	$\{G_n\}, \{H_n\}$	$\{g_k\}, \{h_k\}$	
2.	$\{G_n + H_n\}$	$\{g_k + h_k\}$	Linearity
3.	$\{\alpha G_n\}, \ \alpha \in \mathcal{R}$	$\{\alpha g_k\}$	Linearity
4.	$\{G_{n+rN}\} \ r \in Z$	$\{g_k\}$	Periodicity
5.	$\{g_n\}$	$\frac{1}{N}\{G_{-k}\}$	Symmetry
6.	$\{G_{n-l}\}$	$\{e^{\frac{i2\pi lk}{N}} g_k\}$	Shift in frequency
7.	$\{e^{\frac{-i2\pi ln}{N}} G_n\}$	$\{g_{k-l}\}$	Shift in time
8.	$\{\overline{G_n}\}$	$\{\overline{g_k}\}$	Conjugation
9.	$\{G_n H_n\}$	$\{g * h\}_k$	Convolution
10.	$\{G * H\}_n$	$\{g_k h_k\}$	Convolution
		(b) Sequence pairs	
11.	1	$\delta_{k,0}$	
12.	$e^{\frac{-i2\pi\alpha n}{N}}$	$\delta_{k,\alpha}$	
13.	$\cos\frac{2\pi\alpha n}{N}$	$\frac{1}{2}(\delta_{k,\alpha} + \delta_{k,N-\alpha})$	
14.	$\sin\frac{2\pi\alpha n}{N}$	$\frac{i}{2}(\delta_{k,\alpha} - \delta_{k,N-\alpha})$	
15.	$\delta_{n,0}$	$\frac{1}{N}$	
16.	$\delta_{n,\alpha}$	$\frac{1}{N}e^{\frac{i2\pi\alpha k}{N}}$	
17.	n	$\begin{cases} \frac{N-1}{2}, & k=0 \\ \frac{1}{\omega^k-1}, 1 \le k \le N-1 \end{cases}$	
18.	$n^{(2)}$	$\begin{cases} \frac{(N-1)(N-2)}{3}, & k=0 \\ \frac{N-1}{\omega^k-1} - \frac{2}{(\omega^k-1)^2}, & k=1,2,\ldots \end{cases}$	
19.	$n^{(p)}$	$\begin{cases} g_0, & k=0 \\ \frac{1}{\omega^k-1} \cdot \left[\frac{N^{(p)}}{N} - p\mathcal{F}\{n^{(p-1)}\}\right], \\ k=1,2,\ldots \end{cases}$	
20.	$\begin{cases} 0, & 0 \le n \le \frac{N}{2} \\ 1, & \frac{N}{2} \le n < N \end{cases}$	$\begin{cases} \frac{1}{2}, & k=0 \\ 0, & k \text{ even} \\ \frac{2}{N(1-\omega^k)}, & k \text{ odd} \end{cases}$	

Table 4.1: Operational and sequence pairs of discrete Fourier transform

Exercises 4.5

1. Prove the symmetry relation

$$\mathcal{F}^{-1}\{\mathcal{F}^{-1}\{g_k\}\} = Ng_{-k}.$$

2. Show that

$$\mathcal{F}\{\Delta G_n\} = (\omega^{-k} - 1)g_k$$

and that

$$\mathcal{F}\{\nabla\Delta G_n\} = \mathcal{F}\{G_{n+1} - 2G_n + G_{n-1}\} = 2\left(\cos\frac{2\pi k}{N} - 1\right)g_k.$$

3. In view of problem 2, it is possible to give a more direct alternate (but entirely equivalent) approach to using the DFT for solving difference equations. As we have indicated following Example 2.12, such an approach is more in the direction of operational calculus, or the familiar integral transform method, which we have termed "Operational Sum Calculus." Consider the difference equation

$$G_{n+1} - 2G_n + G_{n-1} = F_n$$

Instead of assuming a solution of the form

$$G_n = \sum_{k=0}^{N-1} g_k \omega^{-nk}$$

one can simply transform the equation directly using the results of problem 2. Apply this method to the above equation and show that the expression for g_k is the same in either case.

4. Prove the shift in time identity

$$\mathcal{F}\{\omega^{-l}G_n\} = g_{k-l}.$$

Hint: See the proof of (4.37).

5. Verify that discrete convolution is commutative, i.e.,

$$G * H = H * G \quad \text{and} \quad g * h = h * g.$$

6. Use the discrete convolution theorem to find

   (a) $\mathcal{F}^{-1}\{\frac{\delta_{k,\alpha}}{\omega^k - 1}\}$, where $0 \leq k \leq N$,

   (b) $\mathcal{F}\{\cos \frac{2\pi\alpha}{N} \sin \frac{2\pi\alpha}{N}\}$.

   Verify these results by a direct calculation using the definition of the DFT and its inverse.

7. Consider the difference equation of problem 3 in G_n with $F_n = n$, $0 < n < N$ and the periodic boundary condition $G_0 = G_N = 1$. Use the DFT method of this section to solve the problem.

4.6   The Fast Fourier Transform

No discussion of the DFT could be complete without at least a mention of a relatively recent and highly significant development in the computation of the DFT. Another look at the definition of the DFT

$$g_k = \frac{1}{N} \sum_{n=0}^{N-1} G_n \omega^{nk}, \quad 0 \leq k \leq N - 1 \tag{4.11}$$

suggests that this operation may be regarded as the multiplication of an $N \times N$ complex-valued matrix by an N-vector. Letting

$$\vec{g} = (g_0, g_1, \cdots, g_{N-1})$$

and

$$\vec{G} = (G_0, G_1, \ldots, G_{N-1})$$

be N-vectors and W be the matrix with elements $W_{nk} = \omega^{nk}$, we may write

$$\vec{g} = W\vec{G}. \tag{4.46}$$

After this clear note regarding the N-vectors $\vec{g}$ and $\vec{G}$, we shall, in what follows, drop the $\rightarrow$ and use g and G for the N-vector notation. If we choose to call a multiplication followed by an addition one operation, then roughly N^2 operations are needed to compute the

DFT of the N-point sequence, assuming that the elements of the matrix W are available. For small N this is not a prohibitive expense. However, in many practical applications of the DFT (for instance, data analysis, signal or image processing), N may be very large and it may be necessary to evaluate many transforms successively. In these situations the DFT can become a very costly calculation. Therefore, the discovery of the fast Fourier transform (FFT), which is generally attributed to a 1965 paper by Cooley and Tukey, was greeted with considerable acclaim and has since become one of the most widely used numerical algorithms (see the general reference for the FFT at the end of the book.) There is a vast amount of literature dealing with the FFT. Our purpose here is not to examine it in detail, but rather to give an idea of why the FFT is such an efficient algorithm. Perhaps the simplest way to do this is by presenting the *splitting algorithm.*

Assume that a real or complex N-vector G is given as data where $N = 2^p$ is an integer power of 2. Splitting G into its odd and even components and letting

$$Y_n = G_{2n}, \quad Z_n = G_{2n+1} \text{ for } 0 \leq n \leq \frac{N}{2} - 1, \qquad (4.47)$$

the transform of G may be written (omitting multiplication by $\frac{1}{N}$)

$$
\begin{aligned}
g_k &= \sum_{n=0}^{N-1} G_n \omega_N^{nk} = \sum_{n=0}^{\frac{N}{2}-1} \left[G_{2n} \omega_N^{2nk} + G_{2n+1} \omega_N^{(2n+1)k} \right], \quad \omega_N = e^{\frac{i2\pi}{N}} \\
&= \sum_{n=0}^{\frac{N}{2}-1} \left[Y_n (\omega_N^2)^{nk} + Z_n \omega_N^k (\omega_N^2)^{nk} \right], \\
&\quad 0 \leq k \leq N - 1.
\end{aligned}
$$

$$(4.48)$$

The Nth root of unity, $e^{\frac{i2\pi}{N}}$, which we have been denoting ω is now denoted ω_N. The crucial symmetry of the complex exponential function enters at this point by noting that $\omega_N^2 = e^{\frac{i4\pi}{N}} = e^{\frac{i2\pi}{N/2}} = \omega_{\frac{N}{2}}$. The transform now appears as

$$g_k = \sum_{n=0}^{\frac{N}{2}-1} Y_n \omega_{\frac{N}{2}}^{nk} + \omega_N^k \sum_{n=0}^{\frac{N}{2}-1} Z_n \omega_{\frac{N}{2}}^{nk}. \qquad (4.49)$$

However, the two sums on the right hand side are simply the $\frac{N}{2}$-point transform of the sequences $\{Y_n\}_{n=0}^{\frac{N}{2}-1}$ and $\{Z_n\}_{n=0}^{\frac{N}{2}-1}$ which we shall call $\{y_k\}_{k=0}^{\frac{N}{2}-1}$ and $\{z_k\}_{k=0}^{\frac{N}{2}-1}$ respectively. Therefore, for $0 \le k \le \frac{N}{2} - 1$,

$$g_k = y_k + \omega_N^k z_k. \tag{4.50}$$

For indices between $\frac{N}{2}$ and $N - 1$, we must note that the sequences $\{y_k\}$ and $\{z_k\}$ have period $\frac{N}{2}$, by the periodicity property of the $\frac{N}{2}$-point transforms. Therefore,

$$g_{k+\frac{N}{2}} = y_{k+\frac{N}{2}} + \omega_N^{k+\frac{N}{2}} z_{k+\frac{N}{2}} = y_k + \omega_N^{k+\frac{N}{2}} z_k. \tag{4.51}$$

However, $\omega_N^{k+\frac{N}{2}} = -\omega_N^k$. Hence, the entire sequence $\{g_k\}$ may be determined from the relations

$$g_k = y_k + \omega_N^k z_k \tag{4.52}$$

$$g_{k+\frac{N}{2}} = y_k - \omega_N^k z_k \tag{4.53}$$

where $0 \le k \le \frac{N}{2} - 1$. In other words, relations (4.52) and (4.53) tells us how to construct the N-point transform $\{g_k\}$ if we know the two $\frac{N}{2}$-point transform $\{y_k\}$ and $\{z_k\}$. This step already offers a savings in computation. If $\{y_k\}$ and $\{z_k\}$ are computed by matrix multiplication the cost is roughly $2(\frac{N}{2})^2 = \frac{N^2}{2}$ operations. The combination step (4.52), (4.53) costs N additions and $\frac{N}{2}$ multiplications which are negligible. Therefore, with one splitting the cost of computing the N-point DFT can be reduced from N^2 to $\frac{N^2}{2}$.

The dramatic improvement of the FFT comes in realizing that this splitting strategy may be applied again to compute the $\frac{N}{2}$-point transforms $\{y_k\}$ and $\{z_k\}$. This will give the $\frac{N}{2}$-point transforms in terms of $\frac{N}{4}$-point transforms. The splitting may be repeated $p = \log_2 N$ times until the problem has been reduced to finding N one-point transforms. The transform of a single number is itself. The full FFT, then, amounts to using the combination step (4.52), (4.53) to produce m-point transforms from $\frac{m}{2}$-point transforms. The operation count is now straightforward. At each of p steps, the $\frac{N}{2}$ pairs $(g_k, g_{k+\frac{N}{2}})$ of (4.52), (4.53) must be computed. Each pair requires one complex multiply (assuming powers of ω_N have been computed) and two complex adds. This gives a total of $\frac{pN}{2}$ multiplications and pN adds where $p = \log_2 N$. Thus the savings of the FFT over the matrix multiplication is very close to $N \log_2 N$ vs. N^2. For even moderate values of N (e.g. $N = 512, 1024$) this savings is significant. Of course, this algorithm can be used to compute the inverse transform simply by replacing ω_N by ω_N^{-1}.

The algorithm described above corresponds essentially to the original Cooley–Tukey algorithm. It has some important features which must be understood before the algorithm can be implemented. The splitting procedure describes how to combine two $\frac{N}{2}$-point transforms to give one N-point transform. However, in the process the original N-point sequence must be divided into two subsequences, one containing odd terms and one containing even terms. Fig. 4.9 illustrates one application of the splitting algorithm.

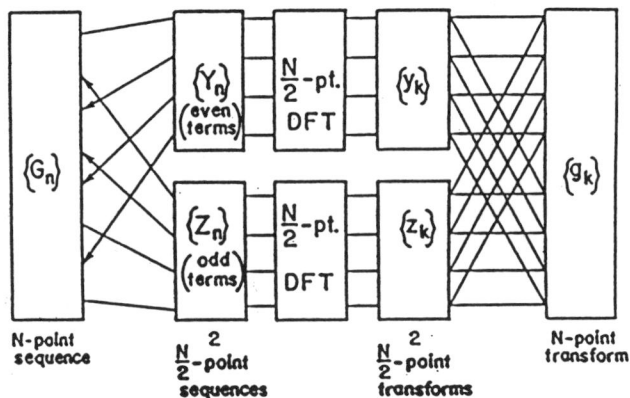

Figure 4.9: An application of the splitting algorithm for the FFT

The first step is the separation of an N-point sequence $\{G_n\}$ into its even subsequence $\{Y_n\}$ and its odd subsequence $\{Z_n\}$. If we now imagine that we have a "box" that can compute $\frac{N}{2}$-point DFT's, then the transforms $\{y_k\}$ and $\{z_k\}$ may be found. The combined step now follows, using (4.52), (4.53), which determines the N-point transform $\{g_k\}$ from $\{y_k\}$ and $\{z_k\}$. This describes just one application of the splitting algorithm. The argument can now be repeated inside of each "box" labeled "$\frac{N}{2}$-point DFT" and again inside of each resulting box labeled "$\frac{N}{4}$-point DFT" until the problem has been reduced to finding 1-point DFT's. This entire procedure is illustrated in Fig. 4.10, which shows the flow of data in an $N = 8$ point transform.

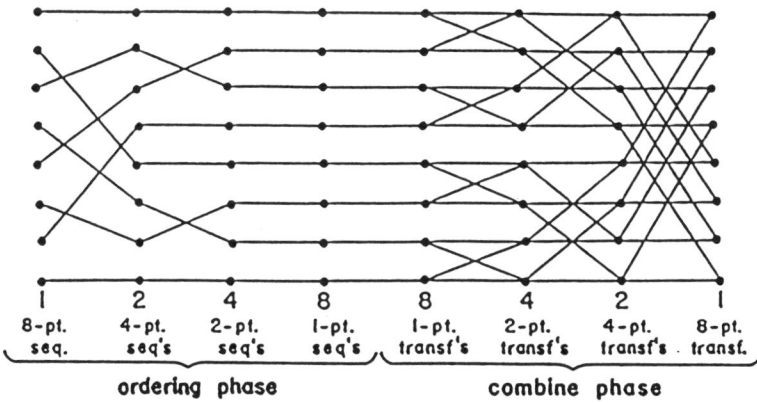

1	2	4	8	8	4	2	1
8-pt. seq.	4-pt. seq's	2-pt. seq's	1-pt. seq's	1-pt. transf's	2-pt. transf's	4-pt. transf's	8-pt. transf.

ordering phase   combine phase

Figure 4.10: The flow of data in an $N = 8$ point transform

From this figure, it is clear that before the combine phase of the Cooley–Tukey algorithm can take place, the original input sequence $\{G_n\}$ must undergo a $\log_2 N$ -stage ordering phase. This ordering phase is often called "bit-reversal", since if the indices of the components are expressed in binary form, the index of each component in the re-ordered sequence is the reverse of its index in the original sequence. Having carried out the ordering phase, the combine phase which also requires $\log_2 N$ stages, can be done. Only in the combine phase does the actual computation of the algorithm take place. Notice that the entire algorithm can be done "in place." No additional arrays of storage are required beyond what is needed to hold the original sequence. In an attempt to clarify the subtleties of the algorithm, Table 4.2 shows the contents of the computational array at each step of the algorithm. The input sequence is $\{G_n\} = \{\cos\frac{2\pi n}{N}\}$ for $N = 8$ which appears in the left column. Columns 2, 3, and 4

show the successive splitting of each subsequence into its own odd
and even subsequences. (Column 4 could be omitted, but it does
show the logical conclusion

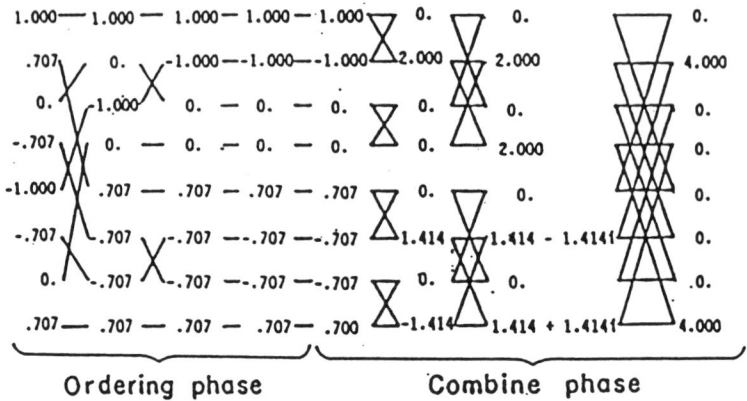

Ordering phase   Combine phase

Table 4.2: Stages of the Cooley–Tukey FFT applied to $\{G_n\}$ =
$\{\cos \frac{2\pi n}{N}\}$, $N = 8$

of the splitting procedure). The step from column 4 to column 5
corresponds to taking the DFT of eight one-point sequences, which
is an identity operation. Columns 6, 7, and 8 show the computation of four 2-point transforms, two four-point transforms and one
eight-point transform respectively. Column 8 gives $\{g_k\}$ without the
multiple of $\frac{1}{N}$.

We shall leave the rest of the details for schematic program of
the Cooley–Tukey and other algorithms for the reference, and where
the FFT algorithm is available on most computers. This very familiar program for the FFT is designed as FFTCC (the Fast Fourier
Transform for Complex–valued Sequence Computation). Among the
more up–to–date version are FFTCF, FFTCB in the IMSL Math.
Library, Ed.1.1.

The Cooley–Tukey algorithm is just one of several versions of
the FFT. Another useful version is generally attributed to Gentleman and Sande. It may be viewed structurally as an inverse of the

Cooley–Tukey algorithm since it begins with a combine phase (which differs from the Cooley–Tukey combine phase) and finishes with the inverse of the Cooley–Tukey ordering phase. The Gentleman–Sande algorithm can also be done in place. Another family of the FFT algorithms are called the autosort algorithms and often have the name Stockham associated with them. These algorithms have the virtue that they do not require an ordering phase. The price which must be paid for this desirable feature is that the algorithm can no longer be done in place and an extra array of storage is needed. Depending on the problem, this may not be prohibitive. It should also be mentioned that all versions of the FFT may be easily adapted to do the inverse transform by replacing ω by ω^{-1} and inserting the factor $\frac{1}{N}$ in the appropriate place.

There are several applications, among them the solution of difference equations, in which the Cooley–Tukey and Gentleman–Sande versions may be used in tandem to eliminate the ordering phase of both algorithms. A boundary value problem such as (4.27), (4.28) could be solved using FFT's without ordering as follows:

1. Use the combine phase of the Gentleman–Sande algorithm to transform the right hand side sequence $\{F_n\}$. This results in the permuted transform $\{f_{p(k)}\}$, where $p(k)$ is known.

2. Do the necessary algebraic calculations in the transform variable to produce the coefficients of the solution $\{g_{p(k)}\}$ which are still in permuted order.

3. Use the combine phase of the Cooley–Tukey algorithm (the ordering phase is no longer needed since the input sequence is already bit-reversed) with ω^{-1} instead of ω. This results in the solution $\{G_n\}$ in natural order.

The complex FFT as discussed has specialized and even more efficient forms which can be used to compute the sine and cosine coefficients of a real input sequence. With proper adaptations, the complex FFT can also be used to compute the coefficients of the sine and cosine series (to be discussed in the next chapter).

Finally, it is worth pointing out that for moderate values of N, the FFT can be used to great advantage in computing the discrete convolution product of two sequences. For example, computing $(g * h)_k$ from the definition (4.42) requires roughly N^2 operations. On the other hand, applying the convolution theorem in the form of (4.42) requires two FFT's to compute $\{G_n\}$ and $\{H_n\}$ from $\{g_k\}$ and $\{h_k\}$, N multiplications to compute the product sequence $\{G_n H_n\}$ and one more FFT to invert $\{G_n H_n\}$. This amounts to roughly $3N \log_2 N + N$ operations.

With these brief remarks on the FFT, we have just touched the surface of an area of computational mathematics which has contributed greatly to many other fields.

Chapter 5

THE DISCRETE SINE (DST) AND COSINE (DCT) TRANSFORMS FOR BOUNDARY VALUE PROBLEMS

In the preceding chapter, the discrete Fourier transform was introduced and was first used in a series–type solution of difference equations with periodic conditions that are the compatible conditions with the series form of the DFT. Then we used the DFT as a direct transform for reducing the difference equation to an algebraic equation as well as involving periodic boundary conditions, or even if there is a jump discontinuity at the end points instead of the zero jump for the (typical) periodic conditions. These series solutions, as well as the direct operational sum method of the DFT, were illustrated in Example 4.8 for solving a traffic flow problem. As we introduced briefly in Section 2.3, integral as well as finite (complex) exponential, sine, and cosine Fourier transforms are all used to algebraize differential operators, but they differ in accommodating different boundary conditions. So, as it may be expected, the discrete sine and cosine Fourier transforms are also compatible with different boundary conditions. This topic was discussed at the end of Section 2.4 in light of their respective operational pairs (2.127), (2.140), and (2.144), (2.143). There, only the important initial step of their algebraizing the difference operator $E+E^{-1}$ in $G_{n+1}+G_{n-1} = (E+E^{-1})G_n$ was illustrated in that discussion, and in Examples 2.12–2.14 for problems to be completed for their final solution in the second section of this chapter. This delay was necessary in order to have more tools of the operational as well as the sequence pairs developed for the discrete sine and cosine transforms. The emphasis in Section 2.4 was on the direct DST and DCT transforms operational sum calculus method of reducing difference equations, of *even* order along with accommodating the boundary conditions that are compatible with either the DST or DCT transforms. As we did in Section 4.4 for the use of the DFT in solving difference equations, with periodic boundary conditions we will, in Section 5.2, use the above DST and DCT for a series solution method as well as a direct operational sum calculus method. But, since the latter method was well discussed in Section 2.4, we will start the same and other problems with the series solution type method of using the DST and the DCT transforms. So in the following section we will establish the discrete sine and cosine transforms, and illustrate them with a few examples. Of course, the reason for beginning our study with the DFT, which seems to apply only to a very specialized kind of problems, is that once the DFT has been presented, the discrete sine and cosine transforms appear with very little additional work.

5.1 Establishing the Discrete Sine and Cosine Transforms

Let us return to the sine-cosine form of the DFT (4.17) where once again we assume that the sequence $\{G_n\}$ has an even number, $N = 2M$, of terms. Recall that this gives the following representation for $\{G_n\}$.

$$G_n = \frac{a_0}{2} + \sum_{k=1}^{M-1}\left[a_k \cos\frac{\pi nk}{M} + b_k \sin\frac{\pi nk}{M}\right] + (-1)^n\frac{a_M}{2} \qquad (4.17)$$

Now consider the situation in which $\{G_n\}$ is an *even sequence* on the interval $0 \le n \le 2M$, that is, $G_{2M-n} = G_n$. With this assumption, the cosine and sine coefficients can be simplified when we realize that $\cos \frac{\pi nk}{M}$ is an even function of n on $[0, N]$ for all k and $\sin \frac{\pi nk}{M}$ is an odd function of n on $[0, N]$ for all k. Using (4.18) and (4.19), this gives us

$$
\begin{aligned}
a_k &= \frac{2}{N} \sum_{n=0}^{N-1} G_n \cos \frac{2\pi nk}{M} \\
&= \frac{1}{M} \left\{ G_0 + \sum_{n=1}^{M-1} G_n \cos \frac{\pi nk}{M} + \right. \\
&\qquad\qquad \left. G_M(-1)^k + \sum_{n=1}^{M-1} G_{2M-n} \cos \frac{\pi(2M-n)k}{M} \right\} \\
&= \frac{1}{M} G_0 + \frac{2}{M} \sum_{n=1}^{M-1} G_n \cos \frac{\pi nk}{M} + \frac{1}{M}(-1)^k G_M, \quad 0 \le k \le M.
\end{aligned}
$$

after using the evenness property ($G_n = G_{2M-n}$) of G_n and that of $\cos \frac{\pi nk}{M}$ in the second sum of the second line.

At the same time, all the sine coefficients vanish, that is, $b_k = 0$ for $0 \le k \le M - 1$, due to the evenness of $\{G_n\}$ and the oddness of the $\sin \frac{\pi nk}{M}$ terms. Thus we have from (4.17) for an *even* sequence $\{G_n\}$ on $0 \le n \le N$,

$$
G_n \equiv \mathcal{C}^{-1}\{a_k\} = \frac{a_0}{2} + \sum_{k=1}^{M-1} a_k \cos \frac{\pi nk}{M} + (-1)^n \frac{a_M}{2}, \quad 0 \le n \le M,
$$

$$(5.1)$$

where

$$
\begin{aligned}
a_k &\equiv \mathcal{C}\{G_n\} \\
&= \frac{1}{M} \left[G_0 + 2 \sum_{n=1}^{M-1} G_n \cos \frac{\pi nk}{M} + (-1)^k G_M \right],
\end{aligned}
$$

$$0 \le k \le M. \qquad (5.2)$$

But now notice what has happened. By incorporating the evenness of the sequence $\{G_n\}$ into the DFT, all of the sine terms have disappeared. Furthermore, the original sequence of length N is now used only for $0 \leq n \leq M = \frac{N}{2}$. Therefore, we may interpret (5.1), (5.2) as a transform pair relating two sequences $\{G_n\}$ and $\{a_k\}$ of length $M+1$ where M may be any integer. This pair constitutes the *discrete cosine transform* (DCT). This result of the DFT reducing to the DCT for even sequences is not surprising, since it parallels with what happened to the Fourier sine–cosine series of the even function $G(x)$ on $(-a, a)$ in becoming a Fourier cosine series as given in (2.83)-(2.84) of Section 2.3.

The important property of the DCT pair is that the sequences $\{G_n\}$ and $\{a_k\}$ to which it relates are no longer periodic in the same way that the DFT is. However, these sequences do have a special symmetry. If the expression (5.1) for $\{G_n\}$ is evaluated outside of the interval $0 \leq n \leq M$, the sequence that is produced is the *even extension* of the sequence $\{G_n\}_{n=0}^{M}$. In particular, this means that $\{G_n\}$ is an even sequence about $n = 0$ (i.e., $G_n = G_{-n}$) and $\{G_n\}$ is an even sequence about $n = M$ (i.e., $G_{M-n} = G_{M+n}$). Fig. 5.1 shows a sequence $\{G_n\}$ defined on $0 \leq n \leq M$ and its even extension outside of this interval.

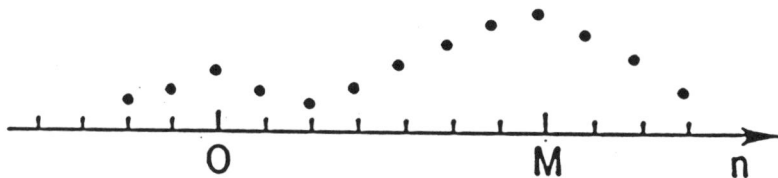

Figure 5.1: The sequence $\{G_n\}$ and its even extension outside its domain $0 \leq n \leq M$

Indeed such an "evenness" boundary condition about the end points $n = 0$ and $n = M$, should remind us of an (approximate) zero slope at the two end points. A zero slope is usually used to model an insulated end, where the absence of a gradient, of the temperature for example, at such ends results in no heat flow. Hence we may conclude that the discrete cosine transform (DCT) is *compatible* with discrete problems that represent (or approximate) a system with two *insulated ends* as the boundary conditions. As we discussed in Section 2.4, such an evenness (or Neumann) boundary conditions can be approximated by $G_1 - G_{-1} = 0$ and $G_{M+1} - G_{M-1} = 0$ at the two end points $n = 0$ and $n = M$, respectively. This is in contrast to the DFT which is compatible with *periodic* boundary condition $G_0 = G_N$ for the two end points $n = 0$ and $n = N$ as was used in (4.28) for Example 4.8 of the traffic flow problem. Compared to these two different boundary conditions, we will soon show, and as we did in Section 2.4, that the discrete sine transform is compatible with vanishing boundary values at the two end points, i.e., $G_0 = G_M = 0$. Hence, the DST is used for zero temperature, for example, at the two end points $n = 0$, and $n = M$.

As we discussed and illustrated in Section 2.4, the above homogeneous boundary conditions G_0, $G_M = 0$, and $G_1 - G_{-1} = 0, G_{M+1} - G_{M-1} = 0$ are accommodated by the DST and DCT, respectively, when these transforms are used as a form of a series solution for the difference equation with these homogeneous boundary conditions. The other, relatively new, operational sum calculus method of the DST and DCT as direct transforms is a bit more general, where it accommodates, instead, nonhomogeneous boundary conditions associated with difference equations of even order. This was shown in (2.140) and (2.143) in Section 2.4 for the DST and DCT, respectively.

A similar procedure may be followed to produce the *discrete sine transform* (DST). We now assume that $\{G_n\}$ is an *odd sequence* on an interval of length $N = 2M$. This means that $G_n = -G_{2M-n}$. The terms $\cos \frac{\pi n k}{M}$ and $\sin \frac{\pi n k}{M}$ have the same symmetry as before. Again we look at the expressions (4.18) and (4.19) for the sine and cosine coefficients in the DFT. We have from (4.19)

$$
\begin{aligned}
b_k &= \frac{2}{N} \sum_{n=0}^{N-1} G_n \sin \frac{2\pi n k}{M} \\
&= \frac{1}{M} \sum_{n=1}^{M-1} G_n \sin \frac{\pi n k}{M} + \frac{1}{M} \sum_{n=1}^{M-1} G_{2M-n} \sin \frac{\pi(2M - n)k}{M} \\
&= \frac{2}{M} \sum_{n=1}^{M-1} G_n \sin \frac{\pi n k}{M}, \quad 0 \le k \le M - 1.
\end{aligned}
$$

after using the oddness property $G_n = -G_{2M-n}$ of G_n and that of $\sin \frac{\pi n k}{M}$ in the second sum of the second line.

At the same time, all of the cosine coefficients vanish, that is $a_k = 0$ for $0 \leq k \leq M$, due to the *oddness* of $\{G_n\}$ and the *evenness* of the $\cos \frac{\pi n k}{M}$ terms. Thus we have for an *odd* sequence $\{G_n\}$ on $0 \leq n \leq N$

$$G_n \equiv \mathcal{S}^{-1}\{b_k\} = \sum_{k=1}^{M-1} b_k \sin \frac{\pi n k}{M}, \quad 1 \leq n \leq M - 1, \qquad (5.3)$$

$$b_k \equiv \mathcal{S}\{G_n\} = \frac{2}{M} \sum_{n=1}^{M-1} G_n \sin \frac{\pi n k}{M}, \quad 1 \leq k \leq M - 1. \qquad (5.4)$$

As in the case of the DCT, the original sequence $\{G_n\}$ of length N is used only for $1 \leq n \leq M - 1$. These $M - 1$ values of the sequence $\{G_n\}$ generate $M - 1$ sine coefficients $\{b_k\}$. Equations (5.3), (5.4) comprise a transform pair relating two sequences $\{G_n\}$ and $\{b_k\}$ of length $M - 1$ where M may be any integer. This pair of relations is the *discrete sine transform (DST)*. As was mentioned after arriving at (5.2) for the DCT, this parallels that of the Fourier sine series for odd functions.

The sequences $\{G_n\}$ and $\{b_k\}$ defined by the DST pair (5.3), (5.4) also have a special symmetry which should be examined. The functions $\sin \frac{\pi n k}{M}$, regarded as functions of either n or k, are odd functions about $n = 0$ (or $k = 0$) and $n = M$ (or $k = M$). Also note that $G_0 = G_M = b_0 = b_M = 0$. Therefore, if the expression (5.3) for G_n, for example, is evaluated outside the interval $0 \leq n \leq M$, the sequence that is produced is the *odd extension* of the sequence $\{G_n\}_{n=0}^{M}$, about both ends $n = 0$ and $n = M$. In particular, this means that $G_n = -G_{-n}$ and $G_{M-n} = -G_{M+n}$. Fig. 5.2 shows a sequence defined on $0 \leq n \leq M$ and its odd extension outside this interval.

The symmetry of the sequences produced by the DST about the endpoints determines that the DST should be chosen in a series solution to solve difference equations with the above homogeneous *Dirichlet* conditions. As we mentioned earlier, this means that the DST is compatible with discrete problems that have, for example, zero temperature at the end points for the two required boundary

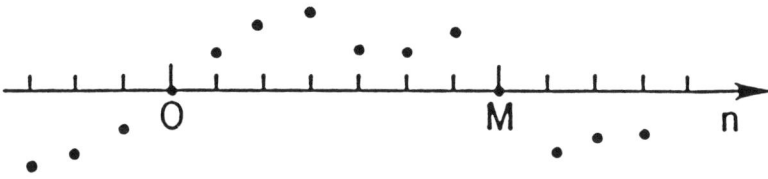

Figure 5.2: The sequence $\{G_n\}$ and its odd extension outside its domain $0 \leq n \leq M$

conditions. As was remarked for the DCT and was established for both the DCT in (2.143) and the DST in (2.140), the more direct method of the operational sum calculus accommodates the more general *nonhomogeneous* boundary conditions associated with the even order difference equations as was established in (2.140) for second order difference equations. The series solution method of the DST was illustrated in Example 2.12, and both of the methods were illustrated in Example 2.13 of Section 2.4. In Example 2.13 we showed how the first series solution method needs more preparations in advance before applying it, while the operational sum calculus method can be applied directly with an apparent shorter computations.

Before turning to the solution of difference equations, some remarks should be made concerning the actual computation of the DCT and DST of sequences. Given a sequence $\{G_n\}_{n=0}^{M}$, its DCT and DST may be computed using (5.2) and (5.4). Inversely, if the coefficients $\{a_k\}_{k=0}^{M}$ or $\{b_k\}_{k=1}^{M-1}$ are known, then the sequence $\{G_n\}$ may be reconstructed using either (5.1) or (5.3) respectively. In situations where the sums of (5.1)-(5.4) cannot be evaluated explicitly, the problem must be done numerically. However, as with the DFT, these calculations can be expedited significantly by using modifications of the FFT.

We now turn to some examples in which the DCT and DST can be computed explicitly. We warn that such explicit expressions in the DCT and the DST can be quite "cumbersome."

Example 5.1   A DCT and DST Expansion of the Sequence
$$\{G_n\} = \{1\}, \ 0 \leq n \leq N$$

Consider the sequence $\{G_n\} = \{1\}$ for $0 \leq n \leq N$. By (5.2) and (5.4), with $M = N$,

$$a_k = \frac{1}{N}\left[1 + 2\sum_{n=1}^{N-1}\cos\frac{\pi nk}{N} + (-1)^k\right], \quad 0 \leq k \leq N \qquad (E.1)$$

and

$$b_k = \frac{2}{N}\sum_{n=1}^{N-1}\sin\frac{\pi nk}{N}, \quad 0 \leq k \leq N - 1. \qquad (E.2)$$

Of course, a_k in (E.1) is good for the DCT expansion of the sequence $G_n = \{1\}$, which gives it its *natural* even extension about the end points $n = 0$ and $n = N \cdot b_k$ in (E.2) on the other hand, is for the DST representation, which gives the same sequence $G_n = \{1\}$ an odd extension about the end points $n = 0$ and $n = N$.

In both expressions the sum may be evaluated by taking the real and imaginary parts of

$$\sum_{n=1}^{N-1} e^{\frac{i\pi nk}{N}} = \sum_{n=1}^{N-1}\left(e^{\frac{i\pi k}{N}}\right)^n = \frac{e^{\frac{i\pi k}{N}} - \left(e^{\frac{i\pi k}{N}}\right)^N}{1 - e^{\frac{i\pi k}{N}}} \qquad (E.3)$$

where the middle sum is clearly a geometric series. This result can now be written as

$$
\begin{aligned}
\sum_{n=1}^{N-1} e^{\frac{i\pi nk}{N}} &= \frac{\cos\frac{\pi k}{N} + i\sin\frac{\pi k}{N} - (-1)^k}{(1 - \cos\frac{\pi k}{N}) - i\sin\frac{\pi k}{N}} \\
&= -\frac{1 + (-1)^k}{2} + i\frac{(1 - (-1)^k)\sin\frac{\pi k}{N}}{2(1 - \cos\frac{\pi k}{N})}
\end{aligned}
\qquad (E.4)
$$

after multiplying numerator and denominator by $(1 - \cos\frac{\pi k}{N}) + i\sin\frac{\pi k}{N}$ and collecting real and imaginary parts. In (E.4), we recognize that its real part $-\frac{1+(-1)^k}{2}$ is the result of the sum in (E.1),

$$\sum_{k=1}^{N-1}\cos\frac{\pi nk}{N} = -\frac{1 + (-1)^k}{2}, \qquad (E.5)$$

while the imaginary part of (E.4) represents the result of the sum in (E.2),

$$\sum_{n=1}^{N-1} \sin \frac{\pi n k}{N} = \frac{(1 - (-1)^k) \sin \frac{\pi k}{N}}{2(1 - \cos \frac{\pi k}{N})} \tag{E.6}$$

Therefore, if we use the results of (E.5) and (E.6) for a_k in (E.1) and b_k in (E.2), respectively, we obtain

$$a_k = \frac{1}{N} \left[1 - (1 + (-1)^k) + (-1)^k \right] = \begin{cases} 2 & k = 0 \\ 0 & k \neq 0 \end{cases} \tag{E.7}$$

$$b_k = \frac{1}{N} \left(\frac{(1 - (-1)^k) \sin \frac{\pi k}{N}}{(1 - \cos \frac{\pi k}{N})}\right). \tag{E.8}$$

The result for a_0 comes from setting $k = 0$ in the definition of a_k. Since the sequence $\{G_n\} = \{1\}$ is even on the interval $0 \leq n \leq N$, only the constant term of the cosine series is needed to represent $\{G_n\}$. Hence, the (simple) even sequence $\{G_n\} = \{1\}$ has the DCT representation

$$G_n = 1 \tag{E.9}$$

as can be seen clearly from (5.1) with the only (surviving) coefficient $a_0 = 2$, since

$$a_0 = \frac{1}{N}[G_0 + 2 \sum_{n=1}^{N-1} G_n + G_M] = \frac{1}{N}[1 + 2(N - 1) + 1]$$

$$= \frac{1}{N}[1 + 2(N - 1) + 1] = \frac{2N}{N} = 2$$

Example 5.2 A DST Expansion of the (Odd) Sequence $\{G_n\} = \{n\}, 0 \leq n \leq N$

A somewhat more involved example is the computation of the DST of the sequence $\{G_n\} = \{n\}, 0 \leq n \leq N$. By (5.4), we have

$$b_k = \frac{2}{N} \sum_{n=0}^{N-1} n \sin \frac{\pi n k}{N}, \quad 1 \leq k \leq N - 1. \tag{E.1}$$

As it soon becomes apparent these computations, and especially their algebraic manipulations, are quite tedious. This point is, of course, in favor of using the FFT as the fast algorithm of computing these DST and DCT finite sums. We will start this example with the basic steps in parallel to what we did in Example 5.1, and leave the rest of the details and algebraic simplifications for the reader as an exercise (see problem 4.)

To evaluate this sum, we must appeal to the summation by parts. Referring to (2.35),

$$\sum_{k=0}^{N-1} v_k \Delta u_k = u_N v_N - u_0 v_0 - \sum_{k=0}^{N-1} u_{k+1} \Delta v_k \qquad (2.35)$$

we let $v_n = n$ and $\Delta u_n = \sin \frac{\pi n k}{N}$. Clearly, $\Delta v_n = 1$. To find u_n, we must compute the anti-difference of $\sin \frac{\pi n k}{N}$. Using the result of problem 5d(ii) of Exercises 1.2 (with $x = k$ and $h = 1$),

$$\Delta(\cos \alpha n) = -2 \sin \frac{\alpha}{2} \cdot \sin [\alpha(n + \frac{1}{2})] \qquad (E.2)$$

it follows that

$$u_n = -\frac{\cos \frac{\pi k}{N}(n - \frac{1}{2})}{2 \sin \frac{\pi k}{2N}}. \qquad (E.3)$$

Applying summation by parts yields

$$b_k = \frac{2}{N} \left\{ \frac{-n \cos \frac{\pi k}{N}(n - \frac{1}{2})}{2 \sin \frac{\pi k}{2N}} \Big|_{n=0}^{n=N} + \sum_{n=1}^{N-1} \frac{\cos \frac{\pi k}{N}(n + \frac{1}{2})}{2 \sin \frac{\pi k}{2N}} \right\}$$

$$b_k = -\frac{N \cos \frac{\pi k}{N}(N - \frac{1}{2})}{N \sin \frac{\pi k}{2N}} + \frac{1}{N \sin \frac{\pi k}{2N}} \sum_{k=1}^{N-1} \frac{\cos \frac{\pi k}{N}(n + \frac{1}{2})}{2 \sin \frac{\pi k}{2N}}$$

Now we try to look at $\cos \frac{\pi k}{N}(n + \frac{1}{2})$ as the real part of $e^{\frac{i\pi kn}{2N}} e^{\frac{i\pi kn}{N}}$. To simplify the notation we let $\theta = \frac{k\pi}{N}$,

$$b_k = \frac{1}{\sin \frac{\theta}{2}} \left[-(\cos \pi k \cos \frac{\theta}{2} - 0) + \frac{1}{N} \text{Re}[e^{\frac{i\theta}{2}} \sum_{n=1}^{N-1} (e^{i\theta})^n] \right]$$

$$= \frac{1}{\sin \frac{\theta}{2}} \left[(-1)^{k+1} \cos \frac{\theta}{2} + \frac{1}{N} \text{Re}[e^{\frac{i\theta}{2}} \frac{e^{i\theta} - e^{iN\theta}}{1 - e^{i\theta}}] \right]$$

$$(E.4)$$

after recognizing the sum as a geometric series. To simplify the complex valued expression in (E.4) before taking its real value, we multiply numerator and denominator by (the complex conjugate) $(1 - e^{-i\theta})$ of the denominator to have

$$
e^{\frac{i\theta}{2}} \cdot \frac{(e^{i\theta} - e^{-iN\theta})(1 - e^{-i\theta})}{(1 - e^{i\theta})(1 - e^{-i\theta})} = e^{\frac{i\theta}{2}} \frac{\left[e^{i\theta} - 1 - e^{iN\theta} + e^{i(N-1)\theta} \right]}{2 - (e^{i\theta} + e^{-i\theta})}
$$

$$
= \frac{e^{i\frac{3}{2}\theta} - e^{i\frac{\theta}{2}} - e^{i(N+\frac{1}{2})\theta} + e^{i(N-\frac{1}{2})\theta}}{2(1 - \cos\theta)}
$$

$$(E.5)$$

Now, we are to write $e^{i\alpha} = \cos\alpha + i\sin\alpha$ for all the complex valued terms above, then collect their real parts to be used as the final result of the sum in (E.4). As it sounds the algebra will be tedious, however, the result can still be simplified, which we leave as exercise 4 with its result to be compared with known results for such finite sine series. (See exercises 4(b) and 5(c)).

This last example suggests that the calculation of discrete sine and cosine transforms of even rather simple sequences may become rather cumbersome. Yet the techniques and results of the previous chapters do allow the calculation to be done. Furthermore the FFT can be adapted to compute the DCT and DST. Having computed a few transforms, we will investigate the ways in which they may be used to solve difference equations in the following section.

Exercises 5.1

1. Verify that the sequences $\{\cos\frac{\pi nk}{M}\}$ and $\{\sin\frac{\pi nk}{M}\}$ are even and odd functions of n respectively.

2. Sketch the first three harmonics of the full sine-cosine DFT ($\cos\frac{2\pi kn}{N}$ and $\sin\frac{2\pi kn}{N}$ for $k = 1, 2, 3$) for $N = 16$. Compare these to sketches of the first three harmonics of the DCT ($\cos\frac{\pi kn}{N}$ for $k = 1, 2, 3$) and the DST ($\sin\frac{\pi kn}{N}$ for $k = 1, 2, 3$) for $N = 16$.

3. Consider the sequence

$$G_n = \begin{cases} n & 0 \le n \le \frac{N}{2} \\ N - n & \frac{N}{2} \le n \le N \end{cases}$$

which defines a *triangular* pulse. Sketch the sequence which is given by the DFT, DST and DCT representations of $\{G_n\}$ on the interval $-2N \le n \le 2N$.

4. (a) Simplify the result in (E.5) to be used in (E.4) for evaluating b_k, the DST of $\{G_n\} = \{n\}$.

   (b) Use the following known finite sine series sum,

$$\sum_{n=1}^{N-1} n \sin n\theta = \frac{\sin N\theta}{4 \sin^2 \frac{\theta}{2}} - \frac{N \cos(\frac{2N-1}{2})\theta}{2 \sin \frac{\theta}{2}}$$

   to compute b_k of part (a), and compare the two answers. *Hint:* Let $\theta = \frac{\pi k}{N}$ in the above series, and note that $\cos N\pi = (-1)^N$.

5. Compute the DCT of the sequences where $-2n \le n \le 2N$

   (a) $G_n = \sin \frac{\pi n}{N}$

   (b) $G_n = n$

   Hint: See the computations of the DST and DCT in Examples 5.1, 5.2, the above problem 4, and part (c).

   (c) Use the following known finite cosine series sum

$$\sum_{n=1}^{N-1} n \cos n\theta = \frac{N \sin(\frac{2N-1}{2}\theta)}{2 \sin \frac{\theta}{2}} - \frac{1 - \cos N\theta}{4 \sin^2 \frac{\theta}{2}}$$

   Hint: Let $\theta = \frac{\pi k}{N}$, and note that $\cos N\pi = (-1)^N$.

5.2 Solution of Difference Equations with Boundary Conditions by DST and DCT

In the preceding chapter, we saw that the DFT pair produces sequences of length N having a period of length N. This periodic property can be used to advantage in solving difference equations with periodic boundary conditions, as was seen in the traffic loop problem of Example 4.8 of Section 4.4 in the previous chapter. There we used the DFT in two ways: a direct transform (or operational sum) method and a series type solution method. In this section we will discuss both methods only for the DST and DCT briefly, since we have already done and illustrated the operational sum method in Examples 2.12, 2.13 for the DST and in Example 2.4 for the DCT. In Example 2.12 we used the series solution method of the DST with homogeneous boundary conditions for the same following boundary value problem in (5.5)-(5.6b). In Example 2.13 we considered the same problem, but with nonhomogeneous Dirichlet boundary conditions $G_0 = A$, $G_N = B$, where we showed very clearly in part (b) of Example 2.13 the advantage of the direct operational sum method of the DST for this problem over the (equivalent of a) series type solution method of part (a) of the same example. We shall start with the series solution method, since the direct method was well covered in Example 2.12 of Section 2.4. The present series solution method of the DST is an exact parallel to that of the DFT that we used in Example 4.8(a) of the last Chapter.

The DST and Dirichlet Boundary Conditions

Consider the second order, constant coefficient difference equation in G_n,

$$\alpha G_{n+1} + \beta G_n + \alpha G_{n-1} = F_n, \quad 1 \le n \le N - 1 \qquad (5.5)$$

$$G_0 = 0 \qquad (5.6a)$$

$$G_N = 0 \qquad (5.6b)$$

along with the homogeneous (Dirichlet) boundary conditions at $n = 0$ and $n = N$, where α, β are given constants and $\{F_n\}_{n=1}^{N-1}$ is a given sequence, and which we have already covered in Section 2.4 with regard to the use of the DST in a direct transform method that was illustrated in Examples 2.12 and 2.13.

(a) The DST for a Sine Series Solution Method .

We wish to choose a representation for the solution $\{G_n\}$ in terms of a discrete transform. The key in making this choice is the form of the boundary conditions. In this particular problem, the boundary conditions specify the value of the solution at the endpoints $n = 0$ and $n = N$. Borrowing terminology from differential equations, we refer to the boundary conditions which specify the value of the solution (at two boundary points) as *Dirichlet Conditions*. We saw in the previous section that a sequence defined by the DST takes a value of zero at the endpoints $n = 0$ and $n = N$. Therefore, if a DST with its sine function kernel as in (5.3)) solution of the form

$$G_n = \sum_{k=1}^{N-1} b_k \sin \frac{\pi n k}{N}, \quad 1 \le n \le N \tag{5.3}$$

is chosen, the homogeneous boundary conditions of this problem are satisfied. It still remains to be seen whether this representation (5.3) simplifies the actual solution of the difference equation. We may point out here that the DST in (5.3) is only good for the homogeneous (vanishing) boundary values of (5.6) at the two end points. As was mentioned in Example 2.13, in case we have nonhomogeneous boundary conditions $G_0 = A$ and $G_N = B$, then we may make a simple change of variable from G_n to, say, H_n, where the boundary conditions become homogeneous $H_0 = H_N = 0$ in the new sequence $\{H_n\}$. The present method can be continued for $\{H_n\}$ as we shall illustrate following this example, and in problem 2 of the exercises.

On the other hand, when the DST is used for a direct transform method or "the operational sum calculus method," a DST operational pair is now derived in (2.140) to accommodate these nonhomogeneous boundary conditions, where we have already illustrated this method for the problem at hand in Example 2.13 of Section 2.4. This is similar to what we see in the variations of the different type Fourier transforms that are derived especially to be compatible with the given boundary conditions, that we introduced briefly in Section 2.3.

As we did in Example 4.8(a) with the DFT, and before substituting our assumed solution (5.3) into the difference equation, it is necessary to represent the right hand side function $\{F_n\}$ of (5.5) in terms of a DST. Since $\{F_n\}$ is known, its DST, which we shall denote $\{f_k\}$ can be computed using (5.4). As we have already noted, in the similar way of using the DFT in a series solution in part (a) of Example 4.8, this step is not so essential when the other method of using the DST as a direct transform is used as illustrated in Example 2.12.

Now substituting for $\{G_n\}$ and $\{F_n\}$ in (5.5), we have

$$\sum_{k=1}^{N-1} b_k \left\{ \alpha \sin \frac{\pi k(n+1)}{N} + \beta \sin \frac{\pi kn}{N} + \alpha \sin \frac{\pi k(n-1)}{N} \right\}$$

$$= \sum_{k=1}^{N-1} f_k \sin \frac{\pi nk}{N}.$$

At this point, to equate the coefficients of the resulting $\sin \frac{\pi kn}{N}$, we may proceed either by appealing to trigonometric identities or by writing the sine as the imaginary part of a complex exponential. The second option is often easier. After combining terms this approach gives

$$Im \left\{ \sum_{k=1}^{N-1} e^{\frac{i\pi nk}{N}} \left[b_k(\alpha e^{\frac{i\pi k}{N}} + \beta + \alpha e^{\frac{-i\pi k}{N}}) - f_k \right] \right\} = 0, \ 1 \leq n \leq N-1$$

or

$$\sum_{k=1}^{N-1} \left\{ \sin \frac{\pi nk}{N} \left[b_k(\alpha \cos \frac{\pi k}{N} + \beta + \alpha \cos \frac{\pi k}{N}) - f_k \right] \right.$$

$$\left. + \cos \frac{\pi nk}{N} b_k \left[\alpha \sin \frac{\pi k}{N} - \alpha \sin \frac{\pi k}{N} \right] \right\} = 0, \ 1 \leq n \leq N-1.$$

This sum can vanish for all $1 \leq n \leq N-1$ only if each term of the sum vanishes independently, which can happen only if the coefficient of each $\sin \frac{\pi nk}{N}$ and $\cos \frac{\pi nk}{N}$ term vanishes independently. Because of the symmetry of the difference equation, all of the cosine terms vanish identically as seen in the second term of the above sum. The condition that the sine terms vanish is

$$b_k \left(\alpha \cos \frac{\pi k}{N} + \beta + \alpha \cos \frac{\pi k}{N} \right) - f_k = 0 \tag{5.7}$$

or

$$b_k = \frac{f_k}{2\alpha \cos \frac{\pi k}{N} + \beta}, \qquad |2\alpha| \leq |\beta|, \qquad 1 \leq k \leq N-1$$

This condition determines the coefficients $\{b_k\}$ in the representation of the solution $\{G_n\}$. The solution may be reconstructed using the DST as given by (5.3).

(b) The Direct Operational Sum Method of the DST and the Nonhomogeneous (Dirichlet) Boundary Conditions .

As we did for part (b) of Example 4.8, we may also employ the more direct operational sum calculus method of using the DST in a direct way, as the compatible discrete transform with the above boundary value problem (5.5)-(5.6) with its zero boundary values of (5.6). This has the added advantage of giving us the result b_k of (5.7) in one simple step, thus avoiding the above complicated algebra of equating the coefficients. This is shown in (E.6) of Example 2.12 (for this same problem) via the DST operational pair in (2.140).

The boundary value problem (5.5), (5.6) is rather specialized. It would be useful to see how the DST can be used to solve more general boundary value problems. We first consider slightly more general boundary conditions. The problem

$$\alpha G_{n+1} + \beta G_n + \alpha G_{n-1} = F_n, \quad 1 \leq n \leq N-1 \qquad (5.7)$$

$$G_0 = A, \quad G_N = B \qquad (5.8)$$

where $\{F_n\}$ is given, and A, B are specified for general Dirichlet boundary conditions. As was discussed in Example 2.13 of Section 2.4, the strategy in solving this problem (5.8)-(5.9) is to convert it to a problem of the form (5.5)-(5.6) with *homogeneous* boundary conditions as in (5.6). One way to do this is to define a *new* variable

$$H_n = G_n - \Theta_n \equiv G_n - A - (B - A)\frac{n}{N}. \qquad (5.9)$$

The sequence $\{\Theta_n\}$ serves to "subtract out" the nonhomogeneous boundary conditions in the following way:

$$H_0 = G_0 - \Theta_0 = A - A = 0$$

$$H_N = G_N - \Theta_N = B - B = 0$$

Substituting $G_n = H_n + \Theta_n$ into the difference equation (5.8) gives

$$
\begin{aligned}
\alpha H_{n+1} + \beta H_n + \alpha H_{n-1} &= F_n - \alpha\Theta_{n+1} - \beta\Theta_n - \alpha\Theta_{n-1} \\
&= F_n - (2\alpha + \beta)\Theta_n
\end{aligned}
$$

$$ \tag{5.10} $$

$$ H_0 = 0 \tag{5.11a} $$

$$ H_N = 0 \tag{5.11b} $$

as the desired homogeneous boundary condition in the new variable H_n. We are now left with a new boundary value problem which has the form of problem (5.5)-(5.6). Once again the boundary conditions are homogeneous. The price which is paid to create homogeneous boundary conditions is that the right hand side sequence of the difference equation (5.10) (in the new variable H_n) is slightly more complicated. However, the problem (5.10)-(5.11) can be solved using methods already discussed. The comparison between using the DST in a series solution method and a direct transform method of solving this problem, is discussed and illustrated in Example 2.13 of Section 2.4.

As we discussed and illustrated in (2.151)-(2.155) of Section 2.4, another way in which we might generalize the boundary value problem solvable by the DST is to consider *non-symmetric* difference equations. Consider the boundary value problem.

$$ \alpha G_{n+1} + \beta G_n + \gamma G_{n-1} = F_n, \quad 1 \leq n \leq N - 1 \tag{5.12} $$

$$ G_0 = G_N = 0 \tag{5.13} $$

Once again the strategy is to convert this problem to a problem of the form of (5.5)-(5.6) which we already know how to solve. If we assume that α and γ have the same sign, then the following change of variable is possible. Let

$$ G_n = \left(\frac{\gamma}{\alpha}\right)^{\frac{n}{2}} H_n. \tag{5.14} $$

Substituting this into (5.12) and simplifying, leads to the new boundary value problem

$$ \sqrt{\alpha\gamma} H_{n+1} + \beta H_n + \sqrt{\alpha\gamma} H_{n-1} = \left(\frac{\gamma}{\alpha}\right)^{-n} F_n, \quad 1 \leq n \leq N - 1 \tag{5.15} $$

$$H_0 = H_N = 0. \qquad\qquad (5.16)$$

This equation for the boundary condition (5.16) has the symmetry of equation (5.6). This symmetry was achieved at the expense of complicating the right hand side sequence in (5.15). Of course, once we have this symmetry in (5.15), the direct DST transform method of Example 2.12 can be used with much ease to solve the boundary value problem (5.15)–(5.16).

We have reached the end of the obvious manipulations that will make a boundary value problem amenable to the DST, however, we have also added a new method, namely, the direct operational sum method of Section 2.4 for the DST as well as the DCT and the DFT to avoid the necessity of such manipulations and advanced planning! In general, difference equations with variable coefficients, nonlinear difference equations and problems with other than Dirichlet boundary conditions cannot be handled by the DST. As we have seen, however, the DST is a compatible transform for (linear) second order, *constant coefficient* problems with Dirichlet conditions. We add that a "modified DST" may very well be developed to accommodate a more general or "mixed" boundary conditions. This will have the same spirit, for example, as the "modified" finite Fourier sine transform of (2.86) that accommodates the boundary condition $G(0)$ as well as $G'(\pi)$ at the two boundary points $x = 0$ and $x = \pi$ as shown in its operational pair (2.87) of algebraizing the second order derivative $\frac{d^2 G}{dx^2}$ and accommodating these boundary conditions, which are away from being both Dirichlet boundary conditions.

As we had indicated towards the end of Section 2.4 in Chapter 2, an interesting illustration for using the DST is that of Example 7.4 in Chapter 7. It models a discrete system with springs and masses.

The DCT and Neumann Boundary Conditions

We now turn to another type of boundary conditions with which the DCT is compatible. Consider the boundary value problem

$$\alpha G_{n+1} + \beta G_n + \alpha G_{n-1} = F_n, \ \ 1 \le n \le N \qquad (5.17)$$

$$\delta G_0 = G_1 - G_{-1} = 0, \qquad\qquad (5.18a)$$

$$\delta G_N = G_{N+1} - G_{N-1} = 0 \qquad\qquad (5.18b)$$

with its Neumann type boundary conditions (5.18a)-(5.18b), which we have already discussed in Example 2.14 with emphasis on the direct operational sum method of the DCT.

The boundary conditions in this problem require that the centered difference of the solution vanish at the endpoints. Physically, this may be interpreted as requiring that the solution have "no flux" into or out of the domain of the problem. For example if $\{G_n\}$ represented temperature, then these boundary conditions require no temperature difference, and hence no heat flux, across the boundaries. Borrowing again from the terminology of differential equations, boundary conditions which specify the difference of the solution at the boundary are called *Neumann conditions*.

At this point, it is necessary to recall the symmetry associated with the DCT. The sequences $\{G_n\}_{n=0}^{N}$ and $\{a_k\}_{k=0}^{N}$ related by the transform pair (5.1), (5.2) have the property that they are even sequences about the endpoints $n = 0$ and $n = N$. In particular, this means that $G_{-1} = G_1$ and $G_{N-1} = G_{N+1}$. Therefore, the DCT would seem to be a good candidate for a compatible transform for the problem (5.17)-(5.18). Before proceeding, it should be mentioned that, unlike the case of Dirichlet boundary conditions, Neumann conditions do not allow the solution itself to be determined at the boundary before solving the difference equation. The difference equation is to be solved for $1 \leq n \leq N$ and the $(N + 1)$ unknowns include G_0 and G_N.

As we did in the case of the DFT transform method in Section 4.4 and its Example 4.8, and the above DST transform method of solving boundary value problems, we shall discuss here both the DCT use in a (cosine) series solution method as well as its use as a direct transform in another special case of the operational sum calculus method.

(a) The DCT for a (Cosine) Series Solution Method .

Here we shall do the first series solution method of the DCT with more details, since the DCT operational sum method was the emphasized one in Example 2.14 of Section 2.4.

If we assume a solution of the form

$$G_n \equiv \frac{a_0}{2} + \sum_{k=1}^{N-1} a_k \cos \frac{\pi nk}{N} + (-1)^n \frac{a_N}{2}, \quad 0 \le n \le N \qquad (5.1)$$

we have verified that the (homogeneous) boundary conditions of (5.18) are immediately satisfied. As with the DST, we must now represent the right hand side sequence $\{F_n\}$ in terms of the DCT. Substituting for $\{G_n\}$ and $\{F_n\}$ in (5.17) gives

$$\alpha \left[\frac{a_0}{2} + \sum_{k=1}^{N-1} a_k \cos \frac{\pi k(n+1)}{N} + (-1)^n \frac{a_N}{2} \right] +$$

$$\beta \left[\frac{a_0}{2} + \sum_{k=1}^{N-1} a_k \cos \frac{\pi kn}{N} + (-1)^n \frac{a_N}{2} \right] +$$

$$\alpha \left[\frac{a_0}{2} + \sum_{k=1}^{N-1} a_k \cos \frac{\pi k(n-1)}{N} + (-1)^{n-1} \frac{a_N}{2} \right] =$$

$$\frac{f_0}{2} + \sum_{k=1}^{N-1} f_k \cos \frac{\pi nk}{N} + (-1)^n \frac{f_N}{2}. \qquad (5.19)$$

Our eventual aim is *to match coefficients* of the various cosine terms in these sums. We may simplify our work if we match the coefficients of the constant terms and the $(-1)^n$ terms right now. This gives us the conditions that

$$a_0 = \frac{f_0}{\beta + 2\alpha}, \quad a_N = \frac{f_N}{\beta - 2\alpha}. \qquad (5.20)$$

This determines two of the unknown coefficients $\{a_k\}$. To find the remaining coefficients, we combine the sum of (5.19)

$$\sum_{k=1}^{N-1} \left[a_k (\alpha \cos \frac{\pi k(n+1)}{N} + \beta \cos \frac{\pi kn}{N} + \alpha \cos \frac{\pi k(n-1)}{N}) \right.$$

$$\left. - f_k \cos \frac{\pi kn}{N} \right] = 0.$$

Again we have a choice of using trigonometric identities or regarding the cosine as the real part of a complex exponential. The latter approach gives us

$$Re\left\{\sum_{k=1}^{N-1}e^{\frac{i\pi nk}{N}}\left[a_k(\alpha e^{\frac{i\pi k}{N}}+\beta+\alpha e^{\frac{-i\pi k}{N}})-f_k\right]\right\}=0,\ 1\le n\le N-1$$

or

$$\sum_{k=1}^{N-1}\left\{\cos\frac{\pi nk}{N}\left[a_k(\alpha\cos\frac{\pi k}{N}+\beta+\alpha\cos\frac{\pi k}{N})-f_k\right]-\right.$$

$$\left.\sin\frac{\pi nk}{N}\left[a_k(\alpha\sin\frac{\pi k}{N}-\alpha\sin\frac{\pi k}{N})\right]\right\}=0,\ 1\le n\le N-1. \quad (5.21)$$

This sum can vanish for all $1\le n\le N-1$ only if each term of the sum vanishes independently, which can happen only if the coefficient of each $\sin\frac{\pi nk}{N}$ and $\cos\frac{\pi nk}{N}$ term vanishes independently. Because of the symmetry of the difference equation, all of the sine terms vanish identically as seen from the second term inside the above sum. The condition that the cosine terms vanish is

$$a_k(\alpha\cos\frac{\pi k}{N}+\beta+\alpha\cos\frac{\pi k}{N})-f_k=0, \quad (5.22)$$

$$a_k=\frac{f_k}{2\alpha\cos\frac{\pi k}{N}+\beta},\ 1\le k\le N-1$$

These conditions together with conditions (5.20) for a_0 and a_N determine the coefficients in the DCT representation of the solution $\{G_n\}$. The solution G_n of (5.17) may now be reconstructed using the DCT as given by (5.1).

(b) The Direct Operational Sum Method of the DCT and the Nonhomogeneous (Neumann) Boundary Conditions .

As we just mentioned for the direct DST method, all the above details in (5.19)–(5.21) can be avoided with the use of the DCT as a direct transform, where we arrive at the desired result of a_k in (5.22) in one simple step via the DCT operational pairs (2.144) or (2.143), the latter pair (2.143) is a more general one that accommodates

nonhomogeneous Neumann boundary conditions such as that of the
following (5.24a)-(5.24b). This was illustrated for this same problem
in Example 2.14, in particular the same a_k of (5.22) is found in (E.6)
of the example.

It is now possible to generalize the class of boundary value problems which may be handled by the DCT. Consider the more general
Neumann boundary conditions of the following problem.

$$\alpha G_{n+1} + \beta G_n + \alpha G_{n-1} = F_n, \quad 1 \leq n \leq N \tag{5.23}$$

$$\delta G_0 = A, \tag{5.24a}$$

$$\delta G_N = B \tag{5.24b}$$

We now look for a change of variable that will restore homogeneous
boundary conditions. Let a new variable $\{H_n\}$ be given by

$$H_n = G_n - \Phi_n \equiv G_n - \left(\frac{nA}{2} + \frac{B - A}{4N}n^2\right).$$

It may now be verified that the new variable $\{H_n\}$ does satisfy

$$\delta H_0 = \delta G_0 - \delta \Phi_0 = A - A = 0$$

$$\delta H_N = \delta G_N - \delta \Phi_N = B - B = 0 \tag{5.25}$$

Introducing the new variable does, of course, change the difference
equation. Substituting $H_n = G_n - \Phi_n$ into (5.23) along with what
we obtained as homogeneous boundary conditions in (5.25), gives the
new boundary value problem in H_n

$$\alpha H_{n+1} + \beta H_n + \alpha H_{n-1} = F_n - \alpha \Phi_{n+1} - \beta \Phi_n - \alpha \Phi_{n-1}$$

$$= F_n - (2\alpha + \beta)\Phi_n + \frac{\alpha(B - A)}{2N} \tag{5.26}$$

$$\delta H_0 = 0, \quad \delta H_N = 0 \tag{5.27}$$

Apart from a slightly more complicated right hand side of (5.25),
this problem has the form of (5.18)-(5.19). The DCT may be used to
find $\{H_n\}$ and then the solution of the original problem $\{G_n\}$ may
be found using $G_n = H_n + \Phi_n$. So all these preparations following
(5.24) have to be done to a problem with nonhomogeneous Neumann

boundary conditions such as (5.24) in order to have it ready for the method of series solution of the DCT. In contrast, the operational sum method of the DCT with its (new) operational pair in (2.143) can take care of the Neumann boundary value problem (5.23)-(5.24) with its nonhomogeneous boundary conditions in a very direct way, and without the need for the above advanced, and may be tedious preparations.

Finally, a non-symmetric difference equation with Newmann conditions may be treated just as it was with Dirichlet conditions. A change of variables restores the symmetry of the equation and then the DCT may be applied. In conclusion, we have found that the DCT is the compatible transform for (linear) second order, *constant coefficient* difference equations with Neumann boundary conditions. However, just as we remarked for the DST transform, a modified DCT may well be developed to accommodate a more general condition than the Neumann condition. A lead for that is to look at a parallel to a modified finite Fourier cosine transform and how it was constructed to accommodate viable boundary conditions away from the usual Neumann condition of the typical finite Fourier cosine transform. More details, on this topic of constructing compatible transforms, are found in the author's book on the subject of integral and discrete transforms, in particular Section 1.6 of Chapter 1.

As we mentioned towards the end of Section 2.4 in Chapter 2, an interesting illustration of using the DCT is that of Example 7.7 in Chapter 7. It models a diffusion process in the altruistic neighborhood relations.

Exercises 5.2

1. A difference equation may be transformed directly if we know the effect of the DCT and DST on a particular difference operator. Let $\mathcal{C}$ and $\mathcal{S}$ denote the operation of performing the DCT and DST respectively. Show the following properties are true

   (a) $\mathcal{S}\{\Delta G_n\} = (\cos\frac{\pi k}{N} - 1)\mathcal{S}\{G_n\} + \sin\frac{\pi k}{N}\mathcal{C}\{G_n\}$

   (b) $\mathcal{S}\{\Delta G_n\} = (\cos\frac{\pi k}{N} - 1)\mathcal{C}\{G_n\} - \sin\frac{\pi k}{N}\mathcal{S}\{G_n\}$

   (c) $\mathcal{S}\{\Delta \triangledown G_n\} = \mathcal{S}\{G_{n+1} - 2G_n + G_{n-1}\}$
   $$= 2(\cos\frac{\pi k}{N} - 1)\mathcal{S}\{G_n\}$$

(d) $C\{\Delta \triangledown G_n\} = C\{G_{n+1} - 2G_n + G_{n-1}\}$
$$= 2(\cos \tfrac{\pi k}{N} - 1)C\{G_n\}$$

The first two properties are generally not too useful since the DCT or DST are generally not applied to first order difference equations. However, the second two properties may be used frequently for second order equations.

2. Transform the boundary value problem

$$2G_{n+1} - 5G_n + 8G_{n-1} = n, \quad 1 \le n \le N - 1$$

$$G_0 = 4, \quad G_N = 3$$

into the form (5.5)-(5.6) so that the DST may be applied.

Hint: See (5.8)-(5.15), in particular, the important step following (5.9) that led to the homogeneous boundary condition (5.11).

3. Transform the boundary value problem

$$2G_{n+1} - 5G_n + 8G_{n-1} = 1, \quad 1 \le n \le N$$

$$\delta G_0 = 1, \quad \delta G_N = 1$$

into the form (5.16)-(5.17) so that the DCT may be applied.

Hint: See (5.23)-(5.27).

4. Solve the resulting boundary value problems of problems 2 and 3. You might try both the method given in the text and the method of direct transformation which used properties (c) and (d) of problem 1.

5. Can you see the significance of the fact that, in the DCT solution, a_0 in (5.20) is undefined if $\beta + 2\alpha = 0$? In general, how do you explain that a_k for the DCT solution and b_k for the DST solution are undefined if $2\alpha \cos \tfrac{\pi k}{N} + \beta = 0$?

6. Use the methods of Section 3.3 to solve the boundary value problem of exercise 2.

7. Use the methods of Section 3.3 to solve the boundary value problem of exercise 3.

8. Use the direct operational sum calculus method of using the DST on the boundary value problem (5.5)-(5.6) to arrive at the result of g_k in (5.7).

   Hint: You may consult problem 1.

9. Similar to problem 7, use the DCT on the boundary value problem (6.17)-(5.18) to arrive at the same result of g_k in (5.22).

10. Consider the boundary value problem (5.5)-(5.6) of exercise 8 for G_n, $0 < n < N - 1$, with its boundary condition $G_0 = G_N = 0$, and where $F_n = n$, $0 < n < N - 1$. Use the methods of Section 3.3 to solve this boundary value problem.

11. Consider the boundary value problem (5.17)-(5.18) with $F_n = 1$, $1 \leq n \leq N$. Use the methods of Section 3.3 to solve the problem.

Chapter 6

THE z-TRANSFORM FOR INITIAL VALUE PROBLEMS

6.1 Discrete Initial Value Problems for the z-Transform

In the previous two chapters we investigated various *boundary value problems* associated with difference equations. In the process, we developed discrete transforms which are compatible with each kind of boundary value problem. This included the discrete Fourier transform (DFT) of Chapter 4 for periodic boundary conditions, the discrete sine transform (DST) and the discrete cosine transform (DCT) of Chapter 5, which are compatible, respectively, with the Dirichlet condition, i.e., when the value of the function is given at the two end points; and the Neumann condition where the slope (or the central difference) is given at the two end points. We now turn our attention to a different sort of problem in which difference equation may appear. This is the *initial value problem* which is often associated with *time-dependent* or *evolutionary* system. As we will see, there is a "hybrid–type" transform that transforms the (discrete) *infinite* sequence $\{u_n\}_{n=0}^{\infty}$ to the *continuous* z-plane. This transform is called the *discrete Laplace* or z-transform, which is defined in (6.4), and where as we shall see in the next section, its properties parallel those of the Laplace transform. In particular, it parallels the operational calculus property of the Laplace transform, in algebraizing (linear) differential operators as well as involving the given constant initial conditions, as was briefly discussed in Section 2.3 with its main such operational pair given in (2.49) for algebraizing the second order derivative. The z-transform does the same operation of algebraizing difference equations and accommodating the given constant initial conditions. We shall develop such pairs in (6.8), (6.9), then apply them to solving a difference equation with constant initial conditions in the last section. For this we add the (appropriate) constant initial conditions, which help in determining the final *unique* solution of the resulting initial value problem associated with the difference equation considered.

Before developing the transform method for solving initial value problems, we begin by giving an example of an initial value problem, and repeat again the familiar method of solving it as we had already discussed in Chapter 3.

Consider the second order difference equation

$$u_{n+2} - u_{n+1} + 6u_n = 0, \quad n \geq 0 \tag{6.1}$$

with the constant initial conditions

$$u_0 = 0, \tag{6.2}$$

$$u_1 = 3. \tag{6.3}$$

The two extra conditions which we have come to expect with second order difference equations are now given, not at the endpoints of the interval on which the problem is to be solved, as in the case of boundary conditions, but rather at the "initial end" of the interval on which the problem is to be solved. The constant initial conditions give the initial configuration of the system which is being modeled, while the difference equation gives the subsequent behavior of the system as it evolves from the constant initial conditions.

The most familiar way to solve this initial value problem is to assume, as we did in Section 3.2 for linear, homogeneous and constant coefficient difference equations, that it has solutions of the form $u_n = $

λ^n, where λ is a constant to be determined. Substituting this trial solution into the difference equation (6.1) gives

$$\lambda^n(\lambda^2 - \lambda - 6) = 0 \tag{6.4}$$

This equation can be satisfied if we choose $\lambda = 0$, which gives very uninteresting (trivial) solutions, or if we require that

$$\lambda^2 - \lambda - 6 = 0 \tag{6.5}$$

This *second* degree polynomial (arising from a *second* order difference equation) is called the *auxiliary* or *characteristic equation* of the difference equation. It can be solved to yield the (distinct) roots $\lambda = 3$ and $\lambda = -2$. Therefore, $u_n = 3^n$ is a solution of the equation and so is $u_n = (-2)^n$. However, as we have already discussed and illustrated in Sections 3.1, 3.2, because the difference equation is linear, it is easy to show that the following linear combination of the two solutions

$$u_n = A3^n + B(-2)^n$$

is also a solution. This last expression is called the *general solution* and is valid for the arbitrary values of A and B. It is now a straightforward matter to find the specific solution $\{u_n\}$ which also satisfies the constant initial conditions (6.2)–(6.3). The condition $u_0 = 0$ requires that

$$u_0 = A + B = 0$$

The condition $u_1 = 3$ requires that

$$u_1 = 3A - 2B = 3.$$

These two conditions on A and B imply that $A = \frac{3}{5}$, $B = -\frac{3}{5}$. Thus the solution to the initial value problem (6.1)–(6.3) is

$$u_n = \frac{3}{5}3^n - \frac{3}{5}(-2)^n. \tag{6.6}$$

As we showed in Chapter 3, this solution technique can be extended to higher order constant coefficient difference equations, to cases in which the characteristic polynomial has multiple real or complex roots, and to nonhomogeneous equations. In this section we will not pursue this method any further and mention it here only for the sake of completeness for the comparison with the more direct "operational sum calculus" of the z–transform. Instead, we will remain with the added feature of this book and investigate the direct use of the z–transform in solving *initial value problems* associated with difference equations that involve *infinite* sequences. This is in contrast to the discrete Fourier transform of the last two chapters that we used for solving, primarily, *boundary value problems* associated with difference equations that involve *finite* sequences.

Exercises 6.1

1. Use the method of Section 3.2 to solve the following initial value problems.

   (a)
   $$u_{n+2} - 4u_n = 0, \quad n \geq 0$$
   $$u_0 = 0, \quad u_1 = 1$$

   (b)
   $$u_{n+1} - u_n - 12u_{n-1} = 0, \quad n \geq 1$$
   $$u_0 = 4, \quad u_1 = 2$$

   (c)
   $$u_{n+2} - 2u_{n+1} - 4u_n - 8u_{n-1} = 0, \quad n \geq 1$$
   $$u_0 = 0, \quad u_1 = 1, \quad u_2 = 0$$

   Hint: Watch for the repeated root $\lambda = 2$.

2. Find the solution of the following initial value problems. Watch for repeated roots of the characteristic equation, where you may consult Section 3.2.

   (a)
   $$u_{n+1} - 4u_n + 4u_{n-1} = 0, \quad n \geq 1$$
   $$u_0 = 0, \quad u_1 = 4$$

   (b)
   $$u_{n+2} + 6u_{n+1} + 9u_n = 0, \quad n \geq 0$$
   $$u_0 = 1, \quad u_1 = 0$$

6.2   The *z*-Transform

The nature of the initial value problems generally requires that we find a solution sequence $\{u_n\}$ over a range of indices $n \geq N$, where N is some initial index. Whenever possible, we will take $N = 0$, so that our sequences can be indexed $n \geq 0$. Given the *infinite* sequence $\{u_n\}_{n=0}^{\infty}$, its *z*-transform is defined and denoted

$$\mathcal{Z}\{u_n\} = U(z) = \sum_{n=0}^{\infty} u_n z^{-n} \qquad (6.7)$$

We will denote sequences by lower case letters, their *z*-transforms by corresponding upper case letters, which is more in line with the notation for the Laplace transform of (2.46). The $\mathcal{Z}$ notation denotes the operation of applying the *z*-transform. The transform variable, z, is now a *complex variable* which may take any value in the complex plane. In this regard, the *z*-transform is different and not "as discrete" as the discrete Fourier transforms in which both the original variable (which we called n) and the transform variable (which we called k) are integers. The fact that the *z*-transform takes us into the complex plane results in that the inversion of the *z*-transform (that is, given $U(z)$, find $\{u_n\}$) becomes rather complicated and requires the techniques of contour integration, a clear parallel to what is done in the case of the Laplace transform inversion. Nevertheless, by building a table of *z*-transforms of various sequences and by collecting properties of the *z*-transform, it will be possible to do the inverse transform by a look-up procedure. Toward this end, we will do some examples that can become part of a table of transforms. It is the place here to remark that even though a good table of the discrete Fourier transforms is most useful, nevertheless it is not as essential as it is the case here for the *z*–transform. The reason lies in the availability of the simple numerical inversion of the discrete Fourier transforms, thanks to their very fast algorithm, the FFT. A numerical inversion for the Laplace and the *z*–transform may be available, but it is complicated, and its computations may be unstable, a subject that we shall not cover in this book. So, we shall limit ourselves to the, hopefully, extensive tables of the *z*–transform. It is then no surprise to find much longer tables of the *z*–transform pairs compared to the tables of the discrete Fourier transforms pairs.

Example 6.1   Computing a z-transform

Find the z-transform of the *unit* step sequence $u_n^{(s)}$

$$u_n^{(s)} = \begin{cases} 0 & n < 0 \\ 1 & n \geq 0. \end{cases} \qquad (E.1)$$

By the definition (6.7) we have

$$\mathcal{Z}\{u_n^{(s)}\} = \sum_{n=0}^{\infty} u_n^{(s)} z^{-n} = \sum_{n=0}^{\infty} z^{-n} = \sum_{n=0}^{\infty} \left(\frac{1}{z}\right)^n. \qquad (E.2)$$

This unit step sequence $u_n^{(s)}$ is very important in facilitating the description of the *causal* C_n sequences, i.e., sequences that vanish for negative integer n,

$$C_n = u_n \cdot u_n^{(s)} = \begin{cases} 0, & n < 0 \\ u_n, & n \geq 0. \end{cases}$$

At this point, we appeal to a result which will be used in all that follows. It is really the extension of the convergence theorem for the *geometric series* to the complex plane. If $g(z)$ is an expression involving a complex variable z, then the infinite series

$$p(z) = \sum_{n=0}^{\infty} (g(z))^n \qquad (E.3)$$

converges for all z for which $|g(z)| < 1$, and it converges to

$$p(z) = \frac{1}{1 - g(z)}. \qquad (E.4)$$

This may be compared to the result for a real number α that

$$\sum_{n=0}^{\infty} \alpha^n = \frac{1}{1 - \alpha}, \quad \text{for } |\alpha| < 1. \qquad (E.5)$$

In the present example, if we compare (E.2) with (E.3) we have: $g(z) = \frac{1}{z}$. Hence, according to (E.4), the desired result of (E.2) becomes:

$$\mathcal{Z}\{u_n^{(s)}\} = \frac{1}{1 - \frac{1}{z}} = \frac{z}{z - 1}, \quad \text{for } |z| > 1.$$

Example 6.2   The z-transform of $\{a^n\}_{n=0}^\infty$

The geometric series may be used to find several other z-transforms. Consider $\{u_n\}_{n=0}^\infty = \{a^n\}_{n=0}^\infty$ where a is a complex constant. By (6.4) we have

$$
\mathcal{Z}\{u_n\} = \mathcal{Z}\{a^n\} = \sum_{n=0}^\infty u_n z^{-n} = \sum_{n=0}^\infty \left(\frac{a}{z}\right)^n
$$

$$
= \frac{1}{1 - \frac{a}{z}} = \frac{z}{z - a}
$$

$$(E.1)$$

provided $|z| > |a|$. From this example, we may deduce two other useful transforms. Letting $a = e^{\pm i\theta}$ where θ is real, we find that

$$
\begin{aligned}
\mathcal{Z}\{\cos n\theta\} &= \mathcal{Z}\{\frac{1}{2}(e^{in\theta} + e^{-in\theta})\} \\
&= \frac{1}{2}\left(\frac{z}{z - e^{in\theta}} + \frac{z}{z - e^{-in\theta}}\right) \\
&= \frac{z - \cos\theta}{z - 2\cos\theta + z^{-1}} \quad \text{for} \quad |z| > \left|e^{\pm i\theta}\right| = 1.
\end{aligned}
$$

$$(E.2)$$

Similarly,

$$
\begin{aligned}
\mathcal{Z}\{\sin n\theta\} &= \mathcal{Z}\{\frac{1}{2i}(e^{in\theta} - e^{-in\theta})\} \\
&= \frac{\sin\theta}{z - 2\cos\theta + z^{-1}}, \quad \text{for} \quad |z| > 1.
\end{aligned}
$$

$$(E.3)$$

Example 6.3   An Inverse z-transform

We will use the geometric series in a slightly different way now to show how an inverse transform may be found. Denoting the inverse z-transform as $\mathcal{Z}^{-1}$, consider the problem of finding $\mathcal{Z}^{-1}\{\frac{1}{z-a}\}$ where a is a complex constant. The idea now is to represent $U(z) = \frac{1}{z-a}$

as a geometric series and then identify the coefficients of the series as the sequence $\{u_n\}$. Before doing this, it is important to note that the z-transform involves negative powers of z, that is $\left(\frac{1}{z}\right)^n$. If such a series is to converge, then it must converge for values of z lying *outside* of a circle or for $|z| > R$ for some R. The preceding examples show that this is the case. Therefore, we look for a geometric series for $U(z)$ which involves negative powers of z. This can usually be done by an algebraic manipulation. We can write

$$
\begin{aligned}
U(z) &= \frac{1}{z-a} = \frac{1}{z}\frac{1}{\left(1-\frac{a}{z}\right)} \\
&= \frac{1}{z}\sum_{n=0}^{\infty}\left(\frac{a}{z}\right)^n = \sum_{n=1}^{\infty}\frac{a^{n-1}}{z^n}
\end{aligned}
\qquad (E.1)
$$

after making a change of variable $m = n+1$ for the index n to suit the definition of the z-transform in (6.7). The above result is valid provided $|z| > |a|$. Comparing this with the definition of the z-transform (6.7) we see that

$$
u_n = \begin{cases} 0, & n = 0 \\ a^{n-1}, & n \geq 1. \end{cases}
\qquad (E.2)
$$

This idea of finding the inverse transform by algebraic manipulation and use of the geometric series may be applied to many rational functions of z. The following example illustrates more of this strategy.

Example 6.4   An Inverse z-transform of a Rational Function

To have the rational function

$$
U(z) = \frac{z^2 + 2z}{z^2 - 2z + 1}
\qquad (E.1)
$$

in the form suitable for the z–transform in (6.7), where it must be written in powers of $\frac{1}{z}$, we should continue the long division to have such an expansion as follows:

$$
U(z) = 1 + \frac{4}{z} + \frac{7}{z^2} + \frac{10}{z^3} + \cdots + \frac{3n+1}{z^n} + \cdots
\qquad (E.2)
$$

from which it is seen from (6.7) that $u_n = 3n + 1$ for $n \geq 0$.

Example 6.5 Partial Fraction for the Inverse z-transform

Another powerful tool in finding the inverse transform of rational functions is the use of partial fraction expansions. For example, the function

$$U(z) = \frac{z + 6}{z^2 - 4} \qquad (E.1)$$

has the partial fraction expansion

$$U(z) = \frac{2}{z - 2} - \frac{1}{z + 2}. \qquad (E.2)$$

Both of these fractions have inverse z-transforms given by Example 6.3. We find that

$$u_n = \begin{cases} 0 & \text{for } n = 0 \\ 2(2)^{n-1} - (-2)^{n-1} & \text{for } n \geq 1. \end{cases} \qquad (E.3)$$

Example 6.6 The Use of Partial Fraction Expansion

We look at the two last examples which are of interest in themselves, but which also allow additional inverse z-transforms to be done by partial fraction expansion. Consider the function

$$U(z) = \frac{1}{z^2 + a^2} \qquad (E.1)$$

where a is a real number. Writing this as

$$
\begin{aligned}
U(z) &= \frac{1}{z^2}\left(1 + (\frac{a}{z})^2\right) = \frac{1}{z^2}\sum_{n=0}^{\infty}\left[-(\frac{a}{z})^2\right]^n \\
&= \sum_{n=0}^{\infty}\frac{(-1)^n a^{2n}}{z^{2n+2}}
\end{aligned}
\qquad (E.2)
$$

provided $|z| > |a|$. It now follows that

$$u_n = \begin{cases} 0, & \text{if } n = 0 \text{ or } n \text{ is odd} \\ (-1)^{\frac{n}{2}+1}a^{n-2}, & \text{if } n \text{ is even,} \end{cases} \qquad (E.3)$$

after letting $2n + 2 = k$ in the last sum, then comparing with (6.7) for u_k as the coefficient of z^{-k}.

Example 6.7   Another Inverse z-transform

Finally, the function

$$U(z) = \frac{z}{z^2 + a^2} \tag{E.1}$$

also arises in partial fraction representations. Letting a be a real number, we have

$$
\begin{aligned}
U(z) &= \frac{z}{z^2}\left(1 + (\frac{a}{z})^2\right) \\
&= \frac{1}{z}\sum_{n=0}^{\infty}\left[-(\frac{a}{z})^2\right]^n \\
&= \sum_{n=1}^{\infty}\frac{(-1)^n a^{2n-2}}{z^{2n-1}} \\
&= \sum_{k=1}^{\infty}(-1)^{\frac{k+1}{2}}\frac{a^{k-1}}{z^k}
\end{aligned}
\tag{E.2}
$$

provided $|z| > |a|$. In the last sum we made a change of variable $m = n + 1$ for the index to suit the expression (6.7) of the z-transform. Identifying the coefficients of z^{-n}, we can recover the inverse z-transform sequence $\{u_n\}$ as

$$u_n = \begin{cases} 0 & \text{if } n \text{ is even} \\ (-1)^{\frac{n+1}{2}} a^{n-1} & \text{if } n \text{ is odd,} \end{cases} \tag{E.3}$$

after another index manipulation $k = 2n - 1$ like that done for $\{u_n\}$ of the above Example 6.6.

Operational Pairs for the z-transform

The above examples illustrate finding the z-transform pairs for infinite sequences, that are of value, especially for the z-transform, which depends entirely on such pairs. This is true, since, for all practical purposes, an efficient numerical inversion in the sense of the

FFT of the discrete Fourier transforms, is not available. In addition, the operational pairs of the z-transform, for example, the pair that transforms the difference operator E in $Eu_n = u_{n+1}$ to an algebraic form in the z-transform $U(z)$,

$$\mathcal{Z}\{u_{n+1}\} = zU(z) - zu_0 \qquad (6.8)$$

is very essential in our attempt of using the z-transform as an "operational sum calculus method" for solving difference equations, that are associated (in the present case of the z-transform) with an initial condition as seen in (6.8). This is in contrast with the boundary conditions associated with the use of the discrete Fourier transforms of Chapters 4 and 5. We shall list the most important operational properties of the z-transform here and in Table 6.1, prove a few of them, and leave establishing the rest for the exercises.

The z-transform pair (6.8) can be established with the following simple computations, with due attention to the change of variables for the index of the infinite sum in the intermediate steps,

$$
\begin{aligned}
\mathcal{Z}\{u_{n+1}\} &= \sum_{n=0}^{\infty} u_{n+1} z^{-n} = z \sum_{n=0}^{\infty} u_{n+1} z^{-n-1} \\
&= z \sum_{n=0}^{\infty} u_{n+1} z^{-(n+1)} = z \sum_{m=1}^{\infty} u_m z^{-m} \\
&= z\left[-u_0 + \sum_{m=0}^{\infty} u_m z^{-m}\right] = z[-u_0 + U(z)], \\
\mathcal{Z}\{u_{n+1}\} &= zU(z) - zu_0
\end{aligned}
\qquad (6.8)
$$

after we changed variable by letting $m = n+1$ in the second line, then added the term u_0 to the sum to make it run from $m = 0$ to ∞ to fit the definition of the z-transform in (6.7), and, of course, we subtracted u_0 as it appears outside the sum. A similar calculation gives

$$\mathcal{Z}\{u_{n+2}\} = z^2 U(z) - z^2 u_0 - z u_1. \qquad (6.9)$$

Here, as we did for the Laplace transform of $\frac{d^2 f}{dx^2}$ in (2.49), we can let $v_n = u_{n+1}$, where the above u_{n+2} becomes v_{n+1}, then use the result of (6.8) for v_{n+1} with $V(z) = \mathcal{Z}\{v_n\} = \mathcal{Z}\{u_{n+1}\} = zU(z) - zu_0$ to have

$$
\begin{aligned}
\mathcal{Z}\{u_{n+2}\} &= \mathcal{Z}\{v_{n+1}\} = zV(z) - zv_0 \\
&= z[zU(z) - zu_0] - zu_1, \\
\mathcal{Z}\{u_{n+2}\} &= z^2U(z) - z^2u_0 - zu_1,
\end{aligned} \tag{6.9}
$$

where in the last step we had $v_0 = u_1$. With the same method we can establish the general result for $\mathcal{Z}\{u_{n+k}\}$,

$$
\begin{aligned}
\mathcal{Z}\{u_{n+k}\} &= z^kU(z) - z^ku_0 - z^{k-1}u_0 \quad - \quad z^{k-1}u_1 \\
&\qquad\qquad - \quad \cdots - zu_{k-1} \\
&= z^k\mathcal{Z}\{u_n\} - \sum_{m=0}^{k-1} u_m z^{k-m}
\end{aligned}
$$

$$\tag{6.10}$$

In (6.8)-(6.10) we note how the z-transform algebraizes the shifted sequence u_{n+k}, and at the same time involves all the initial values $u_0, u_1, u_2, \ldots,$ up to u_{k-1}. This is done where the shifted sequence is with constant coefficient 1, which is usually the case for almost all problems covered by the Laplace transform. On the other hand, we see that the algebraic result in (6.7), for example, have the transform $U(z)$ with variable coefficient z as $zU(z)$. So, we may say that, essentially, multiplying $U(z)$ by z in the z-transform space corresponds to the differencing (or forward shifting) u_{n+1} in the original space. On the other hand, the algebraic operation of variable coefficient in the original space, for example nu_n, will be transformed to a differentiation operation with variable coefficient $-z\frac{d}{dz}U(z)$ in the z-space. Indeed, we can easily show that

$$
\mathcal{Z}\{nu_n\} = -z\frac{dU}{dz}, \tag{6.11}
$$

and

$$
\mathcal{Z}\{n^2u_n\} = z^2\frac{d^2U}{dz^2} + z\frac{dU}{dz} \tag{6.12}
$$

This, of course, will complicate the situation a great deal in the z-space, where we now have to deal with variable coefficient differential equations instead of the simple algebraic equations corresponding to the z-transform of shifted sequences with constant coefficients. This is the reason behind seeing the use of the z-transform, and its (integral) analog, the Laplace transform, being practically limited to constant coefficient difference and differential equations, respectively. Of course, there are occasions where we do some variable coefficient equations with these operational methods, which are due to an essential need, or that we strike a compromise. For example, the second order shifted sequence u_{n+2} with variable coefficient of degree 1 as nu_{n+2}, will, according to (6.11), transform to an expression that involves only a first order derivative $\frac{dU}{dz}$,

$$
\begin{aligned}
\mathcal{Z}\{nu_{n+2}\} &= -z\frac{d}{dz}[z^2 U(z) - z^2 u_0 - zu_1] \\
&= -z[z^2\frac{dU}{dz} + 2zU(z) - 2zu_0 - u_1] \qquad (6.13) \\
&= -z^3\frac{dU}{dz} - 2z^2 U(z) + 2z^2 u_0 + zu_1
\end{aligned}
$$

This is easier to deal with in a first order differential equation than the nu_{n+2} in a difference equation of second order that is with variable coefficients, as we saw in Section 3.7. The result in (6.11) can be established when we differentiate $U(z)$, as given by its definition in (6.7) as a series, and allow the (formal) term by term differentiation inside this infinite series to have

$$
\begin{aligned}
\frac{dU(z)}{dz} &= \frac{d}{dz}\sum_{n=0}^{\infty} u_n z^{-n} = \sum_{n=0}^{\infty} u_n \frac{dz^{-n}}{dz} \\
&= \sum_{n=0}^{\infty} -nu_n z^{-n-1} = -\frac{1}{z}\sum_{n=0}^{\infty} nu_n z^{-n}, \\
-z\frac{dU}{dz} &= \sum_{n=0}^{\infty} (nu_n)z^{-n}
\end{aligned}
$$

which means that

$$
\mathcal{Z}\{nu_n\} = -z\frac{dU}{dz} \qquad (6.11)
$$

The proof of (6.12) can be obtained with the help of (6.11) after letting $v_n = nu_n$, then use (6.11) for nv_n,

$$
\begin{aligned}
\mathcal{Z}\{n^2 u_n\} = \mathcal{Z}\{nv_n\} = -z\frac{dV(z)}{dz} &= -z\frac{d}{dz}[-z\frac{dU}{dz}] \\
&= z^2\frac{d^2 U}{dz^2} + z\frac{dU}{dz}
\end{aligned} \qquad (6.12)
$$

where we used (6.11) again for $\mathcal{Z}\{v_n\} = \mathcal{Z}\{nu_n\} = -z\frac{dU}{dz}$ in the last line. We will have an illustration of solving a difference equation with variable coefficient in the Exercises of Section 6.3 (see problem 9).

The simplest property to prove for the z-transform, as a summation operation, is its *linearity property* of *superposition*. Let $U(z)$ and $V(z)$ be the respective z-transforms of the (infinite) sequences $\{u_n\}_{n=0}^\infty$ and $\{v_n\}_{n=0}^\infty$, then

$$\mathcal{Z}\{u_n + v_n\} = \mathcal{Z}\{u_n\} + \mathcal{Z}\{v_n\} = U(z) + V(z) \qquad (6.14)$$

Indeed we can easily show that for α constant,

$$\mathcal{Z}\{\alpha u_n\} = \alpha U(z), \qquad (6.15)$$

and the general linearity result that embodies both (6.14) and (6.15), namely for c_1 and c_2 constants,

$$\mathcal{Z}\{c_1 u_n + c_2 v_n\} = c_1 U(z) + c_2 V(z) \qquad (6.16)$$

We will prove this result next,

$$
\begin{aligned}
\mathcal{Z}\{c_1 u_n + c_2 v_n\} &= \sum_{n=0}^\infty [c_1 u_n + c_2 v_n] z^{-n} \\
&= \sum_{n=0}^\infty c_1 u_n z^{-n} + \sum_{n=0}^\infty c_2 v_n z^{-n} \\
&= c_1 \sum_{n=0}^\infty u_n z^{-n} + c_2 \sum_{n=0}^\infty v_n z^{-n} \\
&= c_1 U(z) + c_2 V(z)
\end{aligned}
\qquad (6.16)
$$

The proof of (6.14) and (6.15) should be easy, and we leave it for an exercise.

Another very important operational property of the z-transform, especially with regard to finding the inverse z-transform for its operational sum calculus, is its following *convolution theorem*.

The Convolution Theorem of the z-transform

Let $U(z)$ and $V(z)$ be the z-transforms of the (infinite) sequences $\{u_n\}_{n=0}^\infty$ and $\{v_n\}_{n=0}^\infty$, we define the z-transform *convolution product*

$u_n * v_n$ of the two sequences u_n and v_n as

$$u_n * v_n \equiv \sum_{k=0}^{n} u_k v_{n-k}. \tag{6.17}$$

Then, the *convolution theorem* of the z-transform states that

$$\mathcal{Z}\{\{u * v\}_n\} = \mathcal{Z}\{\sum_{k=0}^{n} u_k v_{n-k}\} = U(z)V(z) \tag{6.18}$$

and we shall leave its proof for an exercise (see problem 7(b)). Also, it should be easy to show that the convolution product in (6.17) is communicative, i.e., $u_n * v_n = v_n * u_n$, $\sum_{k=0}^{n} u_k v_{n-k} = \sum_{k=0}^{n} v_k u_{n-k}$ (see problem 7(a)).

Sometimes we may face a sequence $u_n, 0 \leq n \leq N$, that is being truncated or limited to $u_n,\ 0 \leq n \leq k - 1$, which means that all the terms $u_n,\ k \leq n \leq N$ vanished, i.e.,

$$v_n = \begin{cases} u_n, & 0 \leq n \leq k - 1 \\ 0, & k \leq n \leq N \end{cases} \tag{6.19}$$

To simplify the expression for such sequences we use the (shifted) *unit step* sequence $u_n(k)$,

$$u_n(k) = \begin{cases} 0, & 0 \leq n \leq k - 1 \\ 1, & k \leq n \leq N \end{cases} \tag{6.20}$$

With this unit step sequence in (6.20), we can write the special sequence (6.19) as

$$v_n = u_n - u_n \cdot u_n(k) = \begin{cases} u_n, & 0 \leq n \leq k - 1 \\ 0, & k \leq n \leq N \end{cases} \tag{6.21}$$

The definition of the unit step sequence and the convolution theorem will lead us to find what corresponds to dividing the z-transform $U(z)$ by z^k, i.e., $\frac{U(z)}{z^k}$ in comparison to the shifting of u_{n+k} that, essentially, corresponds to $z^k U(z)$. This correspondence is

$$\mathcal{Z}\{u_{n-k} u_n(k)\} = z^{-k} U(z) \tag{6.22}$$

which is established as follows:

$$\mathcal{Z}\{u_{n-k}u_n(k)\} \;=\; \sum_{n=0}^{\infty} u_{n-k}u_n(k)z^{-n}$$

$$=\; \sum_{n=k}^{\infty} u_{n-k}z^{-n}$$

$$=\; \sum_{m=0}^{\infty} u_m z^{-m-k} = z^{-k}\sum_{n=0}^{\infty} u_n z^{-n}$$

$$=\; z^{-k}U(z)$$

after using the definition of the unit step sequence of (6.20) in the first line, then changing variable by letting $m = n - k$ to have the final result in terms of $U(z)$.

It is easy now to establish

$$\mathcal{Z}\{u_n(k)\} = \frac{z^{1-k}}{z-1}. \tag{6.23}$$

when we use $u_n = 1$ in (6.22), where, clearly $u_{n-k} = 1$, and we note that $\mathcal{Z}\{u_n\} = \mathcal{Z}\{1\} = \frac{z}{z-1}$. Another important property is that of scaling in the z-space,

$$\mathcal{Z}\{b^n u_n\} = U(\frac{z}{b}),\; b \neq 0 \tag{6.24}$$

which can be proven easily,

$$\mathcal{Z}\{b^n u_n\} \;=\; \sum_{n=0}^{\infty} b^n u_n z^{-n} = \sum_{n=0}^{\infty} u_n (\frac{z}{b})^{-n}$$

$$=\; U(\frac{z}{b}),\quad b \neq 0$$

after using the definition $U(z)$ in (6.7) for the first line and for $U(\frac{z}{b})$ in the last line.

With these and other basic results for the z-transform of operations and sequences pairs and the few more that are summarized in the following Tables 6.1 and 6.2, we have the tools to begin solving some difference equations with *initial conditions*, as we shall illustrate in the next section.

Table 6.1: Operational pairs of the z-transform

$\{u_n\}_{n=0}^{\infty}$	$\mathcal{Z}\{u_n\}$
(a) Operations Pairs	
$\{\alpha u_n\}, \quad \alpha \in C$	$\alpha U(z)$
$\sum_{n=0}^{n} u_{n-j} v_j$, see Example 6.1	$U(z) + V(z)$
$\{a^n u_n\}$	$U(\frac{z}{a})$
$\{u_{n+j}\}$	$z^j \mathcal{Z}\{u_n\} - \sum_{m=0}^{j-1} u_m z^{j-m}$
$\{u_{n-j}\}$	$z^{-j} \mathcal{Z}\{u_n\}$
$\{\sum_{m=0}^{n} u_m\}$	$\frac{z}{z-1} \mathcal{Z}\{u_n\}$
$\{n u_n\}$	$-z \frac{d}{dz} \mathcal{Z}\{u_n\}$

Table 6.2: The z-transform of (infinite) sequences

(b) Sequences Pairs	
$\{1\}$	$\frac{z}{z-1}, \quad \|z\| > 1$
$\{n\}$	$\frac{z}{(z-1)^2}, \quad \|z\| > 1$
$\{n^2\}$	$\frac{z(z+1)}{(z-1)^3}, \quad \|z\| > 1$
$\{a^n\}, \quad a \in C$	$\frac{z}{z-a}, \quad \|z\| > \|a\|$
$\{\cos n\theta\}$	$\frac{z-\cos\theta}{z-2\cos\theta+z^{-1}}, \quad \|z\| > 1$
$\{\sin n\theta\}$	$\frac{\sin\theta}{z-2\cos\theta+z^{-1}}, \quad \|z\| > 1$
$\{b^n \cos n\theta\}$	$\frac{z^2-bz\cos\theta}{z^2-2bz\cos\theta+b^2}, \quad \|z\| > 1$
$\{b^n \sin n\theta\}$	$\frac{bz\sin\theta}{z^2-2bz\cos\theta+b^2}, \quad \|z\| > 1$
$\begin{cases} 0, & n=0 \\ a^{n-1}, & n \geq 1. \end{cases}, \quad a \in C$	$\frac{1}{z-a}, \quad \|z\| > \|a\|$
$\begin{cases} 0, & n=0 \text{ or } n \text{ is odd} \\ (-1)^{\frac{n}{2}+1} a^{n-2}, & n \text{ is even} \end{cases}$	$\frac{1}{z^2+a^2}, \quad \|z\| > \|a\|$
$\begin{cases} 0, & n \text{ is even} \\ (-1)^{\frac{n-1}{2}} a^{n-1}, & n \text{ is odd} \end{cases}$	$\frac{z}{z^2+a^2}, \quad \|z\| > \|a\|$
$\delta_{n,0} = \begin{cases} 1, & n=0 \\ 0, & n>0 \end{cases}$	1

Exercises 6.2

1. Find the z-transform of the following sequences $\{u_n\}_{n=0}^{\infty}$

   (a) $u_n = e^{-\alpha n}, \quad \alpha \in R$

   (b) $u_n = \frac{1}{2}(1 + (-1)^n) = \begin{cases} 1, & \text{if } n \text{ is even} \\ 0, & \text{if } n \text{ is odd} \end{cases}$

   (c) $u_n = (-1)^n$

   (d) $u_n = \sin n\theta$

   Hint: Use $\sin\theta = \frac{e^{i\theta} - e^{-i\theta}}{2i}$, and note that that here $|z| > 1$.

   (e) $u_n = \begin{cases} 1 & \text{if } n = 0 \\ 3(\frac{1}{2})^{n-1} & \text{if } n \geq 1 \end{cases}$

   (f) $u_n = \{0, 0, 2, 0, 0, 2, 0, 0, 2, \cdots\}$

   Hint: Here $\mathcal{Z}\{u_n\} = \sum_{n=1}^{\infty} 2z^{-3n} = -2 + \sum_{n=0}^{\infty} 2z^{-3n}$.

   For what values of z are the z-transforms of problem 1 valid?

   Hint: See that $\sum_{n=0}^{\infty} \alpha^n = \frac{1}{1-\alpha}$ is valid for $|\alpha| < 1$.

3. Find the inverse transforms of the following functions

   (a) $U(z) = \frac{1}{z+3}$

   Hint: Write $\frac{1}{z+3} = \frac{1}{z} \cdot \frac{1}{1+\frac{3}{z}} = \frac{1}{z} \sum_{n=0}^{\infty} -(\frac{3}{z})^n$

   $= \sum_{n=0}^{\infty} \frac{(-3)^n}{z^{n+1}} = \sum_{m=1}^{\infty} \frac{(-3)^{m-1}}{z^m}$, whence

   $u_m = \begin{cases} 0, & m = 0 \\ (-3)^{m-1}, & m \geq 1. \end{cases}$

   (b) $U(z) = \frac{z^2 + 4z - 2}{z^2 + z}$

   Hint: $\frac{z^2 - 4z - 2}{z^2 + z} = 1 - \frac{2}{z} + \frac{5}{z+1}$.

   (c) $U(z) = \frac{1}{z^2 - 4}$

(d) $U(z) = \frac{z}{z^2+4}$

Hint: Apply long division then see the coefficients of $\frac{1}{z^n}$.

(e) $U(z) = \frac{z^2+9}{z^2-4}$

Hint: $\frac{z^2+9}{z^2-4} = 1 - \frac{13}{4}(\frac{1}{z+2} - \frac{1}{z-2})$.

(f) $U(z) = \frac{1}{z^p}$, $p = 0, 1, 2, 3, \ldots$

4. The case in which the z-transform $U(z)$ has a multiple root in the denominator is more difficult for inversion. Consider the function $U(z) = \frac{1}{(z-2)^2}$. There are at least two ways to proceed in inverting this function: i. Use long division, ii. Use the *Convolution Theorem for the z-transform* in (6.18) which states that

$$\mathcal{Z}^{-1}\{F(z)G(z)\} = \{f * g\}_n = \sum_{k=0}^{n} f_k g_{n-k}$$

In this particular example, letting $F(z) = G(z) = \frac{1}{z-2}$ and noting that $f_n = g_n = 2^{n-1}$ for $n \geq 1$, $f_0 = g_0 = 0$ allows the sequence $\{u_n\} = \mathcal{Z}^{-1}\{U(z)\}$ to be generated.

Use both of these methods to find the inverse z-transform of

(a) $U(z) = \frac{1}{(z-2)^2}$

(b) $U(z) = \frac{z}{(z-2)^2}$

Hint: The convolution theorem method is easier, after noting in part (a) that $f_n = g_n = 2^{n-1} = \mathcal{Z}^{-1}\{\frac{1}{z-2}\}, n \geq 1, g_0 = f_0 = 0$. Also, for part (b) we have $f_n = 2^n = \mathcal{Z}^{-1}\{\frac{z}{z-2}\}, n \geq 1, f_0 = 0$ and $g_n = 2^{n-1} = \mathcal{Z}^{-1}\{\frac{1}{z-2}\}, n \geq 1, g_0 = 0$.

If it is not possible to find a general expansion for $\{u_n\}$, then simply enumerate several terms.

5. (a) Use the convolution theorem to show that

$$\mathcal{Z}\{\sum_{n=0}^{n} u_k\} = \frac{z}{z-1}U(z) \qquad (E.1)$$

Hint: For the convolution theorem (6.18) let $v_k = 1$, and note that $v_{n-k} = 1$, and $\mathcal{Z}\{v_n\} = \mathcal{Z}\{1\} = \frac{z}{z-1}$.

(b) Use the result in part (a) to show that

$$\mathcal{Z}\{\sum_{k=0}^{n} 2^k\} = \frac{z^2}{(z-1)(z-2)} \qquad (E.2)$$

Hint: Note that $\mathcal{Z}\{2^k\} = \frac{z}{z-2}$.

(c) Use the convolution theorem to establish the result in (E.2).

Hint: Let $V(z) = \frac{z}{z-1}$ and $U(z) = \frac{2}{z-2}$ where their inverses are $v_k = 1$ and $u_k = 2^k$, respectively. Then use the convolution theorem (6.18) with $v_k = 1$.

6. Find the z-transform of the following equation with $u_0 = f_0$,

$$u_{n+1} = f_{n+1} + \sum_{n=0}^{n} g_{n-k} u_k, \qquad (E.1)$$

where both f_n and g_n are known, to show that $U(z) = \frac{2F(z)}{z-G(z)}$, where $U(z)$, $F(z)$, and $G(z)$ are the z-transforms of u_n, f_n, and g_n, respectively.

Hint: Note that the sum in (E.1) is in the (z-transform) convolution product $g_n * u_n$, whose z-transform is $G(z)U(z)$.

7. (a) Prove that the convolution product in (6.17) is communicative.

Hint: Let $n - k = m$, and note that the summation from $m = n$ to $m = 0$ is the same as that for $m = 0$ to $m = n$.

(b) Prove the convolution theorem in (6.18).

Hint: Use the identity

$$\sum_{n=0}^{\infty} a_n \sum_{n=0}^{\infty} b_n = \sum_{n=0}^{\infty} \sum_{m=0}^{n} a_{n-m} \cdot b_m$$

6.3 The Operational Sum Calculus Method of the z-Transform for Solving Initial Value Problems

As we explained in the last section, the following operational–type properties of the z–transform, are essential for its use as a direct transform method of solving difference equations with initial conditions .

We recall now the basic properties of the z-transform for solving difference equations. What effect does the z-transform have upon the shifted sequence $\{u_{n+1}\}$? The answer lies with the following results

$$\mathcal{Z}\{u_{n+1}\} = zU(z) - zu_0, \tag{6.8}$$

and

$$\mathcal{Z}\{u_{n+2}\} = z^2U(z) - z^2u_0 - zu_1. \tag{6.9}$$

that we derived in the last section, and which are special cases for $j = 1, 2$ of the following z–transform pair of $\{u_{n+k}\}_{n=0}^{\infty}$,

$$\mathcal{Z}\{u_{n+k}\} = z^kU(z) - z^ku_0 - z^{k-1}u_1 - \cdots - zu_{k-1}, \tag{6.10}$$

which is found as the seventh entry in Table 6.1. This property is precisely what we need to solve difference equations since it relates the transform of a shifted sequence to the transform of the original sequence. In addition, the transform of the shifted sequence introduces the initial elements of the sequence $u_0, u_1, u_2, \cdots$. But if we are solving an initial value problem, then these initial values are given. This, of course, is exactly how the Laplace transform is used for differential equations with constant initial conditions.

The above very important pairs (6.8), (6.9) illustrate how a difference sequence in the n space is transformed via the z-transform to an algebraic expression in the z-space. Thus with this z-transform we add another (hybrid) discrete transform to the "operational sum calculus" method. Here the z-transform is compatible with difference equations, principally with constant coefficients, that are associated with initial values to comprise an initial value problem in the *infinite* sequence $\{u_n\}_{n=0}^{\infty}$. This is in variance with boundary value problems in the *finite* sequence $\{u_n\}_{n=1}^N$, where we used the discrete Fourier transform DFT and DST, DCT for solving them in Chapters 4 and 5, respectively.

We should note again that the other advantage of using such operational methods, besides algebraizing the difference equation, is that the transformation automatically incorporates the auxiliary conditions, which are the constant initial conditions here as seen in (6.8)-(6.9), and (6.10). This is also the case for the discrete Fourier transform, however, the DFT we used needed only the periodicity of the transformed sequence, which is the DFT intrinsic property. In the case of the DST and the DCT of Chapter 5, we had mostly homogeneous boundary conditions, which did not make a difference, and for most such problems, we went the other route of assuming one of these discrete Fourier transforms as a (finite) series solution for the few nonhomogeneous cases. The direct use of these discrete Fourier transforms, on the other hand, as shown in Section 2.4, and Chapters 4 and 5 were parallel to the z-transform method in immediately transforming the difference equation to an algebraic equation as well as accommodating *homogeneous* and *nonhomogeneous* boundary conditions. The involvement of nonhomogeneous boundary conditions at $n = 0$ and $n = N$ is valuable, since it can take care of the jump discontinuity $u_N - u_0$ at the end points of the periodic extension of the DFT sequence as well as the nonhomogeneous Dirichlet and Neumann boundary conditions with the direct use of the DST and DCT, respectively.

Let us return to the initial value problem (6.1)–(6.3) and solve it using the z-transform. This problem is

$$u_{n+2} - u_{n+1} - 6u_n = 0 \tag{6.1}$$

$$u_0 = 0, \tag{6.2}$$

$$u_1 = 3 \tag{6.3}$$

The first step is to take the z-transform of the entire difference equation and use properties (6.8) and (6.9) for the first and second terms of (6.1), respectively,

$$\mathcal{Z}\{u_{n+2} - u_{n+1} - 6u_n\} = \mathcal{Z}\{0\}, \quad n \geq 0$$

$$= \mathcal{Z}\{u_{n+2}\} - \mathcal{Z}\{u_{n+1}\} - 6\mathcal{Z}\{u_n\} = 0$$
$$= z^2 U(z) - z^2 u_0 - z u_1 - (z U(z) - z u_0) - 6 U(z) = 0 \qquad (6.25)$$

Substituting the given initial values $u_0 = 0$, $u_1 = 3$ and solving for $U(z)$ gives

$$U(z) = \frac{3z}{z^2 - z - 6} = \frac{9/5}{z - 3} + \frac{6/5}{z + 2}. \qquad (6.26)$$

In anticipation of taking the inverse transform of $U(z)$, it has been written in terms of partial fractions. The inverse transform comes immediately from the 9th entry in Table 6.2 and we find that

$$
\begin{aligned}
u_n &= \mathcal{Z}^{-1}\{U(z)\} = \mathcal{Z}^{-1}\{\tfrac{9/5}{z-3}\} + \mathcal{Z}^{-1}\{\tfrac{6/5}{z+2}\} \\
&= \tfrac{9}{5}(3)^{n-1} + \tfrac{6}{5}(-2)^{n-1} \qquad\qquad (6.27) \\
&= \tfrac{3}{5}(3)^n - \tfrac{3}{5}(-2)^n
\end{aligned}
$$

This solution is valid for $n \geq 2$, but notice that it also satisfies the constant initial conditions.

Of course this method is used mainly for constant coefficient difference equations with constant initial conditions, and as such it should be a simple matter to verify the above solution by using the familiar method of solving linear difference equations with constant coefficients that we covered in Section 3.2.

The use of z-transform to solve a problem which is this simple hardly offers an improvement over the method of Section 3.2 that uses the characteristic polynomial. However, the z-transform begins to justify itself for nonhomogeneous problems. Consider the same initial value problem with a right hand side sequence.

$$u_{n+2} - u_{n+1} - 6u_n = \sin\frac{\pi n}{2}, \quad n \geq 2 \qquad (6.28)$$

$$u_0 = 0, \qquad\qquad\qquad (6.29)$$

$$u_1 = 3 \qquad\qquad\qquad (6.30)$$

We proceed in much the same way as before. Transforming the entire difference equation and using the shift properties (6.8), (6.9), as we did in (6.25), we have

$$(z^2 - z - 6)U(z) - z u_1 = \mathcal{Z}\{\sin\frac{\pi n}{2}\}$$

$$= \frac{1}{z + z^{-1}} = \frac{z}{z^2 + 1} \qquad (6.31)$$

after using the z-transform of $\{\sin \frac{\pi n}{2}\}$ from the 6th entry in Table 6.2. Solving for $U(z)$

$$U(z) = \frac{3z + \frac{z}{z^2+1}}{z^2 - z - 6} = \frac{3z^3 + 4z}{(z^2 + 1)(z^2 - z - 6)} \qquad (6.32)$$

At this point, $U(z)$ must be decomposed into partial fractions, a direct calculation which gives

$$U(z) = \frac{31}{50} \frac{1}{z - 3} - \frac{14}{25} \frac{1}{z + 2} - \frac{1}{50} \frac{7z + 1}{z^2 + 1} \qquad (6.33)$$

Now each of these terms has an inverse transform from the 9th to 11th entries of Table 6.2. The final solution sequence $\{u_n\}$ may be expressed as

$$u_n = \frac{31}{50}(3)^n - \frac{16}{25}(-2)^n - \frac{1}{50} \begin{cases} (-1)^{\frac{n}{2}-1} & \text{if } n \text{ even} \\ 7(-1)^{\frac{n-1}{2}} & \text{if } n \text{ odd.} \end{cases} \qquad (6.34)$$

This solution is valid for $n \geq 0$. It may be checked by verifying that it produces the sequence $\{0, 3, 3, 20, 48, \cdots\}$ which comes from evaluating the difference equation (6.28) recursively.

We should note that the constant coefficient nonhomogeneous difference equation (6.28) is with a nonhomogeneous term $\sin \frac{\pi n}{2}$ on a UC sequence, where the initial value problem (6.28)-(6.30) can be solved with the UC method of Section 3.3 as was illustrated there with Examples 3.16, 3.12-3.15 in Section 3.3 (see also problem 3 of the exercises of the present section).

A Problem in Combinatorics

We close this chapter by solving a very famous difference equation which appears in many settings. We will introduce the equation by means of a problem in combinatorics. A store manager is putting boxes on a long shelf. He has boxes of two sizes: one type of box occupies one space on the shelf, the other type occupies two spaces on the shelf. The figure shows the shelf loaded with 6 boxes occupying 9 spaces.

There are many different ways in which boxes could be placed on the shelf 9 spaces long (for example, 9 short boxes or 4 long boxes and one short box or 3 short boxes and 3 long boxes). We ask the question: how many different combinations of long and short boxes can be arranged on a shelf which is n spaces long? Let C_n denote the number of combinations on a shelf of length n spaces. In particular we would like to find C_n for any positive integer n and for values of n which may be large. An ingenious argument leads to a difference equation.

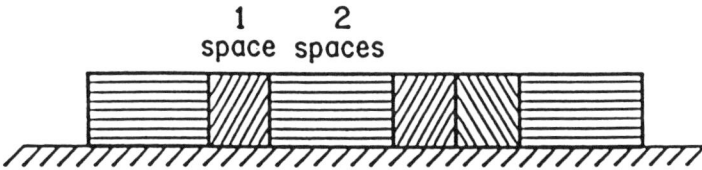

1 space **2 spaces**

Figure 6.1: A shelf loaded with boxes (of two sizes) occupying nine (equal) spaces

Assume we know how many ways the boxes may be arranged on shelves of length n and $n+1$. How can we express C_{n+2}? A shelf of length $n + 2$ can be considered a shelf of length n plus 2 spaces (one large box) or a shelf of length $(n+1)$ plus one space (one small box). In the first instance, there are C_n combinations of boxes on a shelf of length n with one large box added in each case which makes C_n combinations. In the second instance, there are C_{n+1} combinations of boxes on a shelf of length $n + 1$ with one small box added in each case which makes C_{n+1} combinations. Therefore, the total number of combinations on a shelf of length $n + 2$ spaces is

$$C_{n+2} = C_{n+1} + C_n, \quad \text{for } n \geq 1. \tag{6.35}$$

This is a second order difference equation and it is associated with the *Fibonacci sequence*. If C_1 and C_2 are specified as constant initial conditions, then we have an initial value problem which can be solved. The initial values can be easily found. On a shelf which is one space long, there is only one arrangement of boxes, namely, one short box. Thus $C_1 = 1$. On a shelf of length two spaces, there are two arrangements, namely two short boxes or one long box. Thus $C_2 = 2$. To put our initial value problem in a standard form define a new sequence $\{u_n\}$ by letting $u_n = C_{n+1}$ for $n \geq 0$. Then an equivalent initial value problem is

$$u_{n+2} - u_{n+1} - u_n = 0 \quad n \geq 0 \tag{6.36}$$

$$u_0 = 1, \tag{6.37}$$

$$u_1 = 2 \tag{6.38}$$

We may now proceed by taking a z-transform of the difference equation. Letting $U(z) = \mathcal{Z}\{u_n\}$, we have

$$z^2 U(z) - z^2 u_0 - z u_1 - [z U(z) - z u_0] - U(z) = 0 \tag{6.39}$$

Solving for $U(z)$ gives us

$$U(z) = \frac{z^2 + z}{z^2 - z - 1} = 1 + \frac{1 + \frac{2}{\sqrt{5}}}{z - \alpha_1} + \frac{1 - \frac{2}{\sqrt{5}}}{z - \alpha_2} \tag{6.40}$$

where

$$\alpha_1 = \frac{1 + \sqrt{5}}{2}, \quad \alpha_2 = \frac{1 - \sqrt{5}}{2}.$$

Once again, we have expressed $U(z)$ in partial fractions in preparation for taking the inverse transform. Using the 12th and 9th entries of Table 6.2 we find that since

$$\mathcal{Z}^{-1}\{1\} = \delta_{n,0} \tag{6.41}$$

and

$$\mathcal{Z}^{-1}\{\frac{1}{z - \alpha}\} = \begin{cases} 0, & n = 0 \\ \alpha^{n-1}, & n \geq 1 \end{cases} \tag{6.42}$$

the solution is given by

$$u_n = \begin{cases} 1, & n = 0 \\ \left(1 + \frac{2}{\sqrt{5}}\right)\left(\frac{1+\sqrt{5}}{2}\right)^{n-1} + \left(1 - \frac{2}{\sqrt{5}}\right)\left(\frac{1-\sqrt{5}}{2}\right)^{n-1}, & n \geq 1. \end{cases} \tag{6.43}$$

This expression can not be simplified any further. Despite its complexity and the appearance of irrational numbers, it can be checked that this expression generates the sequence of integers $\{u_n\}_{n=0}^{\infty} = \{1, 2, 3, 5, 8, \cdots\}$. The number of combinations of boxes on a shelf of length n can be found by re-adjusting the index such that $C_{n+1} = u_n$ for $n \geq 0$. Clearly, the number of combinations grows rapidly with the length of the shelf since

$$\{C_n\}_{n=1}^{\infty} = \{C_1, C_2, C_3, \cdots\} = \{1, 2, 3, 5, 8, \cdots\}.$$

A number of interesting illustrations that are modeled as initial value problems associated with linear constant coefficients difference equations, which can use the z-transform method, are found in Examples 7.1, 7.2, 7.5 and in (7.20)-(7.21) of Chapter 7. They model, respectively, the chemical concentration in a tank due to the input as drops injected at equal intervals, the compounding of interest on investments, the evaluation of a determinant, and the recurrence relation for the Tchebychev polynomials. Of course, and as we mentioned towards the end of Sections 3.2 and 3.3 of Chapter 3, these problems can be solved by the traditional methods discussed there.

Exercises 6.3

1. Prove the results (6.9) and (6.10) for the z-transform of shifted sequences:

   (a) $\mathcal{Z}\{u_{n+2}\} = z^2 U(z) - z^2 u_0 - z u_1$

   $$(6.9)$$

   (b) $\mathcal{Z}\{u_{n+k}\} = z^k U(z) - z^k u_0 - z^{k-1} u_1 - \cdots - z u_{k-1}$

   $$(6.10)$$

   Hint: See the derivation of (6.9) following (6.8) of Sec. 6.2.

2. Use the (UC) method of Section 3.3 to solve the initial value problem (6.28)-(6.30), and compare your answer with that of the z-transform method as we obtained it in (6.34).

3. Use the method of Section 3.2 to solve the initial value problem (6.36)-(6.38), and compare your answer with that of the z-transform method as in (6.43).

4. Solve the following initial value problems using the z-transform

   (a) $$u_{n+1} + 4u_n = 0, \quad n \geq 0$$

$$u_0 = 2$$

   (b) $$u_{n+1} - 3u_n = 6, \quad n \geq 0$$

$$u_0 = 0$$

   (c) $$u_{n+2} - 9u_n = 0, \quad n \geq 0$$

$$u_0 = 0, \quad u_1 = 2$$

   (d) $$u_{n+2} - 2u_{n+1} - 15u_n = 2^n, \quad n \geq 0$$

$$u_0 = u_1 = 0$$

Hint: For the resulting z-transform $U(z) = \frac{z}{z-2} \frac{1}{z^2-2z-15} = \frac{z}{(z-2)(z-5)(z+3)}$, use its partial fraction $\frac{z}{(z-2)(z-5)(z+3)} = -\frac{2}{15}\frac{1}{z-2} + \frac{5}{24}\frac{1}{z-5} - \frac{3}{40}\frac{1}{z+3}$.

5. Solve the initial value problem in exercise 4 by using the method of Section 3.2, 3.3. Then compare the answers with their corresponding ones obtained via the z-transform in the above problem.

6. The case in which the characteristic polynomial of the difference equation has a multiple root introduces some additional work when using the z-transform. Solve the initial value problem

$$u_{n+2} - 4u_{n+1} + 4u_n = 0, \quad n \geq 0$$

$$u_0 = 0, \quad u_1 = 2$$

using the results of problem 4(b) of Section 6.2. Try to find a general expression for $\{u_n\}$ that is valid for all n.

7. Use the method of Section 3.2 to solve exercise 6, then compare the two answers.

8. Consider the following difference equation of the first order with *variable coefficients*, and with an initial condition

$$(k+1)u_{k+1} - (50-k)u_k = 0 \qquad (E.1)$$

$$u_0 = 1 \qquad (E.2)$$

(a) Find the z-transform of this initial value problem (E.1)-(E.2).

   Hint: For this variable coefficient equation (E.1), you need the z- transform operational pairs (6.11) and (6.8) for the z-transform of ku_k and u_{k+1} respectively. In addition, watch for

$$\mathcal{Z}\{ku_{k+1}\} = -z\frac{d}{dz}[zU(z) - zu_0]$$
$$= -z^2\frac{dU}{dz} - zU(z) + z^2.$$

(b) Find the solution u_n.

   Hint: You may need the pair

$$\mathcal{Z}\left\{\begin{pmatrix} m \\ n \end{pmatrix}\right\} = \frac{(z+1)^m}{z^m},$$

where $\begin{pmatrix} m \\ n \end{pmatrix} = \frac{m!}{(m-n)!n!}$.

Chapter 7

MODELING WITH DIFFERENCE EQUATIONS

In this chapter, we return to the questions of how difference equations are formulated and how difference equations arise. We will proceed by example and attempt to give only a broad sample of the many uses of difference equations. Whenever appropriate, the methods of the previous chapters will be applied to solve the various equations. Of course, we have added here a new feature for this book, namely, the use of the discrete Fourier transforms for solving difference equations associated with boundary conditions, especially their direct use in algebraizing the difference equation and at the same time accommodating the boundary conditions. The z-transform has long been used in difference equations associated with initial conditions. Thus, the direct use of the discrete Fourier transforms in this book plus the use of the z-transform constitute examples of what we term, the "operational sum calculus" method. We may add that more "modified" discrete Fourier and the more general "Sturm–Liouville" – type discrete transforms may be constructed in parallel to what was done for their parallels of the general finite transforms that were discussed briefly in Section 2.3.

To use these examples of this chapter as illustrations for the methods of solutions in the previous chapters, we have made a clear reference to them at the end of the section that is most appropriate for the method of solution of the particular example or examples.

7.1 Various Discrete Problems Modeled as Difference Equations

Example 7.1   Stirred Tank Problems

This is a problem in which the necessity of using a difference equation is particularly clear, because the process which is being modeled has an unavoidably discrete nature. Imagine a large tank which contains a volume V liters of solution. At the beginning of the experiment, the solution has a concentration of solute of K gm/liter. At regular intervals, an input pipe injects one drop of volume v into the tank (Fig. 7.1) with a concentration of solute of K_I gm/liter. At the same time, an outlet pipe emits one drop of volume v from the tank. We assume that the tank is stirred continuously so that each new drop of solution is thoroughly mixed in a short period of time. The problem is to find the concentration of the solution in the tank after n drops have been added.

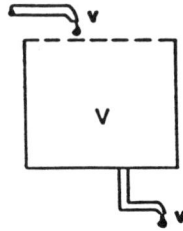

Figure 7.1: A chemical concentration problem

Before formulating the problem, it will pay to make a few qualitative observations. First, it is clear that the volume of the solution in the tank remains a constant V throughout the experiment. Secondly, as the experiment proceeds, the solution which was initially in the tank is replaced by the incoming solution. After a sufficiently long time, we would expect most of the original solution to have been replaced by solution of concentration K_I. Therefore, our solution to the problem should have the property that as the number of drops, n becomes large the concentration in the tank approaches K_I.

Let C_n denote the concentration of the solution in the tank after n drops have been exchanged. Assuming we can find C_n, how can C_{n+1} be determined? The amount of solute (in grams) in the tank after the $(n+1)$st drops are exchanged is given by

$$
\begin{array}{ccc}
C_n V & + \quad K_I v & - \quad C_n v \\
\text{previous} & \text{incoming} & \text{outgoing} \\
\text{solute} & \text{solute} & \text{solute}
\end{array}
$$

Therefore, the concentration of the solution after $n+1$ drops is

$$
C_{n+1} = \frac{C_n V + K_I v - C_n v}{V} = C_n (1 - \frac{v}{V}) + K_I \frac{v}{V}. \tag{7.1}
$$

Letting $B = \frac{v}{V}$, we recognize this as a first order, constant coefficient, nonhomogeneous difference equation with an initial condition $C_0 = K$ and a constant nonhomogeneous term $K_I B$.

$$
C_{n+1} - (1 - B)C_n = K_I B \tag{7.2}
$$

$$
C_0 = K \tag{7.3}
$$

This problem is an initial value problem associated with first order, linear, constant coefficient equation in C_n, hence it is suited to a solution by z-transform. Perhaps simpler in this case, we may follow the method of Section 3.3 and recognize that a complementary solution is given by $C_n^{(c)} = (1 - B)^n$ and the *particular* solution is $C_n^{(p)} = K_I$. Therefore, the general solution is

$$
C_n = A(1 - B)^n + K_I \tag{7.4}
$$

The initial condition $C_0 = K$, requires that $A = K - K_I$. The final solution to the problem is now given by

$$
C_n = (K - K_I)(1 - \frac{v}{V})^n + K_I. \tag{7.5}
$$

Notice that since $1 - \frac{v}{V} < 1$, $\lim_{n\to\infty} C_n = K_I$ as we had anticipated. Furthermore, if the initial concentration in the tank is less
than the incoming drop concentration, $(K - K_I < 0)$, then $\{C_n\}$ is
an increasing sequence converging to K_I. Otherwise, $(K - K_I > 0)$,
$\{C_n\}$ decreases to K_I. See also problem 1 of the Exercises for solving the initial value problem (7.2)-(7.3) by the z-transform method
of Section 6.3, and also by the above UC method of Section 3.3.

   This problem could be made more difficult, but still solvable by
the method of this book, by allowing the incoming drop concentration
to vary. For example, to represent increasingly dilute drops, we might
replace K_I by $K_I e^{-\alpha n}$, where α determines the rate of dilution.

   A significantly more difficult problem is obtained by plugging the
outlet pipe in the tank. The volume of the solution in the tank now
increases from its initial volume V by an amount v with each drop.
The governing difference equation is now

$$C_{n+1} = \frac{C_n(V + nv) + K_I v}{V + (n + 1)v} \tag{7.6}$$

This equation is still first order and nonhomogeneous, but no longer
constant coefficient. It can be solved by the methods discussed in
Section 3.6 in Chapter 3. We, definitely, would not want to use the
z-transform on it because of its variable coefficient.

Example 7.2   Compounding of Interest on Investments

We will briefly mention some other processes that lead to difference
equations very much like (7.2). One common problem is the compounding of interest on an investment. Assume that an amount $\$P$
is put in an account and that at regular intervals (say monthly or
quarterly) interest is added at a rate i for that period. In addition,
during each period a deposit of $\$d$ is made into the account. How
much is in the account after n interest payments have been made?

   The account increases its value from period n to period $(n + 1)$
due to the addition of i times the current balance in the account and
also the deposit of $\$d$ to the account. If we let P_n be the balance in
the account after n interest payments, we have that

$$P_{n+1} = P_n + iP_n + d = (1 + i)P_n + d \tag{7.7}$$

$$P_0 = P \tag{7.8}$$

This is the same initial value problem which governs the stirred tank problem. A very similar equation may be used to describe the payments on a loan. Here we have an initial value problem in (7.7)-(7.8), associated with a first order, nonhomogeneous difference equation (7.7), where the z-transform method or the usual method of undetermined coefficients of Section 3.3 can be used.

Example 7.3 Population Growth

Finally, one of the simplest models describing population growth also takes the form of (7.1). If a population is measured at regular intervals and during the intervening period is assumed to increase by a fraction b due to births and decrease by a fraction c due to deaths, then we have

$$\begin{aligned}
P_{n+1} &= P_n + bP_n - cP_n \\
P_{n+1} &= (1 + b - c)P_n
\end{aligned} \tag{7.9}$$

where P_n is the population after n intervals, and P_0 is the initial population at the initial time $n = 0$. This difference equation also takes the form of the preceding examples. It is first order, homogeneous and constant coefficient, where the method of Section 3.2 applies. With the initial condition P_0 given, we can also use the z-transform method. Solving this difference equation (see exercises) shows that the population grows exponentially if $b > c$ (births outnumber deaths) and decays exponentially if $b < c$.

Example 7.4 Mechanical Systems–Coupled Spring and Mass System

Very often, Mechanical Systems consisting of several distinct components can be modeled in terms of a difference equation. The following problem is instructive because it can be formulated first as a *differential-difference equation*, which in turn can be reduced to a difference equation. Consider a system of $N-1$ identical masses resting on a horizontal surface and connected by identical springs. The first and last springs are attached to fixed supports. When the system is at equilibrium (all springs neither stretched nor compressed), the masses are all separated by the same distance as shown in Fig. 7.2.

From this equilibrium configuration, each mass is given its own displacement and velocity along the line of the springs and the entire system is set into motion. The problem is to describe the ensuing motion of each mass as it evolves in time.

Figure 7.2: Coupled spring and mass system at equilibrium, $t = 0$

The system as it might appear sometime after the beginning of the experiment is shown in Fig. 7.3 for $t > 0$.

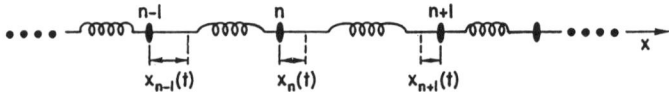

Figure 7.3: Coupled spring and mass system in motion, $t > 0$

The distance between the nth mass and its equilibrium position is marked $x_n(t)$ which is a function of time. In order to formulate the problem, it is necessary to write an equation of motion for each mass. This equation must be Newton's second law in which all forces acting on the particle have been included. We shall assume that the only forces acting on a mass are the restoring forces of the springs to which the mass is attached. In so doing, we neglect friction, gravity and other external forces. To model the restoring force of a spring, we may use Hooke's Law which says that the force is proportional to the amount that the spring is stretched or compressed and acts in the direction opposite to the extension of the spring. The constant of proportionality that gives the restoring force is the spring constant

$K > 0$ which we shall take to be the same for all springs in the system.

Consider the nth mass in the system. It experiences a force $-K(x_n - x_{n-1})$ due to the spring which attached it to the $(n-1)$st mass and a force $-K(x_n - x_{n+1})$ due to the spring which attaches it to the $(n+1)$th mass. To verify this several cases must be examined. In the case shown in Fig. 7.3, $x_n - x_{n-1} > 0$ and $-K(x_n - x_{n-1}) < 0$ which means the force due to the spring $(n-1, n)$ acts in the negative x-direction or to the left. This is consistent with the figure in which this spring is stretched. Similarly, $-K(x_n - x_{n+1}) > 0$ which means that spring $(n, n+1)$ pulls mass n to the right.) Therefore, the equation of motion for the nth mass is

$$
\begin{aligned}
m\frac{d^2 x_n(t)}{dt^2} &= -K(x_n - x_{n-1}) - K(x_n - x_{n+1}) \\
&= K(x_{n-1} - 2x_n + x_{n+1}). \ 1 \le n \le N - 1
\end{aligned}
\tag{7.10}
$$

which is called a "differential–difference" equation, since it involves differentiation with respect to t, the time variable, and differencing in x_n, the spacial variable. We may remark here that the discretization or differencing process is done for the spacial variable x_n, hence it is very likely that we expect a boundary value problem in x_n, especially in anticipation of the following two *boundary conditions* that describe what happens to the two masses that are supported (fixed) at the two end points $n = 0$ and $n = N$.

If the end supports are denoted mass 0 and mass N, then we may include the boundary conditions

$$
x_0 = x_N = 0
\tag{7.11}
$$

Equation (7.10) constitute a set of coupled, ordinary differential equations. The next step is to reduce these equations to a set of difference equations. This can be done if we note that the motion of the masses about their respective equilibrium positions will be oscillations of periodic motions. Therefore it is not unreasonable to assume that the unknown displacement function x_n can be expressed as periodic function. One way to do this is to assume solutions of the form

$$
x_n(t) = A_n \cos(\omega t + \phi)
\tag{7.12}
$$

where A_n is the amplitude of the oscillation of the nth mass, ω is its frequency and ϕ is a phase which essentially allows for the fact that different masses may pass through their equilibrium at different times. Substituting this solution into (7.10) and canceling like terms on both sides gives now a difference equation in A_n

$$-m\omega^2 A_n = K(A_{n-1} - 2A_n + A_{n+1}), \quad 1 \le n \le N - 1,$$

with boundary conditions

$$A_0 = A_N = 0$$

or the following *boundary value problem* in A_n,

$$A_{n+1} - \left(2 - \frac{m\omega^2}{K}\right) A_n + A_{n-1} = 0, \qquad (7.13)$$

$$A_0 = 0$$

$$A_N = 0. \qquad (7.14)$$

We have arrived at a second order, homogeneous difference equation for the amplitude A_n. This boundary value problem appears to be suited to a discrete sine transform because of the form of the boundary conditions. However, the equation itself is also homogeneous, which makes it easier to solve by the familiar methods of Chapter 3, in particular Section 3.2. Nevertheless, we will proceed by assuming a solution of the form given by the DST (5.3)

$$A_n = \sum_{k=1}^{N-1} b_k \sin \frac{\pi n k}{N}, \quad 1 \le n \le N - 1$$

This solution satisfies the boundary conditions already. Substituting this expression into the difference equation of (7.13) and simplifying eventually results in

$$\sum_{k=1}^{N-1} b_k [2 \cos \frac{\pi k}{N} - (2 - \frac{m\omega^2}{K})] \sin \frac{\pi n k}{N} = 0., \quad 1 \le n \le N - 1$$

This sum vanishes only if each coefficient of $\sin \frac{\pi n k}{N}$ vanishes, which requires that either $b_k = 0$ for $1 \leq k \leq N - 1$ or

$$\frac{m\omega^2}{K} - 2\left(1 - \cos \frac{\pi k}{N}\right) = 0, \quad 1 \leq k \leq N - 1$$

The first alternative ($b_k = 0$) leads nowhere as it would imply that all of the amplitudes A_n are also zero. Therefore, we are left with the second condition. But on what variable is this a condition? The quantities m, K and N are given. Only the frequency ω was never specified when it was includes in the trial solution for x_n. We have now determined the admissible values of ω. They are

$$\omega_k^2 = \frac{2K}{m}\left(1 - \cos \frac{\pi k}{N}\right), \quad 1 \leq k \leq N - 1 \tag{7.15}$$

and we see that, in fact, there are $N - 1$ possible values of ω.

But now we are in a curious position. We began by trying to solve the difference equation (7.13) for the amplitude $\{A_n\}$. However, instead of finding the coefficients $\{b_k\}$, from which the amplitudes could be computed, we found a set of values for the frequency ω. This state of affairs occurs whenever there is a linear, *homogeneous* problem in which an *unspecified parameter* (in this case ω) appears. The parameter is called an *eigenvalue*. The calculations above tell us that the solution to the difference equation (7.13) is $A_n = 0$ except when ω takes one of the $N - 1$ special values given by (7.15). For these values of ω there is a non-trivial solution $\{A_n\}$. At this point, however, this non-trivial solution is still undetermined. Briefly, the problem is completed as follows. A similar problem (with homogeneous difference equation), which is the only boundary value problem done in Chapter 3 is found in problem 15 of Section 3.2 with very detailed hints. Boundary value problems associated with nonhomogeneous difference equations are solved in Chapters 4, 5 and Section 2.4 of Chapter 2.

We assumed a solution for the motion of the nth mass of the form

$$x_n(t) = A_n \cos(\omega t + \phi_n), \quad 1 \leq n \leq N - 1 \tag{7.16}$$

We have now discussed that for each mass, there are actually $(N-1)$ different frequencies ω that give a non-trivial amplitude. Therefore, for each mass, there are $(N-1)$ different modes which are solutions. We could write the modes for the nth mass as

$$x_{n,k}(t) = A_{n,k} \cos(\omega_k t + \phi_k), \quad 1 \le k \le N-1$$

We do have some additional information about the modes of the amplitudes. Recall that a solution for the nth amplitude was assumed to have the form

$$A_n = \sum_{k=1}^{N-1} b_k \sin \frac{\pi n k}{N}$$

Therefore the kth mode of A_n can be written

$$A_{n,k} = b_k \sin \frac{\pi n k}{N}$$

Since each of the modes

$$x_{n,k}(t) = b_k \sin \frac{\pi n k}{N} \cos(\omega_k t + \phi_k), \quad 1 \le k \le N-1$$

is a solution for the nth mass, the sum of all of these modes is also a solution. We finally have the motion of the nth mass given by

$$x_n(t) = \sum_{k=1}^{N-1} x_{n,k}(t) = \sum_{k=1}^{N-1} b_k \sin \frac{\pi n k}{N} \cos(\omega_k t + \phi_k),$$

$$1 \le n \le N-1 \tag{7.17}$$

Since the frequencies ω_k have been determined only the $2(N-1)$ constants $b_k, \phi_k, 1 \le k \le N-1$ are unspecified. Once the initial displacement and velocity of each mass are given, the amplitudes and the phases, b_k and ϕ_k can be found for the final solution of (7.17).

Example 7.5   Mathematical Problems–Evaluating Determinants

Difference equations often arise when mathematical problems are reformulated. Consider the problem of evaluating the determinant of the $n \times n$ tridiagonal matrix

$$A_n = \begin{vmatrix} 2a & b & & & & \\ b & 2a & b & & 0 & \\ & b & 2a & b & & \\ & & & \ddots & & \\ & 0 & & & \ddots & b \\ & & & & b & 2a \end{vmatrix}$$

Using the cofactor expansion about the last column gives

$$A_n = 2aA_{n-1} - b^2 A_{n-2}, \quad n \geq 2$$

which is a second order difference equation. It can be solved if initial conditions are given. We can compute A_1 and A_2 of the above matrix with $n = 1$ and $n = 2$ to be

$$A_1 = 2a, \quad A_2 = 4a^2 - b^2$$

Working "backwards" with the difference equation also gives

$$A_0 = \frac{2aA_1 - A_2}{b^2} = 1$$

This leads to the following initial value problem which, when solved, gives the determinant of all orders.

$$A_n - 2aA_{n-1} + b^2 A_{n-2} = 0, \quad n \geq 2 \tag{7.18}$$

$$A_0 = 1, \quad A_1 = 2a \tag{7.19}$$

This *initial value problem* may be solved using either z-transform as described in Chapter 6, or the usual method of Section 3.2 that involves the characteristic polynomial equation (see exercise 5).

The *initial value problem* (7.18)-(7.19) is very closely related to the recurrence relation for the Tchebyshev polynomials which are worthy of mention in passing, here, although we had it as the subject of problem 3 in Exercises 1.1. The initial value problem

$$T_{n+1}(x) - 2xT_n(x) + T_{n-1}(x) = 0, \quad n \geq 1 \tag{7.20}$$

$$T_0(x) = 1, \quad T_1(x) = x \tag{7.21}$$

can be shown to have the solution

$$T_n(x) = \cos(n\theta) \tag{7.22}$$

where $\theta = \cos^{-1}(x)$.

In solving this initial value problem, x may be regarded as a parameter in the interval $-1 \leq x \leq 1$. The sequence of polynomial may be generated either directly from the difference equation or from its solution. For example, from (7.22) we have

$$T_2(x) = \cos(2\theta) = 2\cos^2\theta - 1 = 2x^2 - 1$$

$$T_3(x) = \cos(3\theta) = 4\cos^3\theta - 3\cos\theta = 4x^3 - 3x.$$

At this point it may be easier to use the difference equation to produce the higher degree polynomials. Tchebyshev polynomials are an example of a large and important class of functions called *orthogonal polynomials* such as the Legendre polynomial $P_n(x)$, $-1 \leq x \leq 1$, which was discussed along with its main recurrence relation in Example 1.4, in addition to its other recurrence relations that were the subject of problem 2 of Exercises 1.1. Orthogonal polynomials can generally be defined by a difference equation (or recurrence relation) such as (7.20). Orthogonal polynomials may also serve as the kernels of other discrete transforms.

Example 7.6   Evaluating Integrals

Difference equations often arise in evaluating families of integrals. Consider the integral

$$I_n = \int_0^\infty x^n e^{-x} dx, \quad n \geq 0. \tag{7.23}$$

For large n, the evaluation of this integral would be quite labourous. However, one integration by parts gives us

$$I_n = nI_{n-1} \tag{7.24}$$

We may also quickly find that $I_0 = 1$ from evaluating the simple improper integral in (7.23) for $n = 0$. This leaves us with a first order initial value problem. However, it is a variable coefficient problem, which can be solved by the methods of Section 3.6. Fortunately, for this very special form of (7.24), a solution may be found by inspection since (7.24) is one defining relation of the *factorial function* as a special case of the gamma function $\Gamma(\mu)$ of (3.82) for $\mu = n + 1$. The solution is simply $I_n = n!$ (where the convention $0! = 1$ satisfies the initial condition $I_0 = 1$). This example demonstrates how significantly a problem may be simplified when it is reformulated, in this case, in terms of a difference equation.

Exercises 7.1

1. Use the UC method of Section 3.3, and also the z-transform method of Section 6.3 to solve the initial value problem (7.2)-(7.3) of the stirred tank, then compare your solutions with the solution we obtained in (7.5).

2. A large tank holds 10 liters of solution which has a concentration of 0.1 gm of salt per liter. One milliliter drops of solution with a concentration of 0.5 gm of salt per liter are added to the tank each second and at the same time one milliliter drops leave the tank. Assuming the tank is always thoroughly stirred,

   (a) what is the eventual steady state concentration of the solution in the tank? Note that the incoming drops are measured in millimeter (10^{-3} liter).

   (b) how long does it take the solution in the tank to reach 90% of its steady state concentration?

   Hint: Note here that *one* millimeter drops per second are added, hence time coincides with the index n, the number of drops.

3. An amount of $500 is put in an account that compounds interest monthly at an annual rate of 8%. Find the difference equation that describes the change in the balance in the account. What is the balance after four years? If a $10 deposit

is added to the account each month, what is the balance after four years?

4. (a) Use the method of Section 3.2 to solve the initial value problem of the population forecasting in (7.9) with initial population P_0.

   (b) Use the form of the solution in part (a) to verify the comments made that the population grows exponentially if $b > c$ in (7.9) and decays exponentially if $b < c$.

   (c) Use the method of the z-transform to solve the population problem in part (a), then compare your two answers.

5. (a) Use the method of Section 3.2 to solve the initial value problem (7.18)-(7.19).

   (b) Find the determinant of the 20×20 matrix whose elements are zero except for $a_{ii} = 2$, $a_{i,i+1} = -1$, $a_{i+1,i} = -1$.

6. One of the earliest population models assumed the following rules. Each individual can reproduce for two generations and produces one offspring each generation. If the first generation has one individual, find the difference equation that describes how many offspring are produced in the nth generation.

   Hint: Make a list of the number of *offspring* P_n produced by each generation n starting with $n = 2$, since this is the first offspring produced by the first generation $n = 1$. Note that $P_0 = 0, P_1 = 1, P_2 = P_1 + P_0 = 1$, and $P_3 = P_2 + P_1 = 2$ as the two offspring, one from the first individual and the other from the first offspring. $P_4 = P_3 + P_2 = 2 + 1 = 3$. Also, look for a very familiar sequence.

7.2 Solution of the Altruistic Neighborhood Modeled by DCT

Example 7.7   Diffusion Process–The Altruistic Neighborhood Model

A model of extreme altruism could be formulated as follows. A neighborhood consists of $N + 1$ families that live along one side of street as show in Fig. 7.4. The rule that governs the economy of the neighborhood is that each family gives money to its nearest neighbors when it has more wealth than those neighbors. A family also receives from its nearest neighbors when those neighbors are wealthier. In either case, the amount exchanged between any two neighbors is proportional to the difference in wealth between the neighbors, with a proportionality constant α. We assume that this rule has been in effect for a long time, and we would like to determine the final, "steady state" distribution of wealth in the neighborhood.

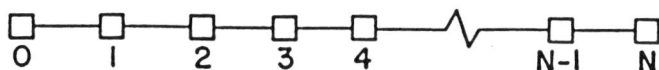

Figure 7.4: A model for $N + 1$ neighbors

This particular problem has been chosen, not for its realism, but rather because the process is rather easy to visualize. However, the arguments which are used to model this system are much like the arguments that a physicist would use to describe the conduction of heat, that an environmental biologist would use to describe the effect of a toxic substance in a pool or that a sociologist would use to describe the spread of rumors. In all of these problems, there is a process of *diffusion* which must be modeled.

To model the altruistic neighborhood, consider the exchanges of wealth that involve family n of $1 \leq n \leq N - 1$. Let G_n be the "steady state" amount of wealth of family n. This family exchanges an amount $\alpha(G_{n+1} - G_n)$ with family $n+1$. Notice that if $G_{n+1} > G_n$, then this amount is positive which means that family n receives, whereas if $G_{n+1} < G_n$, family n gives. In a similar way, family n exchanges an amount $\alpha(G_{n-1} - G_n)$ with family $n - 1$. When this system has reached a steady state, then the wealth of each family remains unchanged, which means that each family has a net transaction of zero with its neighbors. This means that

$$\alpha(G_{n+1} - G_n) + \alpha(G_{n-1} - G_n) = 0$$

or

$$G_{n+1} - 2G_n + G_{n-1} = 0. \tag{7.25}$$

This gives us a difference equation. In order to arrive at some boundary conditions, we need to have some additional information about the neighborhood. In formulating these boundary conditions, care must be taken to insure that a steady state can exist. For example, if we specified boundary conditions of a Neumann type that allowed a new wealth to come into the neighborhood, then a steady state could never be reached.

One boundary condition which does allow for a steady state solution is that $\delta G_0 = \delta G_N = 0$ which implies that no wealth enters or leaves through the $n = 0$ and $n = N$ families. It is easily verified that $G_n = c$, where c is any constant is a solution of (7.25) with this boundary condition. This says that eventually, the total wealth of the neighborhood distributes itself evenly. Extreme altruism eliminates all differences in the absence of other factors.

A slightly more interesting problem can be posed if we assume that family N brings wealth into the neighborhood at the same rate that family 0 loses wealth. Note that a steady state is possible under these assumptions. The boundary conditions which describe this situation are $\delta G_0 = \delta G_N = A$, where $A > 0$ is a constant. Now equation (7.25) together with this boundary condition presents a boundary value problem.

The model may be made somewhat more realistic if we suppose that the families have external sources of wealth. The difference equation (7.25) becomes nonhomogeneous in this case

$$G_{n+1} - 2G_n + G_{n-1} + F_n = 0, \;\; 0 \le n \le N \tag{7.26}$$

Once again, if we are to have a steady state solution, then sequence $\{F_n\}$ which accounts for the external sources (and sinks) of wealth must average to zero over the interval $0 \le n \le N$. There are many ways that external sources of the wealth might be specified. For example, if family p is a creditor and family q is a debtor, then

$F_n = W\delta_{n,p} - W\delta_{n,q}$ as shown in Fig. 7.5. If the left half of the neighborhood is providing wealth and the right half is losing wealth, then a distribution sequence $F_n = W \cos\frac{\pi n}{N}$ might be appropriate, as shown in Fig. 7.6. Either of these choices in Fig. 7.5, 7.6 leads to a boundary value problem which may be solved using the DCT.

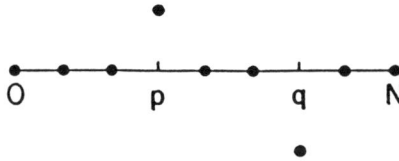

Figure 7.5: Single creditor – single debtor distribution

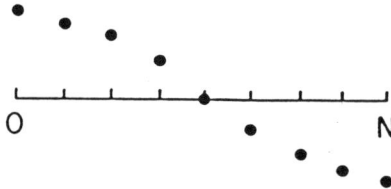

Figure 7.6: Left – right distribution

Let us carry out the solution of the single creditor–single debtor problem. It is modeled by the boundary value problem

$$G_{n+1} - 2G_n + G_{n-1} = -(W\delta_{n,p} - W\delta_{n,q}) = -F_n, \quad 0 \le n \le N \quad (7.27)$$

$$\delta G_0 = \delta G_N = 0. \quad (7.28)$$

This problem is recognized as special case of the boundary value problem (5.17)-(5.18) (with $\alpha = 1$ and $\beta = -2$) with its Neumman type boundary condition (5.18), which was analyzed in (5.17)-(5.22), via the typical method of substituting the DCT as a series–type solution. Following the procedure outlined in Chapter 5, we must first transform the right hand side sequence $\{F_n\}$ using the DCT. As a sub-problem, it is necessary to find the DCT of the sequence $\{\delta_{n,\alpha}\}$

where $1 \leq \alpha \leq N - 1$ is an integer. If this sequence is represented in the form (5.1)

$$\delta_{n,\alpha} = \frac{a_0}{2} + \sum_{k=1}^{N-1} a_k \cos \frac{\pi n k}{N} + \frac{a_N}{2}(-1)^n, \quad 0 \leq n \leq N \qquad (7.29)$$

then the coefficients $\{a_k\}$ are given by (5.2)

$$a_k = \frac{1}{N} \left\{ \delta_{0,\alpha} + 2 \sum_{n=1}^{N-1} \delta_{n,\alpha} \cos \frac{\pi n k}{N} + \delta_{N,\alpha}(-1)^k \right\}, \quad 0 \leq k \leq N$$

Assuming $\alpha \neq 0$ and $\alpha \neq N$, we have

$$a_k = \frac{2}{N} \cos \frac{\pi k \alpha}{N}, \quad 0 \leq k \leq N \qquad (7.30)$$

Therefore, the right hand side sequence $\{F_n\}$ has the representation

$$F_n = \frac{f_0}{2} + \sum_{k=1}^{N-1} f_k \cos \frac{\pi n k}{N} + \frac{f_N}{2}(-1)^n, \quad 0 \leq n \leq N \qquad (7.31)$$

where

$$f_k = \frac{2W}{N} \left\{ \cos \frac{\pi k p}{N} - \cos \frac{\pi k q}{N} \right\}, \quad 0 \leq k \leq N \qquad (7.32)$$

The method of solution described in Section 5.2, specially in (5.17)-(5.22) for the same type Neumann boundary value problem (5.17)-(5.18), may be followed where the coefficients of the difference equation are $\alpha = 1$, $\beta = -2$. The solution to the boundary value problem (5.17)-(5.18) has the form

$$G_n = \frac{a_0}{2} + \sum_{k=1}^{N-1} a_k \cos \frac{\pi n k}{N} + \frac{a_N}{2}(-1)^n, \quad 0 \leq n \leq N \qquad (7.33)$$

where the coefficients $\{a_k\}$ are given by (5.22), with its nonhomogeneous term F_n in the right hand side instead of $-F_n$ in the above (7.27)

$$a_k = \frac{f_k}{2\alpha \cos \frac{\pi k}{N} + \beta}, \quad 0 \leq k \leq N, \ |2\alpha| < |\beta| \qquad (7.34)$$

Another interesting feature emerges at this point. Notice that for our particular difference equation (7.27) with $\alpha = 1$, $\beta = -2$, the zeroth coefficient a_0 is undefined in (7.34). Since a_0 is the constant term in the representation for G_n, this implies that G_n is determined only up to an additive constant. In the absence of any further information that might allow this constant to be determined, it can be chosen arbitrarily (or set to zero). However the remaining coefficients are well defined and given by

$$a_k = \frac{W}{N} \frac{\cos\frac{\pi k p}{N} - \cos\frac{\pi k q}{N}}{\cos(\frac{\pi k}{N}) - 1}, \quad 1 \leq k \leq N \tag{7.35}$$

This solution $\{G_n\}$ may then be reconstructed as

$$G_n = \frac{a_0}{2} + \frac{W}{N} \left\{ \sum_{k=1}^{N-1} \frac{\cos\frac{\pi k p}{N} - \cos\frac{\pi k q}{N}}{\cos\frac{\pi k}{N} - 1} \cos\frac{\pi k n}{N} - \frac{1}{4}((-1)^{p+n} - (-1)^{q+n}) \right\}, \quad 0 \leq k \leq N. \tag{7.36}$$

Further evaluation of this expression seems difficult unless it is done numerically. Fig. 7.7 shows the solution sequence $\{G_n\}$ that gives the steady state distribution of wealth in an $N = 8$ family neighborhood. These values were determined by direct evaluation of the above expression for $\{G_n\}$ with $W = 1$, $p = 2$, $q = 6$ and $a_0 = 0$.

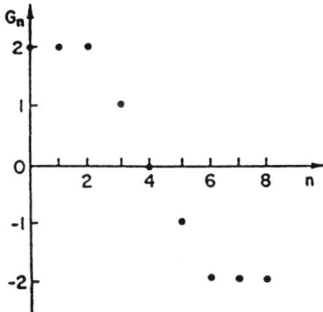

Figure 7.7: Distribution of wealth in an altruistic neighborhood of 9 families ($N = 8$) where family 2 is a creditor and family 6 is a debtor. $G_n = \{2, 2, 2, 1, 0, -1, -2, -2, -2\}$

Exercises 7.2

1. Use the operational sum method of the DCT, as discussed and illustrated in the last part of Section 2.4, and also in Chapter 5, to solve the boundary value problem (7.27)-(7.28). Compare your answer with that found in (7.34) by using the DCT series solution method, and the final solution G_n in (7.36).

REFERENCES

Difference Equations

There are very few books devoted entirely to difference equations. We list here six books which treat difference equation as principal theme, and with a level close to what is found in this book, except for the absence of the present discrete Fourier transforms method, The last three are more advanced in certain topics, especially the more detailed applications.

1. Goldberg, S., Introduction to Difference Equations, Wiley, 1958.

2. Spiegel, M., Finite Differences and Difference Equations, McGraw-Hill, 1971.

3. Hildebrand, F. B., Finite Difference Equations and Simulation, Prentice Hall, 1968.

4. Mickens, R. E., Difference Equations, Van Nostrand Reinhold, 1990.

5. Kelly, W.G. and A. C. Peterson, Difference Equations; An Introduction with Applications, Academic Press, 1991.

6. Elaydi, S., An Introduction to Difference Equations, Springer-Verlag, 1996.

Operational Calculus

Integral transforms techniques for continuous problems can be found in countless sources. The first reference is one of the first books to touch upon using the discrete Fourier transforms for solving difference equations associated with boundary value problems to start the "operational sum calculus" method. It, along with the other two references represent a useful background reading for the topics of the integral, finite and discrete Fourier transforms, needed for the tranforms method of this book.

7. Jerri, A. J., Integral and Discrete Transforms with Applications and Error Analysis, Marcel Dekker Inc., 1992.

8. Weaver, H. J., Theory of Discrete and Continuous Fourier Analysis, Wiley, 1989.

9. Briggs, W. L. and V. E. Henson, The DFT–An Owner's Manual for the Discrete Fourier Transform, SIAM, 1995.

10. Berg, L., Introduction to Operational Calculus, North Holland, 1967.

11. Bracewell, R. N., The Fourier Transform and its Application, McGraw-Hill, 1965.

12. Jolley, L.B.W., Summation of Series, Dover, 1961.

Fast Fourier Transforms

Much has been written on the FFT in thirty years. The original papers are of interest from a historical point of view. Several recent surveys are also valuable.

13. Brigham, E. O., The Fast Fourier Transforms, Prentice Hall,1988.

14. Bergland, G. D., A Guided Tour of Fast Fourier Transform, IEEE spectrum, No.7 (1969), pp41-52.

15. Cooley, J. W. and J. W. Tukey, An algorithm for the Machine Calculation of Complex Fourier Series, Math. of Computation Vol.19, No.90 (1965), pp.297-301.

16. Cooley, T. W., P. A. W. Lewis and P. D. Welch, The Finite Fourier Transform, IEEE Trans. on Audio and Electroacoustics, Vol.AV-17, No.2(1969), pp.77-85.

17. Gentleman, W. M. and G. Sande, Fast Fourier Transforms–
    For Fun and Profit, AFIPS Proc. 1966 Fall Joint Computer
    Conference, Vol.29(1966), pp.563-578.

18. Swarztrauber, P. N., Vectorizing the FFT's, in Parallel Com-
    putations, Academic Press, 1982.

z-**Transforms**

The most complete coverage of the z-transform seems to be in
the engineering literature, particularly in the field of systems analysis. The first of the following three references has the most detailed
coverage of basic theory, varied applications, and one of the earliest
extensive tables of the z-Transform.

19. Jury, E. I., Theory and Application of the z-transform Method,
    Wiley, 1964.

20. Freeman, H., Discrete Time Systems, Wiley 1965.

21. Luenberger, D. G., Introduction to Dynamical Systems, Wiley,
    1979.

ANSWERS TO EXERCISES

Exercises 1.1 p.9

1. $\{\sin \frac{2\pi k}{3}\} = \{\sin 0, \sin \frac{2\pi}{3}, \sin \frac{4\pi}{3}, \cdots, \sin \frac{2\pi N}{3}\}$

   $\{\sin \frac{4\pi k}{3}\} = \{\sin 0, \sin \frac{4\pi}{3}, \sin \frac{8\pi}{3}, \cdots, \sin \frac{4\pi N}{3}\}$

   The second sequence has double the frequency of the first sequence.

2. (b) $P_2(x) = \frac{3}{2}x^2 - \frac{1}{2}$

   (d) "difference – differential" equations.

3. $R_{n+1} = n + 1 + R_n$

Exercise 1.2 p.26

1. (a) $\Delta(\sin x) = \sin(x + h) - \sin x = \sin x \cos h + \cos x \sin h - \sin x$ or use the simple trigonometric identity $\sin A - \sin B = 2\cos \frac{A+B}{2} \sin \frac{A-B}{2}$ for the first line to have $\Delta(\sin x) = 2\cos(x + \frac{h}{2})\sin \frac{h}{2}$.

   (b) $\lim_{h \to 0} \frac{\Delta(\sin x)}{h} = \lim_{h \to 0} \frac{2\cos(x + \frac{h}{2})\sin \frac{h}{2}}{h} = \lim_{h \to 0} \cos(x + \frac{h}{2})\frac{\sin \frac{h}{2}}{\frac{h}{2}} = \lim_{h \to 0} \cos(x + \frac{h}{2}) \cdot \lim_{h \to 0} \frac{\sin \frac{h}{2}}{\frac{h}{2}} = \cos x$,

   after using $\lim_{\theta \to 0} \frac{\sin \theta}{\theta} = 1$.

409

(c) From (1.35) $\delta^2(u_k) = u_{k+1} - 2u_k + u_{k-1}$, so $\delta^2(\sin x) =$
$\sin(x+h) - 2\sin x + \sin(x-h) = 2\sin x \cos h - 2\sin x =$
$2\sin x[\cos h - 1]$

(d) $\lim_{h\to 0} \frac{\delta^2 \sin x}{h^2} = \lim_{h\to 0} 2\sin x \frac{(\cos h - 1)}{h^2} = -\frac{1}{2}\sin x \lim_{h\to 0}$
$4\frac{1-\cos h}{h^2} = -\frac{1}{2}\sin x$,

after using the result of part (c) and $\lim_{h\to 0} 4\frac{1-\cos h}{h^2} = 1$.

3. (a) $\Delta(c) = c - c = 0$

(b) $\Delta k^{(4)} = 4k^{(3)} = 4k(k-1)(k-2)$ after using (1.47)

(c) $\Delta k^{(-2)} = -2k^{(-3)} = -\frac{2}{(k+3)(k+2)(k+1)}$ after using (1.51)
and the definition of $k^{(-3)}$ in (1.50).

(d) $\Delta(k^2+3k+1) = (k+1)^2+3(k+1)+1-(k^2+3k+1) = 2k+4$

(e) $\Delta \sin \frac{\pi k}{4} = 2\cos \frac{\pi}{4}(k+\frac{1}{2})\sin \frac{\pi}{8}$ from Example 1.11 with
$\alpha = \frac{\pi}{4}$.

(f) $\Delta \cos \alpha k = \cos \alpha(k+1) - \cos \alpha k = -2\sin \frac{\alpha}{2}(2k+1)\sin \frac{\alpha}{2}$
after using the trigonometric identity $\cos C - \cos D =$
$2\sin \frac{C+D}{2}\sin \frac{D-C}{2}$

(g) $\Delta \ln k = \ln(k+1) - \ln k = \ln \frac{k+1}{k}$

(h) $\Delta\left((-1)^k\right) = (-1)^{k+1} - (-1)^k = (-1)^{k+1}[1+1] = 2(-1)^{k+1}$

(i) $\Delta\left((-1)^k k\right) = (-1)^{k+1}(k+1) - (-1)^k k =$
$(-1)^{k+1}[k+1+k] = (-1)^{k+1}(2k+1)$

5. (a) $x^{(1)} = x$, $x^{(2)} = x(x-h) = x^2 - xh$, $x^{(3)} =$
$x(x-h)(x-2h) = x^3 - 3x^2h + 2xh^2$

(b) $[f(x)]^{(0)} = 1$, $[f(x)]^{(-m)} = \frac{1}{f(x+h)f(x+2h)\cdots f(x+mh)}$,
$m = 1, 2, 3, \cdots$.

6. (a) (i) $4hx + 2h^2 + 3h$ (ii) $5h^2 + 2hx + 2h - x^2 - 2x - 1$ (iii)
$6h^2x + 6h^3 - 2h^2$

(b) (i) $4x + 4h - x^2 - 2hx - h^2$ (ii) $3x + 9h - 2 = 3(x + 3h) - 2$
(iii) $-h$

8. (a) $28hx^{(3)} = 28hx(x-h)(x-2h)$

(b) $-18hx^{(-3)} = -\frac{18h}{(x+h)(x+2h)(x+3h)}$

(c) $[-6x^{(-3)} - 4x^{(-1)} + 5x^{(-2)}]h$

9. (b) (i) $2hx + 2h + h^2$
   (ii) $2hx + 2h$

14. (c) $k^{(4)} = k^4 - 6k^3 + 11k^2 - 6k$
    (d) $k^3 = k^{(3)} + 3k^{(2)} + k^{(1)}$

Exercises 1.3   p.39

1. (a) C
   (b) $k + C$
   (c) $\frac{1}{2}k^{(2)} = \frac{1}{2}k(k-1) + C$
   (d) $\frac{k^{(3)}}{3} + \frac{k^{(2)}}{2} + C = \frac{1}{6}k(k+1)(2k+1) + C$
   (e) $k^2 + 3k + C$
   (f) $\frac{2^k}{2-1} + C = 2^k + C$
   (g) $(-1)^k + C$
   (h) $(-1)^k k + C$
   (i) $\ln k + C$
   (j) $\cos \alpha k + C$
   (k) $\sin \frac{\pi k}{4} + C$
   (l) $\sin \alpha k + C$

Exercises 2.1   p.53

1. (a) Third order, linear, nonhomogeneous, constant coefficient.
   (b) Second order, linear, nonhomogeneous, variable coefficient (see the k factor in the ku_k term).
   (c) First order, nonlinear (see the term u_k^2), homogeneous, constant coefficient.
   (d) Third order, linear, nonhomogeneous, constant coefficient.

(e) Second order, linear, nonhomogeneous, variable coefficient.

(f) First order, nonlinear, nonhomogeneous (see the nonlinear factor $u_{k+1}u_k$ in the first term $(2k+1)u_{k+1}u_k$).

(g) First order (see the largest difference in the indices if u_k: $|(k+2)-(k+1)| = 1$), linear, nonhomogeneous, constant coefficient.

2. (a) $[E^3 - 2E + 4]u_k = k^2$

(b) $[E^2 + k]u_k = 1$

(d) $[E^3 - 5E^2 + 6E + 8]u_k = k + 3$

(e) $[(2k+1)E^2 - 3kE + 4]u_k = 5k^2 - 3k$

(g) $[E^2 + E]u_k = k$

3. (a) $F(k, u_k, Eu_k, E^2 u_k, \cdots, E^n u_k) = 0$

(b)-1(c) $Eu_k - u_k^2 = 0$

(b)-1(f) $[(2k+1)u_k E - 3kE + 4]u_k = 1$

6. (a) Second order, linear, homogeneous, variable coefficient.

(b) First order, nonlinear, homogeneous.

(c) Second order, linear, homogeneous, constant coefficient.

(d) First order, linear, nonhomogeneous, variable coefficient.

7. -4(a) $[E^2 - 4E + 3 + \frac{4}{k}]u_k = k^2$

-4(c) $[E - 2\cos\alpha + E^{-1}]u_k = 0$ where $E^{-1}u_k \equiv u_{k-1}$.

-4(d) $[E - 1 - \frac{k}{2}]u_k = 1$

9. The auxiliary equation is $(\lambda^2 - \lambda - 6) = 0$, $\lambda_1 = 3$, $\lambda_2 = -2$, $u_n = A3^n + B(-2)^n$, and find $A = \frac{3}{5}$ and $B = -\frac{3}{5}$ from using the given initial conditions on the general solution, $u_n = \frac{3}{5}3^n - \frac{3}{5}(-2)^n$.

10. (a) $\lambda^2 - 3\lambda + 2 = 0$, $\lambda_1 = 1$, $\lambda_2 = 2$.

(b) $u_k = A(1)^k + B(2)^k = A + B \cdot 2^k$

11. (a) $\lambda^2 + 3\lambda + 1 = 0$, $\lambda_1 = \frac{-3+\sqrt{5}}{2}$, $\lambda_2 = \frac{-3-\sqrt{5}}{2}$

(b) $u_k = A\left(\frac{-3+\sqrt{5}}{2}\right)^k + B\left(\frac{-3-\sqrt{5}}{2}\right)^k$

12. $\lambda^2 + 1 = 0$, $\lambda = \mp\sqrt{-1} = \mp i$

13. (a) $\lambda^2 - 2\lambda + 1 = (\lambda - 1)^2 = 0$, $\lambda_1 = \lambda_2 = 1$

    (b) $u_k = c_1(1)^k + c_2 k(1)^k = c_1 + c_2 k$

14. $(2\lambda + 1)^2 = 0$, $\lambda_1 = \lambda_2 = -\frac{1}{2}$, $u_k = c_1(-\frac{1}{2})^k + c_2 k(-\frac{1}{2})^k = (c_1 + c_2 k)(-\frac{1}{2})^k$

15. $\lambda^2 - 4\lambda + 4 = (\lambda - 2)^2$, $\lambda_1 = \lambda_2 = 2$, $u_k = A2^k + Bk2^k = (A + Bk)2^k$

Exercises 2.2   p.63

2. (a) $\frac{e^{-20}-1}{e^{-1}-1}$

   (b) $\frac{3}{2} - \frac{3^{10}}{2}$

   (c) $82^{-10} + 2$

   (d) $-10317 \cdot (3)^{24}$

3. (a) $\frac{1}{6}n(n+1)(2n+1)$

   (b) $-\frac{1}{n+2} + \frac{1}{2}$

   (c) $\frac{1}{4}n(n+1)(n^2 + n + 2)$

   (d) $\frac{-2n-9}{2(n+3)(n+2)} + \frac{9}{4}$

4. (a) 1

   (b) $\frac{3}{4}$

5. $\sum_{k=0}^{N-1} \sin\frac{2\pi kn}{N} = \text{Im}\left\{\sum_{k=0}^{N-1} e^{i\frac{2\pi kn}{N}}\right\} = 0$

   $\sum_{k=0}^{N-1} \cos\frac{2\pi kn}{N} = \text{Re}\left\{\sum_{k=0}^{N-1} e^{i\frac{2\pi kn}{N}}\right\} = \begin{cases} 0, & k \neq 0 \\ N, & k = 0 \end{cases}$

Exercises 2.3 p.91

4(b) The Fourier sine and cosine transforms algebraize *even* order derivatives.

Exercises 3.1 p.153

1. (b) $u_k = \frac{3^k}{2} - \frac{1}{2}2^k + \frac{1}{2}$

2. (c) $u_k = c_1 + c_2 k + c_3 k^2$

3. (a) $u_k = c_1(-1)^k + c_2(-1)^k k + c_3$

   (b) $u_k = c_1 + c_2 k + c_3 k^2$

4. $u_k = c_1 + c_2(-1)^k + c_3 \cos \frac{k\pi}{2} + c_4 \sin \frac{k\pi}{2}$

5. The difference equation is linear.

6. (a) $u_{k+1} - u_k = 0$

   (b) $u_{k+2} - 7u_{k+1} + 10u_k = 0$

   (c) After eliminating the arbitrary constants in (E.1)-(E.2), we expect a relation between k, u_k, u_{k+1}, u_{k+2}, $\cdots$, and u_{k+n}, which represents a general nth order difference equation in u_k.

7. (a) $u_{k+2} - 4u_{k+1} + 4u_k = 0$

   (b) The characteristic equation $\lambda^2 - 4\lambda + 4 = (\lambda - 2)^2$ having repeated roots $\lambda_1 = \lambda_2 = 2$.

8. $u_k = 2k - 1$

9. (a) $u_k = (c_1 - 1) + c_2 2^k + 12 \cdot 5^{k-1}$

   (b) $u_k = \frac{1}{5}(26 - 28 \cdot 2^k + 12 \cdot 5^k)$

Exercises 3.2 p. 164

1. (a) $\lambda^2 - 3\lambda + 2 = (\lambda - 2)(\lambda - 1)$, $\lambda_1 = 2$, $\lambda_2 = 1$

(b) $u_k^{(1)} = 2^k$, $u_k^{(2)} = 1^k = 1$, $u_k = c_1 2^k + c_2$

2. $\lambda^3 - 3\lambda^2 + 3\lambda - 1 = (\lambda - 1)^3 = 0$ with three real repeated roots $\lambda_1 = \lambda_2 = \lambda_3 = 1$. The corresponding (modified) linearly independent solutions are $u_k^{(1)} = 1$, $u_k^{(2)} = k \cdot 1 = k$ and $u_k^{(3)} = k^2 \cdot 1 = k^2$, $u_k^{(c)} = c_1 + c_2 k + c_3 k^2$.

3. $\lambda^3 + \lambda^2 - \lambda - 1 = (\lambda + 1)^2 (\lambda - 1) = 0$, $\lambda_1 = \lambda_2 = -1$, $\lambda_3 = 1$, $u_k^{(1)} = (-1)^k$, $u_k^{(2)} = (-1)^k k$, $u_k^{(3)} = 1$, $u_k^{(c)} = c_1(-1)^k + c_2(-1)^k k + c_3$

4. $u_k = c_1 2^k + c_2 3^k$

5. $u_k = c_1 3^k + c_2 k 3^k + c_3 2^k$

6. $\lambda^4 - 16 = (\lambda^2 - 4)(\lambda^2 + 4) = (\lambda - 2)(\lambda + 2)(\lambda - 2i)(\lambda + 2i)$, $\lambda_1 = 2$, $\lambda_2 = -2$, $\lambda_3 = 2i$, $\lambda_4 = -2i$, $u_k^{(1)} = 2^k$, $u_k^{(2)} = (-2)^k = (-1)^k 2^k$, $u_k^{(3)} = (2i)^k = 2^k i^k = 2^k(\cos \frac{k\pi}{2} + i \sin \frac{k\pi}{2})$, $u_k^{(4)} = 2^k[\cos \frac{k\pi}{2} - i \sin \frac{k\pi}{2}]$, $u_k^{(c)} = c_1 2^k + c_2(-1)^k 2^k + c_3 2^k[\cos \frac{k\pi}{2} + i \sin \frac{k\pi}{2}] + c_4 2^k[\cos \frac{k\pi}{2} - i \sin \frac{k\pi}{2}] = c_1 2^k + c_2(-1)^k 2^k + A 2^k \cos \frac{k\pi}{2} + B 2^k \sin \frac{k\pi}{2}$, $A = c_3 + c_4$, $B = i(c_3 - c_4)$.

7. $\lambda^2 - \lambda + 1 = 0$, $\lambda = \frac{1 \mp i\sqrt{3}}{2}$, $u_k^{(1)} = \left(\frac{1+i\sqrt{3}}{2}\right)^k = \cos \frac{2\pi k}{3} + i \sin \frac{2k\pi}{3}$, $u_k^{(2)} = \left(\frac{1-i\sqrt{3}}{2}\right)^k = \cos \frac{2k\pi}{3} - i \sin \frac{2k\pi}{3}$; $u_k = c_1 u_k^{(1)} + c_2 u_k^{(2)} = A \cos \frac{2k\pi}{3} + B \sin \frac{2k\pi}{3}$.

8. $\lambda^4 + 12\lambda^2 - 64 = (\lambda^2 - 4)(\lambda^2 + 4) = (\lambda - 2)(\lambda + 2)(\lambda - 2i)(\lambda + 2i)$, $\lambda_1 = 2$, $\lambda_2 = -2$, $\lambda_3 = 2i$, $\lambda_4 = -2i$, $u_k = c_1 2^k + c_2(-1)^k 2^k + 4^k(A \cos \frac{k\pi}{2} + B \sin \frac{k\pi}{2})$

9.

$$\lambda^4 - 2\lambda^3 + 3\lambda^2 - 2\lambda + 1 = (\lambda^2 - \lambda + 1)^2$$

$$= \left(\lambda - \frac{1 + i\sqrt{3}}{2}\right)^2 \left(\lambda - \frac{1 - i\sqrt{3}}{2}\right)^2 = 0,$$

$$\lambda_1 = \lambda_2 = \frac{1 + i\sqrt{3}}{2}, \quad \lambda_3 = \lambda_4 = \frac{1 - i\sqrt{3}}{2},$$

$$u_k = A\cos\frac{2k\pi}{3} + B\sin\frac{2k\pi}{3} + Ck\cos\frac{2k\pi}{3} + Dk\sin\frac{2k\pi}{3}.$$

10. $\lambda^3 - 8 = (\lambda - 2)(\lambda^2 + 4\lambda + 4) = 0, \lambda_1 = 2, \quad \lambda_2 = \frac{-1+i\sqrt{3}}{2},$
$\lambda_3 = \frac{-1-i\sqrt{3}}{2}$
$u_k = c_1 2^k + A\cos\frac{2k\pi}{3} + B\sin\frac{2k\pi}{3}.$

11. $u_k = 5\cdot 2^k - 2\cdot 4^k$

Exercises 3.3 p.193

2. (a) $u_k^{(p)} = \frac{1}{6}4^k - k^3 - \frac{3}{2}k^2 - \frac{13}{2}k$

   (b) $u_k = u_k^{(p)} + u_k^{(c)} = u_k^{(p)} + c_1 2^k + c_2$ where $u_k^{(p)}$ is as in part (a).

3. $u_k = u_k^{(p)} + u_k^{(c)} = -3 + c_1(-1)^k + c_2 2^k$

4. $u_k = k3^{k-1} + c_1 2^k + c_2 3^k$

5. $u_k = k^4 + 2k^3 + c_1 + c_2 k + c_3 k^2$

6. $u_k = \frac{3}{2}[k^2 - k]\cdot 4^{k-2}$

7. $u_k = \frac{1}{2}k^2 - \frac{1}{2}k + 1$

8. $u_k = -\frac{7}{9}k - \frac{k^2}{6} + \frac{k2^k}{8} - c_1 + c_2 2^k + c_3(-1)^k 2^k$

9. $u_k = 3k^2\cdot 2^{k-3} + 5\cdot 4^{k-1} + c_1 2^k + c_2 k\cdot 2^k$

10. (a) $u_k = 2[\cos 2k + \cos(2k - 4)]/(1+\cos 4) + c_1\cos\frac{k\pi}{2} + c_2\sin\frac{k\pi}{2}$

    (b) $u_k = \frac{2^k}{65+15\cos 9}[8\cos(3k - 9) + \cos 3k]$

13. See the answer to problem 10.

14. $u_k = k^2 + \frac{8}{3}k + \frac{44}{9} + c_1 2^k + c_2 4^k$

15. $u_k = \frac{3}{8}k^2\cdot 2^k - \frac{3}{8}k\cdot 2^k + c_1 2^k + c_2 k\cdot 2^k = \frac{3}{8}k^2 2^k$
    $+c_1 2^k + (c_2 - \frac{3}{8})k2^k = \frac{3}{8}k^2 2^k + \frac{5}{4}4^k + c_1 2^k + c_3 k2^k, c_3 = c_2 - \frac{3}{8}.$
    Also see the answer to problem 9.

Exercises 3.4   p.202

1.   (c) $v_k = b_1 2^k + b_2 6^k - 4^{k-1} - \frac{3}{5}k - \frac{34}{25}$

(d) $b_1 = c_1, \ b_2 = -3c_2$

(e) $c_1 = b_1 = \frac{133}{100}, \ c_2 = -\frac{1}{60}, \ b_2 = -3c_2 = \frac{1}{20}, \ u_k = \frac{133}{100}2^k - \frac{1}{160}6^k + 4^{k-1} - \frac{4}{5}k - \frac{19}{25}, \ v_k = \frac{133}{100}2^k + \frac{1}{20}6^k - 4^{k-1} - \frac{3}{5}k - \frac{34}{25}.$

2. $u_k = c_1\left(-\frac{1}{2}\right)^k + c_2 4^k + \frac{37}{9}$ and from (E.1)

$v_k = -\frac{1}{5}(2E - 3)u_k + \frac{2}{5} = \frac{4}{5}c_1(-1)^k - c_2 4^k + \frac{11}{9}$

3. $v_k = c_1(-1)^k + c_2 k(-1)^k + \frac{a-1}{(a+1)^2}a^k$

$u_k = -\frac{1}{2}[(c_2 + 2c_1) + 2c_2 k](-1)^k - \frac{2a^{k+1}}{(a+1)^2}$

Exercises 3.5   p.219

5.   (a) $w_{k,l} = 2^{2k} \cdot \mu^{k+l}$

(b) $w_{k,l} = 2^{2k} \int C(\mu)\mu^{k+l}d\mu$

8. $w_{k,l} = (2\mu + 3)^k \mu^l$

9. $w_{k,l}^{(1)} = 2^k \mu^{k+l}$ for $\lambda_1 = 2\mu^{\frac{1}{2}}, \ w_{k,l}^{(2)} = (-2)^k \mu^{k+l}$ for $\lambda_2 = -2\mu^{\frac{1}{2}}$

10. $w_{k,l}^{(1)} = H(k - l), \ w_{k,l}^{(2)} = 2^k G(k - l)$ where $H(j)$ and $G(j)$ are arbitrary function of j. See also $2^l F(k - l)$.

11.   (a) $w_{k,l} = \lambda^k \mu^l = \mu^{l-2k}$ and $w_{k,l} = \int C(\mu)\mu^{l-2k}d\mu \equiv G(l-2k)$ where $G(j)$ is an arbitrary function of the single variable j.

(b) $R_{k,l} = 2^{k-l}$ is a UC sequence, since all its differences end up with a constant multiple of 2^{k-l}, for example $R_{k+1,l} = 2^{k+1-l} = (2)2^{k-l}, \ R_{k+2,l} = 2^{k+2-l} = (4)2^{k-l}, \ \cdots,$ etc.

(c) $w_{k,l}^{(p)} = (-2)(2)^{k-l} \ w_{k,l} = -2^{k-l+1} + G(l - 2k)$, where $G(j)$ is as in the answer of part (a).

Exercises 3.6 p.235

1. $u_k = (-1)^{k-1}(k-1)! \sum_{j=2}^{n-1} \frac{(-1)^j}{(j-2)!} + C_1(-1)^{k-1}(k-1)!$

4. $u_k = (k-1)! \sum_{j=1}^{k-1} \frac{(-1)^j}{j} + C \cdot (k-1)!$

Exercises 3.7 p.266

1. $u_k = \frac{1}{2}k^2 + \frac{3}{2}k + \frac{5}{2} + c_1 2^k + c_2 3^k$

2. $u_k = c_1(k-1) \sum_{j=1}^{k-1} \frac{1}{j} + c_2(k-1)$

3. (a.) $(E+k)(E-k)u_k = 0$

   (b.) $u_k = c_1 u_k^{(1)} + c_2 u_k^{(2)} = c_1(k-1)! + c_2 \sum_{j=1}^{k-1} \frac{(-1)^j}{j}$

4. $u_k = u_k^{(p)} + u_k^{(c)} = \frac{1}{6}4^k - k^3 - \frac{3}{2}k^2 - \frac{13}{2}k + c_1 2^k + c_2$

5. $u_k = u_k^{(p)} + u_k^{(c)} = k^2 + \frac{1}{6}k^3 + \frac{1}{9}4^k + c_1 + c_2 k$

6. (a.) $u_k = \frac{2}{\sqrt{3}} \sum_{j=0}^{k} \frac{1}{(k-j-2)!} \sin \frac{\pi(j+1)}{3} + c_1 \cos \frac{\pi k}{3} + c_2 \sin \frac{k\pi}{3}$

   (b.) The variation of parameters and the reduction of order methods.

Exercises 3.8 p.278

1. $\lambda_1 = i, \lambda_2 = -i; |\lambda_1| = |\lambda_2| = 1, \max(|\lambda_1|, |\lambda_2|) = \max(1,1) = 1$

2. (a.) $\lambda_1 = \frac{1}{2}, \lambda_2 = \frac{1}{4}; \ \max(|\lambda_1|, |\lambda_2|) = \max(\frac{1}{2}, \frac{1}{4}) = \frac{1}{2} < 1$

3. (b) $\lambda_1 = 1, \lambda_2 = 2; \ \max(|\lambda_1|, |\lambda_2|) = \max(1,2) = 2 > 1$

4. $\lambda_1 = -\frac{1}{2} + \frac{i}{2}, \lambda_2 = -\frac{1}{2} - \frac{i}{2}, |\lambda_1| = \sqrt{\frac{1}{4} + \frac{1}{4}} = \frac{1}{\sqrt{2}}, |\lambda_2| = \frac{1}{\sqrt{2}}; \max(|\lambda_1|, |\lambda_2|) = \max(\frac{1}{\sqrt{2}}, \frac{1}{\sqrt{2}}) = \frac{1}{\sqrt{2}} < 1$

5. (a.) $\lambda_1 = 0.4 + 0.58i$, $\lambda_2 = 0.4 - 0.58i$, $|\lambda_1| = \sqrt{(0.4)^2 + (0.58)^2}$
$\sim 0.71 = |\lambda_2|$, $\max(|\lambda_1|, |\lambda_2|) = \max(0.71, 0.71) = 0.71 < 1$ and $u_k^{(c)}$ converges.

(b.) $u_k = u_k^{(c)} + u_k^{(p)} = (0.71)^k[c_1 \cos \alpha k + c_2 \sin \alpha k] + 1.414$, where $\alpha = \tan^{-1} \frac{0.58}{0.4}$. So $u_k^{(c)}$ converges to zero and u_k converges to $u_k^{(p)} = 1.414$ as $k \to \infty$.

6. (a.) $\lambda_1 = 1.2 + 0.4i$, $\lambda_2 = 1.2 - 0.4i$, $|\lambda_1| = \sqrt{(1.2)^2 + (0.4)^2} = 1.265 = |\lambda_2|$; $\max(|\lambda_1|, |\lambda_2|) = 1.265 > 0$. Hence $u_k^{(c)}$ diverges as $k \to \infty$.
$u_k^{(c)} = (1.265)^k[c_1 \cos \alpha k + c_2 \sin \alpha k]$, $\alpha = \tan^{-1} \frac{0.4}{0.2}$
$u_k^{(p)} = 5$, $\lim_{k \to \infty} u_k = \lim_{k \to \infty}[u_k^{(c)} + 5] = 0 + 5 = 5$

(b.) There is no stable equilibrium solution, since $u_k^{(c)}$ diverges as $k \to \infty$.

(c.) $a_1 = -2.4, a_2 = 1.6$
i) $1 + a_1 + a_2 = 1 - 2.4 + 1.6 = 0.2 > 0$
ii) $1 - a_1 + a_2 = 1 + 2.4 + 1.6 = 5 > 0$
iii) $1 - a_2 = 1 - 1.6 = -0.6 < 0$
Here condition iii) of Theorem 3.17 is not satisfied; since $1 - a_2 = -0.6 < 0$. This supports the observation in part(b), that the difference equation has no stable equilibrium solution.

Exercises 4.2 p. 293

1. (a.) $\frac{1}{N}e^{\frac{2\pi i p k}{N}} = \frac{1}{N}\omega^{pk}$

(b.) $\frac{N-1}{\omega^k - 1}\frac{2\omega^k}{(\omega^k - 1)^2}$

(c.) $\frac{(N-1)(N-2)}{\omega^k - 1} - \frac{3(N-1)\omega^k}{(\omega^k - 1)^2} + \frac{6\omega^{2k}}{(\omega^k - 1)^3} = \frac{1}{\omega^k - 1}\left[\frac{N^{(3)}}{N} - 3\omega^k \mathcal{F}\{n^{(2)}\}\right]$

(d.) $\frac{1}{\omega^k - 1}\left[\frac{N^{(p)}}{N} - p\omega^k \mathcal{F}\{n^{(p-1)}\}\right]$

3. (a.) $g_k = \begin{cases} \frac{1}{2}, & k = 0 \\ \frac{1}{N}\frac{-2}{1-\omega^k}, & k \text{ odd} \end{cases}$

420

(b.) $g_k = \frac{\omega^k}{(\omega^k-1)^2}\left[\frac{1}{2}(-1)^k(1-\omega^{-k}) + \frac{1}{N}(-1)^{k+1} + 1\right]$

$$+ \begin{cases} \frac{1}{2}, & k = 0 \\ \frac{2}{1-\omega^k}, & k \text{ odd} \\ 0, & k \text{ even} \end{cases}$$

Exercises 4.3   p. 302

1. (a.) $b_k = \frac{-\sin\frac{2\pi k}{N}}{1-\cos\frac{2\pi k}{N}}, \quad a_k = -1$

(b.) $a_k = \begin{cases} 1, k = 0 \\ 0, k \neq 0 \end{cases}$; $b_k = \frac{1}{N}\left[\frac{(1-(-1)^k)\sin\frac{\pi k}{N}}{2(1-\cos\frac{\pi k}{N})}\right]$

(c.) $a_k = \frac{N-1}{\omega^k-1} - \frac{2\omega^k}{(\omega^k-1)^2} + \frac{N-1}{\omega^{-k}-1} - \frac{2\omega^{-k}}{(\omega^{-k}-1)^2}$,

$b_k = i\left[\frac{N-1}{\omega^{-k}-1} - \frac{2\omega^{-k}}{(\omega^{-k}-1)^2} - \frac{N-1}{\omega^k-1} + \frac{2\omega^k}{(\omega^k-1)^2}\right]$

Exercises 4.4   p. 311

1. $G_n = -\frac{V(\sqrt{3}-i)}{2}e^{-\frac{4i\pi n}{3}} - \frac{V(1-i\sqrt{3})}{2}e^{-\frac{2i\pi n}{3}}$

2. $G_n = -\frac{iV(1+i\sqrt{3})}{2}e^{\frac{i\pi n}{3}} - \frac{iV(1-i\sqrt{3})}{2}e^{-\frac{5i\pi n}{3}}$

3. (a.) $\mathcal{F}\{G_{n+1} + G_n\} = \mathcal{F}\{n\} = \frac{\omega^{-k}}{N}(G_n - G_0) + \omega^{-k}g_k + g_k$

$= \begin{cases} \frac{1}{\omega^k-1}, & 1 \leq k \leq N-1 \\ \frac{N-1}{2}, & k = 0 \end{cases}$

(b.) $g_k = \begin{cases} \frac{1}{(\omega^k-1)(\omega^{-k}-1)}, & 1 \leq k \leq N-1 \\ \frac{N-1}{2(\omega^{-k}+1)}, & k = 0 \end{cases}$

$$G_n = \frac{N-1}{4} + \sum_{k=1}^{N-1}\left(\frac{\omega^k}{\omega^{2k}-1}\right)\omega^{-nk}$$

5. $G_n = \frac{1}{N}\sum_{k=0}^{N-1}(\omega^{-k} + \omega^k)\omega^{-nk}$

Exercises 4.5 p. 320

3. $g_k = \dfrac{f_k}{2(\cos\frac{2\pi k}{N}-1)}$

5. (a) (i) $\mathcal{F}^{-1}\{g_k h_k\} = \sum_{k=0}^{N-1} g_k h_k \omega^{-nk} = \frac{1}{N}\sum_{l=0}^{N-1} G_l H_{n-l}$
   $= (G*H)_n$
   (ii) $\mathcal{F}^{-1}\{g_k h_k\} = \sum_{k=0}^{N-1} g_k h_k \omega^{-nk} = \frac{1}{N}\sum_{l=0}^{N-1} H_l G_{n-l}$
   $= (H*G)_n$

   (b) (i) $\mathcal{F}\{G_n H_n\} = \frac{1}{N}\sum_{n=0}^{N-1} G_n H_n \omega^{nk} = \sum_{l=0}^{N-1} g_l h_{k-l}$
   $= g*h$

   (ii) $\mathcal{F}\{G_n H_n\} = \frac{1}{N}\sum_{n=0}^{N-1} G_n H_n \omega^{nk} = \sum_{l=0}^{N-1} h_l g_{k-l}$
   $= h*g$

6. Let $g_k = \delta_{k,\alpha}, h_k = \frac{1}{\omega^k-1}, G_n = \mathcal{F}^{-1}\{g_k\} = e^{\frac{-2i\pi\alpha n}{N}}, H_n =$
   $\mathcal{F}^{-1}\{h_k\} = n, \mathcal{F}^{-1}\{\frac{\delta_{k,\alpha}}{\omega^k-1}\} = n\delta_{k,\alpha} - \frac{i}{\omega^\alpha-1}$

7. $(\omega^k + \omega^{-k} - 2)g_k = \begin{cases} \frac{N-1}{2}, & k=0 \\ \frac{1}{\omega^k-1}, & 1\le k\le N-1 \end{cases}$

   $G_n = 1 - \frac{1}{6}N^2 n + \frac{1}{6}n^3$

Exercises 5.1 p. 339

5. (a.) $a_k = \frac{1}{N}\left[\dfrac{(1-(-1)^{k+1})\sin\frac{\pi(k+1)}{N}}{2(1-\cos\frac{\pi(k+1)}{N})} + \dfrac{(1+(-1)^{k+1})\sin\frac{\pi(1-k)}{N}}{2(1-\cos\frac{\pi(1-k)}{N})}\right]$
   (b.) See the details of computing the DST of $G_n = n$ in Example 5.2 and problem 4.

Exercises 5.2 p. 351

2. Let $G_n = (\frac{\gamma}{\alpha})^{n/2} H_n, \alpha = 2, \beta = -5, \delta = 8,$
   $4H_{n+1} - 5H_n + 4H_{n-1} = n(4)^{-n/2}, H_0 = 4, H_N = 3(4)^{N/2},$

Let $Y_n = H_n - \left[4 + (3(4)^{N/2} - 4)\frac{n}{N}\right]$,

$4Y_{n+1} - 5Y_n + 4Y_{n-1} = (n)(4)^{-N/2} - 3\phi_n$,

where $\phi_n = [4 + (3(4)^{N/2} - 4)\frac{n}{N}]$, $Y_0 = Y_N = 0$.

Exercises 6.1   p. 358

1. (a) $\lambda_1 = 2$, $\lambda_2 = -2$; $u_n = 2^{n-2}[1 + (-1)^{n+1}]$

$\lambda_1 = 4$, $\lambda_2 = -3$; $u_n = -14(4)^n + 18(-3)^n$

$\lambda_1 = \lambda_2 = 2$, $\lambda_3 = -2$; $u_n = \frac{1}{4}(2)^n - \frac{1}{4}(-2)^n$

2. (a) $\lambda_1 = \lambda_2 = 2$; $u_n = n2^{n+1}$

(b) $\lambda_1 = \lambda_2 = -3$; $u_n = (-3)^n[1 - n]$

Exercises 6.2   p. 372

1. (a) $U(z) = \frac{z}{z - e^{-\alpha}}$

(b) $U(z) = \frac{z^2}{z^2 - 1}$

(c) $U(z) = \frac{z}{z+1}$

(d) $U(z) = \frac{\sin\theta}{z - 2\cos\theta + z^{-1}}$, $|z| > 1$

(e) $U(z) = -2 + \frac{6}{2z-1}$

(f) $U(z) = -2 + \frac{2z^3}{z^3 - 1}$

2. (a) $|z| > |e^{-\alpha}|$

(b) $|z| > 1$

(c) $|z| > 1$

(d) $|z| > 1$

(e) $|z| > \frac{1}{2}$

(f) $|z| > 1$

3. (a) $u_n = \begin{cases} 0, n = 0 \\ (-3)^{n-1}, n \geq 1 \end{cases}$

(b) $u_n = \delta_{n,0} + \delta_{n,1} + \begin{cases} 0, n = 0 \\ 2(2)^{n-1}, n \geq 1 \end{cases}$

(c) $u_n = \begin{cases} 0, n = 0 \\ \frac{1}{4}[(-2)^{n-1} + (2)^{n-1}], n \geq 1 \end{cases}$

(d) $u_n = \begin{cases} (-1)^n (2)^{2n}, n \text{ odd} \\ 0, n \text{ even} \end{cases}$

(e) $u_n = \delta_{n,0} - \frac{13}{4} \begin{cases} 0, n = 0 \\ (-2)^{n-1} + (2)^{n-1}, n \geq 1 \end{cases}$

(f) $u_n = \delta_{n,p} = \begin{cases} 1, n = p \\ 0, n \neq p \end{cases}$

4. (a) $u_n = \sum_{k=0}^{n} f_k g_{n-k} = \sum_{k=1}^{n} 2^{k-1} \cdot 2^{n-k-1} = 2^n \sum_{k=0}^{n} \frac{1}{4}$
   $= 2^n (\frac{1}{4}) \sum_{k=0}^{n} 1 = 2^{n-2}(n), n \geq 1$, noting that we had
   $f_0 = g_0 = 0$

   (b) $u_n = n(2)^{n-1}$

Exercises 6.3 p. 381

4. (a) $u_n = 2(-4)^n, n \geq 0$
   (b) $u_n = 9(3)^{n-1} - 3$
   (c) $u_n = (3)^{n-1} + (-3)^{n-1}$
   (d) $u_n = -\frac{1}{15}(2)^n + \frac{1}{24}(5)^n + \frac{1}{40}(-3)^n$

6. $u_n = n(2)^n$

7. (a) $U(z) = (\frac{z+1}{z})^{50}$

   (b) $u_n = \begin{pmatrix} 50 \\ n \end{pmatrix}$

Exercises 7.1 p. 397

2. (a) $C_n = -0.4(0.9999)^n + 0.5$

(b) The steady state $C_\infty = 0.5$

(c) 20794 seconds

3. (a) $P_{n+1} = P_n + \frac{0.08}{12} P_n, P_0 = 500$

$P_{n+1} - \frac{1.08}{12} P_n = 0$

$P_n = 500(\frac{1.08}{12})^{n \times 12}$

The balance after 4 years $= \$ 698.80$

(b) $P_n - 1.00666 P_n = 10$

$P_n = (500 - \frac{10}{0.00666})(1.00666)^{n \times 12} + \frac{10}{0.00666}$

$P_4 = \$702.80$

4. (a) $P_n = P_0(1 + b - c)^n$

(b) If $b > c, 1 + b - d > 1, (1 + b - c)^n \to \infty$ as $n \to \infty$ which shows that the population *grows* exponentially if $b > c$

If $b < c, 0 < 1 + b - c < 1, (1 + b - c)^n \to 0$ as $n \to \infty$ which shows that the population *decays* exponentially if $b < d$.

5. (a) Let $A_n = \lambda^n, \lambda = a \pm \sqrt{a^2 - b^2}$, so we have 3 cases;

(i) $|a| = |b| \to \lambda_1 = a = \lambda_2$ (repeated root) $A_n = a^n + na^n$

(ii) $|b| > |a|$ let $a = b \cos \theta \to \lambda = b \cos \theta \pm i \sin \theta, A_n = \frac{b^{n+1} \sin((n+1) \cos^{-1}(\frac{a}{b}))}{\sqrt{b^2 - a^2}}$

(iii) $|a| > |b| A_n = \frac{b^{n+1} \sin((n+1) \cosh^{-1}(\frac{a}{b}))}{\sqrt{b^2 - a^2}}$

(b) $A_n - 2A_{n-1} + A_{n-2} = 0, A_0 = 1, A_1 = 2, A_n = 1 + n, A_{20} = \det(A_{20 \times 20}) = 21$

6. $P_{n+1} = P_n + P_{n-1}, n \geq 2$

$P_0 = 0, P_1 = 1$

P_n is the *Fibonacci* sequence of Example 1.3 and (2.24)-(2.26),

$P_n = \frac{1}{\sqrt{5}}(\frac{1+\sqrt{5}}{2})^n - \frac{1}{\sqrt{5}}(\frac{1-\sqrt{5}}{2})^n, n \geq 0.$

INDEX OF NOTATIONS

428

$\mathcal{S}^{-1}$ Inverse Discrete Fourier Sine transform, 110

T

$T_n(x)$ Tchebychev polynomial of the first kind, 10

U

$u_n^{(s)}$ Unit step sequence, 360

$u_n(k)$ Shifted unit step sequence, 369

$u_k^{(c)}$ Complementary solution, 140–141

$u_k^{(p)}$ Particular solution, 141

$u_k^{(e)}$ Equilibrium solution, 276

UC Undetermined coefficients, 172

$U(z)$ The z-transform of u_n, 359

W

w Nth root of unity, 97 (W is also used in other books)

$W = e^{\frac{i2\pi(k-l)}{N}} = w^{k-l}$, 98 ($W$ is also used for the $N \times N$ matrix with elements $W_{nk} = w^{nk}$, 321)

$w_{k,l}$ Complementary solution of partial difference equations, 214

$w_{k,l}^{(p)}$ Particular solution of partial difference equations, 213

Z

$\mathcal{Z}$ z-transform, 359

$\mathcal{Z}^{-1}$ Inverse z-transform, 361

Miscellaneous Notations

$G_n * H_n \equiv \{G * H\}_n$ Discrete (Fourier) convolution product of G_n and H_n, 316

$(g * h)_k$ Discrete (Fourier) convolution product of g_k and H_k, 317

$u_n * v_n$ Discrete (z-transform) convolution product, 368

$g(kT)$ Discrete exponential Fourier transform (DFT) (sometimes used for the inverse DFT), 286

$G\left(\frac{n}{NT}\right)$ Inverse DFT (sometimes used for the DFT), 286

$f_c(kT) = a_k$ Discrete Fourier cosine transform (sometimes used for the inverse transform), 298,

SUBJECT INDEX

438

Other *Mathematics and Its Applications* titles of interest:

G. Gaeta: *Nonlinear Symmetries and Nonlinear Equations.* 1994, 258 pp.
ISBN 0-7923-3048-X

V.A. Vassiliev: *Ramified Integrals, Singularities and Lacunas.* 1995, 289 pp.
ISBN 0-7923-3193-1

N.Ja. Vilenkin and A.U. Klimyk: *Representation of Lie Groups and Special Functions. Recent Advances.* 1995, 497 pp.   ISBN 0-7923-3210-5

Yu. A. Mitropolsky and A.K. Lopatin: *Nonlinear Mechanics, Groups and Symmetry.* 1995, 388 pp.   ISBN 0-7923-3339-X

R.P. Agarwal and P.Y.H. Pang: *Opial Inequalities with Applications in Differential and Difference Equations.* 1995, 393 pp.   ISBN 0-7923-3365-9

A.G. Kusraev and S.S. Kutateladze: *Subdifferentials: Theory and Applications.* 1995, 408 pp.   ISBN 0-7923-3389-6

M. Cheng, D.-G. Deng, S. Gong and C.-C. Yang (eds.): *Harmonic Analysis in China.* 1995, 318 pp.   ISBN 0-7923-3566-X

M.S. Livšic, N. Kravitsky, A.S. Markus and V. Vinnikov: *Theory of Commuting Nonselfadjoint Operators.* 1995, 314 pp.   ISBN 0-7923-3588-0

A.I. Stepanets: *Classification and Approximation of Periodic Functions.* 1995, 360 pp.   ISBN 0-7923-3603-8

C.-G. Ambrozie and F.-H. Vasilescu: *Banach Space Complexes.* 1995, 205 pp.
ISBN 0-7923-3630-5

E. Pap: *Null-Additive Set Functions.* 1995, 312 pp.   ISBN 0-7923-3658-5

C.J. Colbourn and E.S. Mahmoodian (eds.): *Combinatorics Advances.* 1995, 338 pp.   ISBN 0-7923-3574-0

V.G. Danilov, V.P. Maslov and K.A. Volosov: *Mathematical Modelling of Heat and Mass Transfer Processes.* 1995, 330 pp.   ISBN 0-7923-3789-1

A. Laurinčikas: *Limit Theorems for the Riemann Zeta-Function.* 1996, 312 pp.
ISBN 0-7923-3824-3

A. Kuzhel: *Characteristic Functions and Models of Nonself-Adjoint Operators.* 1996, 283 pp.   ISBN 0-7923-3879-0

G.A. Leonov, I.M. Burkin and A.I. Shepeljavyi: *Frequency Methods in Oscillation Theory.* 1996, 415 pp.   ISBN 0-7923-3896-0

B. Li, S. Wang, S. Yan and C.-C. Yang (eds.): *Functional Analysis in China.* 1996, 390 pp.   ISBN 0-7923-3880-4

P.S. Landa: *Nonlinear Oscillations and Waves in Dynamical Systems.* 1996, 554 pp.   ISBN 0-7923-3931-2

Other *Mathematics and Its Applications* titles of interest:

A.J. Jerri: *Linear Difference Equations with Discrete Transform Methods.* 1996, 462 pp. ISBN 0-7923-3940-1